Building Pathology and Rehabilitation

Volume 2

Series Editors

Vasco Peixoto de Freitas
Aníbal Costa
J. M. P. Q. Delgado

For further volumes:
http://www.springer.com/series/10019

Aníbal Costa · João Miranda Guedes
Humberto Varum
Editors

Structural Rehabilitation
of Old Buildings

 Springer

Editors
Aníbal Costa
Humberto Varum
Department of Civil Engineering
University of Aveiro
Aveiro
Portugal

João Miranda Guedes
Department of Civil Engineering
Faculty of Engineering of Porto University
Porto
Portugal

ISSN 2194-9832   ISSN 2194-9840   (electronic)
ISBN 978-3-662-50899-2   ISBN 978-3-642-39686-1   (eBook)
DOI 10.1007/978-3-642-39686-1
Springer Heidelberg New York Dordrecht London

Printed on acid-free paper

Springer is part of Springer Science+Business Media (www.springer.com)

Preface

During the end of the nineteenth and the beginning of the twentieth centuries, the centers of many urban areas suffered important transformations that were responsible for the demolition of a considerable number of constructions, giving place to wider open areas and avenues. Also, new agglomerates of buildings were created outside the city centers, leading to the abandon and degradation of a considerable number of constructions. In reality, the way constructions are understood has modified over the years, following the changes in peoples' lifestyle and demands. In some cases this made people to move to new and modern areas, in other cases to intervene on the existing buildings, or to demolish and substitute them with new ones.

Intervening on an old building is, therefore, a matter that concerns social, economic, and cultural issues, which may assume different weights depending on the available funding, the knowledge and sensibility of the owners and technicians involved, the location and importance of the construction, the perspective of the authorities, among many other issues. The gathering of these data will constrain the procedures and techniques involved on the intervention of an old building, which, however, should always aim, in parallel with other concerns, the accomplishment of structural safety and the usage or service requirements, but without ignoring the particular value of the building.

Unfortunately, in the past many interventions on existent buildings have been inadequate in terms of the protection of their materials and constructive systems. An extreme, but paradigmatic example is the demolishment and substitution of the whole interior of the buildings, substituting it with new structural systems, just preserving the façades. Such types of interventions are quite invasive and ignore the constructive typologies and techniques that characterize the buildings. Moreover, they may introduce different materials and systems, not always sufficiently tested and known and that can create physical, chemical, and structural incompatibilities with those already existing in the buildings. In some countries, the lack of specific codes for the rehabilitation, enforcing the use of codes aimed for the design of new constructions, has strongly contributed to these results.

Actually, there are international recommendations and charts describing principles that should be respected when intervening on old constructions. They refer to some characteristics the interventions should aim; in particular, they should be low intrusive, reversible, and compatible with the preexistences. Although they are not always easy or possible to be fully respected, a growing effort has been made to converge to interventions closer to these principles, which can be only achieved with a proper knowledge of the materials, structural systems, and construction techniques used in the constructions. In fact, the lack of knowledge is, probably, the most important aspect that leads to the disrespect of the built heritage. It leads to the lack of confidence on old materials and induces technicians to look to old constructions not as something capable of sustaining the current needs of people, providing that proper interventions are made, but as something meant to be replaced by a new construction made of materials they know and rely on better, namely concrete and steel. Such perception often makes technicians to propose solutions on old constructions that lead to invasive and barely reversible interventions, and eventually to the near total destruction of the existing building.

Following this purpose, the book highlights the most important aspects involved in the characterization and understanding of the behavior of the most common structural systems and materials that are part of old constructions. It starts with the description of the structural systems of traditional buildings, referring to the most common structural elements and to their influence on the overall behavior of the construction. The subsequent chapters go more in detail in the structural elements and characterize, separately, the main elements that constitute an old building, namely: masonry walls either made of earth, bricks, or stone, timber and composite walls made of timber and infill material, and timber structural floors and roofs. In the book is also included a chapter dedicated to reinforced concrete structures, probably the most important structural material that, by the beginning of the twentieth century, progressively substituted the previous materials, being used in many constructions from that period on. Each of these chapters describes, with different levels of detailing, the materials, the construction procedures, the mechanical properties, the mechanical behavior, the damage patterns, and the most probable collapse mechanisms. Some of the chapters also present common or pioneering intervention measures applied to the repair and/or strengthening of structural elements, referring to their applicability and expected results.

To conclude, the editors believe that the book gives important information about the characterization of old buildings, helping the reader to have a better understanding of the behavior of these constructions and facilitating information that may help in the development of more precise and correct interventions, more in agreement with their original characteristics and cultural value.

Aníbal Costa
João Miranda Guedes
Humberto Varum

Contents

Construction Systems

Alice Tavares, Dina D'Ayala, Aníbal Costa and Humberto Varum

Abstract Understanding the character of a construction system is the base of any pre-evaluation process to support correct, sustainable rehabilitation decisions. For the uniqueness of a building lies partly in the preservation of its construction system, this testifies to the history or culture of a region with its environmental approaches. This chapter presents an overview of important influences as through treatises and a sample of traditional construction systems including other engineered solutions from the eighteenth century. The evolution of system characteristics and the systems' relationships with seismic regions or routes of dissemination is discussed, with archaeological and published examples. The wall-to-wall connections of antique systems are also emphasised to interpret the links between different traditional construction systems appearing all over the world, for the improvement of box behaviour. The debate around the definition of construction systems and their division in categories is also included to emphasise the particular understanding of the vernacular architecture.

KeyWords Constructions systems • Written sources • Treatises • Structural historical evolution • Structural vulnerabilities • Connections • Vernacular architecture

A. Tavares (✉)
Faculty of Civil Engineering, University of Porto, Porto, Portugal
e-mail: tavares.c.alice@gmail.com

D. D'Ayala
Department of Civil Environmental Geomatic Engineering, University College London, London, UK
e-mail: d.d'ayala@ucl.ac.uk

A. Costa · H. Varum
Department of Civil Engineering, University of Aveiro, Aveiro, Portugal
e-mail: agc@ua.pt

H. Varum
e-mail: hvarum@ua.pt

A. Costa et al. (eds.), *Structural Rehabilitation of Old Buildings*,
Building Pathology and Rehabilitation 2, DOI: 10.1007/978-3-642-39686-1_1,
© Springer-Verlag Berlin Heidelberg 2014

1

1 Introduction

Understanding how a construction system is assembled is the first step towards an insight of how it works, how it behave not only for the loadings and actions it was designed or conceived to withstand, but more importantly towards the ones that it needs to endure in time and for which no provision were made at its inception. This is particularly the case for traditional construction systems exposed to seismic action. The first requirement is a robust definition of what is intended for traditional construction in this context *versus* industrialized systems, as the former and its structural behaviour is the object of this book. The history of architecture and the built environment is often portrayed as a two-track path where formal and grander architecture has followed a route separate from the ordinary and vernacular construction with modest reciprocal influence. A review of historic treatises, from Vitruvius' and Alberti's to the some of the eighteenth century authors more specifically interested in the seismic resistance of contemporary construction, proves not only the deep contamination of the two areas of architecture (the courtly and the ordinary) in seismic prone regions, but also the continuity of thought between architectural design solutions and technical solutions, resulting in concerted choices in each part of the building resulting in a more resilient construction system. The enhanced seismic performance of masonry construction in which perpendicular walls are well connected to provide a box behaviour is a well proven concept in seismic engineering and one that informs most repair and retrofitting solutions proposed in modern design seismic codes and guidelines. A comparative analysis of a number of historic and vernacular construction systems, from diverse seismic prone regions and diverse age shows that this is a fundamental construction detail that found robust solutions well before the development of seismic engineering. The necessity to provide a construction system with sturdiness together with flexural capacity and ductility has led in the past to several more or less "vernacular" composite solutions.

Finally, in this chapter are discussed historical and architectural aspects related to the different construction systems studied in this book, with particular emphasis on traditional systems, but retrofitting solutions for the strengthening of buildings from the Modernism style are also presented.

2 Definitions of Construction System

The evolution of a construction system over the centuries is the result of a process of adaptation to climate, to geographical location and soil conditions, but is also influenced by past and present cultural background, economic considerations, taste and fashion. However, the progressive industrialization of methods of construction and the growing requirement from society for quality controls, assurance in the construction practice and in building codes has fostered increased control over the characteristics of materials and components used and over the structural and environmental building performance. The concept of construction system has gradually become more closely linked to the definition of an industrial process, as shown by the following set of statements stretching over the past 40 years. A construction system is:

- a "combination of structures involving organisation, technology and design process" [1] (Schmid T. and Testa C.);
- a "combination of production technologies, component design and construction organisation" [2] (Warszawski A.);
- a concept that "must only be applied to identify advanced industrialised processes of construction, which can be divided into three categories: (i) the design process and the management and control of construction methods; (ii) the technical subsystems such as structure, roof, walls, etc.; and (iii) the full range of activities involved in the production, construction and maintenance of all the specific components" [3] (Sebestyén G.);
- a concept that "encompasses the activities required to build and validate a new system to the point that it can be turned over for acceptance. This presumes an emphasis on the design process to ensure that technical solutions are based on the functional and operational requirements captured during the analysis phase, which includes a series of tests of each component to verify the entire system" [4, pp.129] (NYS ITS).

These definitions reveal the underlying assumption that a construction can be considered a system only if the foreseen performance of each of its materials and components follows a continuous process of appraisal and control from the planning through to the design and the construction phases, and eventually its use.

As correctly pointed out by Sebestyén statement [3], difficulties may arise when such definitions are applied to pre-industrial or vernacular architecture. Indeed the production processes involving these construction systems were conditioned by diverse social structures and cultural background. The production of construction well into the twentieth century was still based in many regions of Europe on the organisation, delivery and application of different crafts within the building site; crafts learned by apprenticeship and oral communication and whose quality control relied conspicuously upon the pride, skill and sense of ownership of the process by the craftsmen. A process which entailed the repeated application of "rules of thumbs" and procedures with well-established performance, and which saw over the years relatively modest variations and improvements to adapt it to different environmental and economic condition and client demand.

Not withstanding the differences in the mode of production of traditional versus industrialised construction, the requirements that both classes of buildings are expected to fulfil are the same, as first formally stated by Vitruvius: environmental comfort, aesthetic comfort, and durability through robustness. Any system is by definition made up of different components with different shapes, functions, materials and crafting. However, for their optimal performance, it is important to guarantee not only the quality of all materials and their compatibility but also the correct design and dimensioning for each component to fulfil its function and the correct type of connections between elements and components to ensure that the system works as a whole.

The connectivity between elements and components fulfilling different functions is a critical aspect for the ability of the system to withstand the actions a building is designed for, and most importantly, for its resilience against unforeseen environmental demands. In particular such connectivity can be identified as follows:

1. The foundations and their relation with the characteristics of the soil;
2. The connection between the upper structure and its footing, including waterproof layers;
3. The connection between vertical elements;
4. The connection between vertical and horizontal elements;
5. The connection between vertical elements and the roof; and.
6. The connection and correct position of non-structural elements such as chimneys, balconies, windows, doors and other elements with respect to the overall layout and the position of the structural elements.

These concepts have been codified since Roman times in the western world (Vitruvius, De architectura libri decem [5, 6]) and since a similar time in the Asian world (although Yangzi Fashi, the earliest surviving treatise of Chinese architecture was written by Li Jie during the mid-Song dynasty 1097–1100, it is a re-visitation of older pre-existing texts [7]) and represent the basis of any good construction, be it vernacular or formal architecture. To accomplish the Vitruvian "firmitas", knowledge of the mechanical behaviour of materials and components and of their expected as opposed to the achievable performance is essential. For traditional construction this knowledge was developed empirically and through the act of building and it was transmitted from generation to generation by the systems of apprenticeship. It constituted nonetheless a system of knowledge to be applied to a complex system. While a minority of sources exist documenting the process of knowledge transmission in the pre-industrial construction yard (one for all Villard de Honnecurt), more information on the development of structural resilience against seismic action in historical construction and its dissemination through ages and different regions can be obtained by a comparative reading of historic architectural treatises, as outlined in the next section.

3 Knowledge Dissemination Through Treatises

As it is well known western architectural theory and treatises have their archetype in the *De Architettura libri decem* of the roman Vitruvius (30–20 B.C.) [5, 6]. According to Vitruvius two fundamental concepts relate directly to robustness: the concept of proportion as the correspondence of members to one another and to the whole, measured by means of a "fixed part" [5, pp. 196] and the concept of symmetry taken as the fundamental condition for "coherence", the result of calculated relationships where each part bears measurable relation to every other part as well as the configuration of the whole [5]. In 1452 the *De re aedificatoria* [8, 9] of Leon

Battista Alberti (1404–1472), fashioned along the same structure of Vitruvius' treatise, traces back the current construction knowledge to the observation of roman archaeological sites and defines the aesthetic value in architecture as "the unity of all the parts founded upon a precise law and in such a way that nothing can be added, diminished, or altered but for worse" [8, p. 240]. More interestingly to the present discussion, however, in the third book Alberti explains this concept in terms of construction, considering that "the whole method of construction is summed up and accomplished in one principle: the ordered and skilful composition of various materials, be they squared stones, aggregate, timber, or whatever, to form a solid and as far as possible, integral and unified structure" [9, p. 61]. Moreover such structure shall be considered integral and unified only "when the parts it contains are not to be separate or displaced, but their every line joins and matches" [9, p. 61].

This concept that unity and integrity are fundamental to structural robustness and resilience is, to the authors knowledge, further developed and applied in practice in at least two instances following destructive earthquakes in the eighteenth century: the case of the *Pombalino* cage construction system of Lisbon (capital of Portugal) used in the reconstruction of Lisbon after the devastating earthquake of 1755, and in the so called *Casa Baraccata* theorised by Milizia in 1781 and extensively used through Calabria region in the reconstruction post the destructive seismic events of 1783.

The authors of the Pombalino cage (to which a whole chapter is dedicated in this book),were military architects and civil engineers (Manuel da Maya, Eugénio dos Santos and Carlos Mardel) emphasized the needs for proportion and symmetry, but also introduced regularity and progressive standardization as essential principles to design a timber frame structure with infill materials, using the idea of a unified construction system, where the connections had a particular importance to balance and guarantee the distribution of loads in a seismic event and to prevent the collapse of the timber cage.

In 1781, 2 years before the strong earthquake of Calabria (Italy) Francesco Milizia (1725–1798) published his *Principij di architettura civile* of where he proposed a composite timber framed building infilled with masonry where every piece needed to be well connected and embedded with the others explicitly to resist seismic actions [10].

In the following sections a detailed reading of these and other contemporary sources is carried out with relevance to the list of connectivity introduced in Sect. 2. The aim is to identify the specific advice and provision contained in the treatises, on the role of each component including size and relationship among parts, and compare these with the details of historic traditional composite construction as it can be observed in seismic region nowadays. The latter will be carried out in Sect. 4.

3.1 Foundations

Alberti's in his third book (1452) provides detailed advice on soil characteristics requirements to withstand the weight of the building, including the importance of underground inspections by trench digging and wells [9] with the aim of identifying the most suitable stratum to implant the foundations the presence of underground water courses or other instances that could hinder the erection of the building [9]. For construction on marshy ground with poor load bearing capacity, he recommends the use of inverted stakes and piles covering twice the surface footprint of the proposed wall, establishing the length of the piles no less than one eight of the planned height of the building and with a diameter no less than a twelfth of their length. Moreover he advises on ventilation of basements and foundations to prevent rotting [9]. Very specifically he indicated that the foundation should be made of solid stones and that this should be built up to a level of 0.30 m off the ground to prevent rising damp and rain erosion. This assumption is later present in other treatises as in the Vivenzio's, who advocated the use of a lower stone platform bigger than the perimeter of walls, considering the separation between foundation and base soil [10]. These concerns were wider disseminated for earth constructions and even timber-framed constructions in many European regions.

Pirro Ligorio (1513–1583) in his report of the long Ferrara earthquake (1570–1572)—*Libri di diversi terremoti*—specifically stresses the importance of sound foundation to guarantee the stability of the building in a seismic event [11].

Milizia in his treatise voices wider concerns, more akin to a comprehensive hazard assessment from the salubrity of the area to its seismic hazard from exposure to floods, to ground depressions and landslides, soft soil and other forms of unstable ground [12].

These particular concerns involved the shape and desired characteristics of the foundations of walls or of columns. Alberti considered a detachment between structure and foundation despite include the plinth as an element of the foundation [9]. Also, the discontinuity of the wall foundations through the use of arches was addressed for pillars or columns for particular grounds characteristics [9]. The capacity of the ground to withstand the intended load of the building was also highlighted [9]. For the stability of the structure he recommended foundations wider than the thickness of the wall [9], this is also present in the 18th century treatises as of Bélidor (1754), discussing also the correct depth of foundations [13].

3.2 Walls and Openings

Alberti treatise (1452) emphasized the needs of guarantee that the walls connect perpendicularly and complete, as much as possible unbroken from the ground to the roof and the placement of the openings in a way that would maintain the strength of the structure. For this reason he also proposed to keep windows away from the corners, mentioning that in ancient architecture it was custom "never allow openings of any kind to occupy more than a seventh or less than a ninth of walls surface" [9, p. 27]. The proportion of the openings was dependent on the distance between columns. Moreover he recommended the use of arched openings considering this the most suitable and durable form, however taking in account to "avoid having an arch of less than a semicircle with one seventh of the radius added"[9, p. 30], i.e. a raised profile. He stated that this was the only one that does not require ties or other means of support, mentioning that "all the others, when on their own and without the restraint of ties and opposing weights, seemed to crack and give way" [9, p. 31]. Alberti considered that attention should be applied to the wall around the openings, which should be "strengthened according to the size of the load that should bear" [9, p. 71].

Recommendations about corners, arches and openings were also present in the treatise *Trattato del Terremoto* of 1571 by Stefano Breventano (1502–1577) [10], while Antonio Buoni in his book *Dialogo del Terremoto* reported detailed observation of damages to arches after the Bologna and Ferrara earthquake [11].

In the *Pombalino* constructions built from 1756 special attention was given to the regular distribution and dimensions of the openings and the corners, obeying to official design plans of the façades. In 1758 a regulation was published limiting the height of the building and imposing restriction on element jutting out of the façades [14]. The design plans of the interior proposed also regular dimensions with standardized measures for the timber, stone and iron elements. These allowed the prefabrication of the elements in the outskirts of Lisbon, which were transported to the Baixa's construction sites only when needed it [14]. Such procedure allowed cost reduction for materials, production and workmanship [14]. The military civil engineer Manuel da Maya in a first proposal considered a restriction on the height of the buildings just for two floors. However, due to the social pressure imposed by necessity of housing and construction profit, the allowable height was extended to 4–5 floors, making the need for a seismic-proof solution all the more pressing [14]. The allowable height of the building was also correlated to the width of the street, to reduce the risk of people being injured by debris falling in the street from the buildings during an earthquake and to maintain a safe area to rescue people and to allow circulation in the immediate aftermath of a destructive shock.

Also Milizia (1781) establishes similar correlations between the height of the building and the width of the street. He proposed that dwellings built along principal streets could have three floors while along secondary streets no more than two floors [12]. While already Scamozzi (1548–1616) in the 16th century had advised that a building should be no taller than the width of the street is built along, Milizia considered that it was more appropriate to adopt a proportional measure [12].

After the 1783 Calabria earthquake an seismic-proof solution for residential buildings conceived by the architect Vincenzio Ferraresi was shown in the treatise of Giovanni Vivenzio *Istoria e teoria de tremuoti* 1783 [15]. This had a specific urban connotation as it foresee a two storey building flanked by two single storey buildings with the role of propping the taller building. Vivenzio argued that such solution was good because in this way at street corners the height of the construction will be smaller, less vulnerable and with lower risk of obstructing the two streets [16]. His consideration were purely static, and he overlooked the possibility that buildings of different height would have diverse natural periods and stiffness's and could damage each other through pounding. In 1784 the Istruzioni Reali of the Borboni's Government of the reconstruction of Reggio Calabria, included the proposals of Vivenzio and established the maximum height of buildings at approximately 30 palms [17, 18]. In large streets and squares were allowed buildings with the addition of a mezzanine floor with no more than 9–10 palms [17].

3.3 Corners and Wall Materials

Alberti considered each corner as "half of the whole Structure" and the point of the building where any damage or decay would start [9]. So he stated that corners throughout the building need to be exceptionally strong and be solidly constructed [9]. Alberti emphasized the ancient practice of considerably thickening the walls at the corner by adding pilasters to reinforce that area "to keep the wall up to its duty and hinder it from leaning any way from its perpendicular" [9]. In addition he underlined the need to use a system of quoins, stones longer and of the same thickness of the wall so as to avoid filling, that would extend into each of the walls at the corner in alternate courses that could support the remaining panelling [9]. Most importantly, in relation to the connections of façades of adjacent buildings along a street he states: "Stones left every other Row jutting out at the Ends of the Wall, like Teeth, for the Stones of the other Front of the Wall to fasten and catch into" [9]. This attention for the construction of the corners was also present in the treatises of Bélidor [13] (Bélidor B. 1754) and Milizia (1781), who highlighted the importance of this due to the effect of loads on the corners in a seismic event [12].

Recommendations for the most appropriate use of materials in diverse parts of the construction were already present in Vitruvius and cited by Alberti. Several other treatises and books had specific chapter dedicated to materials, as the L'Encyclopédie [19] coordinated by Denis Diderot (1713–1784) and Jean Le Rond D'Alembert (1717–1783) whose several books were published during 1751 until 1772. This work intended to record different fields of knowledge, including information about construction practices, catalogues of construction elements and their organization into construction systems, taking into account their regional variations.

Another encyclopaedia related to the existing building typologies seismically deficient, the World Housing Encyclopaedia (www.worldhousing-net.com) is been developed by EERI with the contribution of many researchers from different countries over the past 15 years. This resource, available online, contains important information concerning vernacular traditional and modern housing construction systems in seismic prone regions of the World. The aim is to identify the specific construction elements and construction practices that render a particular system more or less prone to earth-quake damage, classifying them with respect to the EMS' 98 vulnerability scale [20].

The introduction of tie rods in specific locations is one of the measures adopted traditionally in many regions to guarantee better connection between walls and between walls and the roof or floors [21]. This measure would have a very significant effect on the vulnerability, most outstandingly for taller buildings, 4–6 storeys high, as demonstrated in a study carried out on the traditional constructions of Alfama, a district of Lisbon [22].

From the information of the treatises analysed in this chapter is interesting some specific to walls. Alberti considered in his treatise (1452) that one of the most important rules was to build the wall in level and uniform considering that any side could had larger stones and the other small ones [9]. The explanation for such measure was associated with the assumption that imposed weight put irregular pressure on the structure, in addition to the less grip of the drying mortar leaded to cracks in the wall [9]. After the Calabria earthquake (1783), La Vega presented proposals with similar concerns, mentioning references to the sizing of the stones to be included in rubble stone masonry, highlighting also the need of using a quality blend for the mortar [18]. Also Milizia (1781) emphasized the need of uniform distribution of the weight for the structural equilibrium, considering that the materials used should be of the same quality to ensure such purpose [12].

Observing the ancient constructions Alberti concluded that the infilling of the walls was based on the rule that imposed every single section of infill with no more than 1.52 m approximately without being bonded in some areas with a course of long and broad squared stones [9]. He considered this squared stones as acting as "ligatures or muscles, girding and holding the structure together and also ensured that should subsidence occur in any part of the infill, either by accident or as the result of poor workmanship, it would have a form of fresh base on which to rest" [9, pp. 72]. This construction detail can be observed in many Roman constructions, in some cases of the stone being replaced by brick layers along the wall made of thin ceramic elements. In vernacular constructions in seismic regions this form of lacing is achieved by the use of timber elements laid along and across the wall, as it will be discussed in the next section. Milizia also recommends the use of "a succession of ties made of charred olive wood, binding the two faces of the wall together like pins, to give it lasting endurance" [23, p. 45].

Another interesting statement related with different types of stone masonry is the recommendation to improve the durability of the structure. Alberti emphasized the need of each course of the whole wall be composed entirely by squared stone [9]. Nevertheless, if it was necessary to fill the gaps between the two vertical plans of the wall, must be ensured that the courses on either side were bonded together and level [9]. In addition, he recommended the use of spaced block stones, spanning across the wall connecting both vertical plans "to prevent the two outer surfaces that frame the work from bulging out when the infill is poured in" [9, p. 73]. This recommendation is used even in vernacular construction made of stone masonry, as for public buildings, dwellings and walls. It is also associated to the stability of the structure and the need to improve the mechanical behaviour of the entire wall unifying as much as possible its elements.

In relation to the infill materials, Alberti emphasized again the ancient knowledge, considering that small stones joint and bond together better than the bigger ones. For this reason he also recommended that the infill did not contain stones weighing more than 327.45 g approximately and all the materials should be carefully bound together and filled in [9], again a concern to maintain the stability of the wall.

The same concern is pointed out for the construction of the cornice, that Alberti stated again as important to bind the wall tightly together [9]. For this reason a special care should be given to the stone characteristics used in this area. Considering that the blocks should be extremely long and wide, the jointing continuous and well made, the courses perfectly level and squared [9]. The care in this particular component of the construction is justified by Alberti assuming that is a potential vulnerable area of the construction where "it binds the work together at a point where it is most likely to give way"[9, p.74], besides its function of upper protection of the wall to prevent damages by the rain. This particular aspect can be observed in vernacular architecture with stone masonry and in earth constructions with the use of stone or layers of thin tile bricks on that area, in addition to the protection of the eaves.

3.4 Roof and Protection Against Fire

The connections between walls and roof structure received particular attention in the treatise and practice of reconstruction in seismic region in the eighteenth century. The *Pombalino* cage considered specific connections involving the structure of the roof. Lisbon regulations at the time (from 1756) also forbid any element protruding from the roof, allowing in a first stage only the kitchen chimney. Similar restriction was imposed in Calabria (Italy) in 1784, forbidding the construction of cupolas and steeples in churches [16].

A very extensive proportion of the damage experienced in the events of Lisbon (1755) and Reggio Calabria (1783) was the consequence of the fires that developed after the earthquake in adjacent houses.

The regulation applied for the reconstruction of London after the Great Fire of 1666 [24] was known by the Portuguese civil engineers responsible for the reconstruction of Lisbon. To prevent fire from spreading from house to house the civil engineer Manuel da Maya (1756) proposed that each wall dividing the properties within an urban block should be built above the level of the roof [24] as presented in the drawings of Eugénio dos Santos. A similar rule was introduced in Istanbul in the reconstruction of the Fener-Balat area damage by earthquake and following fire in 1894 [25].

The enthusiasm and admiration for classical architecture developed in the post medieval period and the Renaissance, fostered among others by Alberti's treatises, led to the perception of stone masonry construction, as the most durable and robust form of architecture, the only worthy material for formal and celebrative Architecture, royal and nobles palaces and religious buildings, while brickwork and timberwork was relegated to ordinary construction. However, from its inception earthquake engineering identified as essential attributes of earthquake resistant constructions, redundancy and deformability, without failure, or with controlled failure, i.e. ductility. These were achieved by the coupling of masonry and timber elements.

4 Seismic Positive Performance of Traditional Composite Structures

As shown in the previous section the treatises of the Enlightenment and the early engineering solutions for the reconstruction of building following major destructive earthquakes in the second half of the 18th century, all converged towards mixed construction system with timber frames infilled with masonry as the most suitable construction in seismic regions. Very early examples of composite timber-masonry structure are mentioned by Boethius [26] in the reconstruction of ancient Etruscan civilisation settlement in central Italy, dating back to 600 B.C. This are described as made of sun-dried brick and half-timber on stone foundations. Many more examples of composite systems exist still in several regions of the world prone to seismic hazard. These systems are usually the results of the organic development through time of vernacular and traditional architecture altered following destructive earthquake to withstand the next event. Their actual seismic resilience is a direct function of the relative capacity of timber and masonry as primary and secondary bearing structure or vice versa. A detailed analysis of their structural role in several vernacular examples, *bahareque* in El Salvador, *quincha* in Peru, *taquezal* in Nicaragua, *pontelarisma* in Greece, *taq* or *dhajji dewari* in Kashmir, *hatil and hımış* in Turkey is presented wherein. Their construction details and seismic performance are compared with the corresponding elements of the *Pombalino Cage* and the *Baraccata house*, the two early examples of engineered seismic construction introduced in the previous section.

In the case of the *Pombalino* and *Baraccata* systems, the solutions chosen to ensure robustness were also conditioned by the need to rebuild many dwellings in a short time with local available material. A review of the progress of reconstruction by the responsible in Reggio Calabria, noted that in practice less timber than the original design was employed in the reconstruction due to shortages of the material in the immediate region. In Lisbon and a more expedite and economic rebuilt was achieved by using also debris materials from the collapsed constructions. Attempts were made to standardise elements configuration and connections to allow the necessary regularity and a quicker production of the framing elements.

Construction systems that used timber as a framing structural material with masonry in-fill have two important characteristics: first, the use of a lightweight construction system, easy to build with the capability to be built higher and with more available space due to thinner walls; and second the improvement of the ductility of the whole system to withstand seismic loads.

However, mixed systems with timber with no anti-seismic capacity can be observed in many other regions of Europe, some even showing similarities with

Oriental systems considered to have those characteristics, such as the *hımış*. Nevertheless, many European systems, despite the ductility of the timber, do not have the necessary regularity or number of bracing diagonal elements or specific connections to hold up well during seismic events.

It was mentioned in the previous section how important is the connection among orthogonal walls to ensure a global behaviour of the system and how such importance has been emphasised through the ages in treatises and in practice. Such connection in masonry structures was typically ensured by the insertion of quoins and stone keys and later reinforced through iron ties and anchors. The connection among walls is fundamental not only for the robustness of the system and to keep the walls in plumb, but is essential to resist lateral loading as the ones produced by wind or earthquake by the ensemble of the walls in a box like behaviour, transferring the action from out-of-plane to in-plane behaviour. Hence particular construction solutions aimed at strengthening such connection, implicitly testify of an awareness, if not a knowledge on the part of their builders, of how building behave under seismic loading and of efficient ways to resist such loading. In many of the traditional and historic timber masonry composite systems found in seismic prone areas, the role of the timber rather than one of framing is one of reinforcing.

Tracing the presence and evolution through history of these construction solutions and identifying the conceptual links between a construction system made mainly of masonry reinforced with timber to one of timber framing stabilised with masonry will help us understanding the degree of seismic performance awareness on the basis of which these traditional systems were constructed and how they have influenced if any the early engineered solutions.

The basic composite system is made of series of masonry courses interlaced by timber logs laid horizontally along the sides of the wall and meeting at corner with another couple of timber logs laid in a similar manner along the perpendicular wall. The timber forming the couple of runners are connected to each other by transversal elements at regular intervals, while the two couples are sometimes connected at the corner, otherwise simply lay on top of each other. The connections among the timbers are sometime by scarf joint or simple cut, others by nails.

Palyvou K. (1988) describes an ancient mixed structure with timber that belongs to a construction in the Akrotiri settlement (1650 B.C.) on Santorini Island [27] (Fig. 1).

The construction had exterior walls of stone masonry and interior walls reinforced by a timber structure, load-bearing frames around the door and window openings [28]. According to Touliatos's interpretation the aim of this construction is to improve the tensile capacity of the stone walls by using timber grids embodied in them connected to vertical studs. Touliatos assumes this to be an ancient anti-seismic solution [28]. The ductility of the walls, the type of confinement of the material in-fill and the regularity seem to be relevant characteristics in this evolutionary process.

Buildings with laced bearing walls on the ground floor level and in-fill-frame used for the upper stories were commonly used during the Ottoman Empire and can be found in many variations throughout Greek regions such as Pelion Epirus,

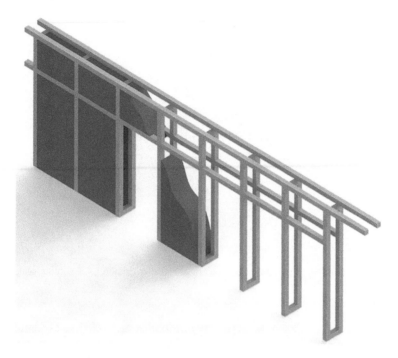

Fig. 1 Timber structure based on the drawings of Palyvou K. [27] of the ground floor construction "Xesti 3", without scale (credit A. Tavares)

Macedonia and Central Greece and also in some Aegean islands such as Thasos and Lesvos that experienced significant urban growth [29].This variation on the use of horizontal timber elements is present in the exterior walls of a traditional Greek construction system (Fig. 2), studied by Sakellaropoulou [30]. Adobe masonry, in many regions of Greece, were built as confined bearing masonry with horizontal timber ties, which were either visible at the façade of the wall or not. The timber ties were spaced at 0.70–1.00 m inside the masonry laid at the level of floor and/or at the openings [31]. The main characteristic of this solution, however, is the use of thinner horizontal elements laid on both sides of the wall and regularly spaced through its whole height. These longitudinal elements are transversally connected also with timber elements (see Fig. 2 for details). The introduction of these horizontal elements improves the overall behaviour of the system, namely and the confinement of the infill material.

As a main concern, the horizontal timber elements were placed at the base of the structure, at window sill and window header height, at the level where the floor beams were fixed to the masonry and at the coronation of the walls [30, 32].

Another relevant characteristic is the solution of the corner. The couples of horizontal timber elements belonging to two orthogonal walls at the same level are connected at the corner and sometimes extend beyond the exterior face of the wall. This connection of double timber elements can improve the stiffness of the walls

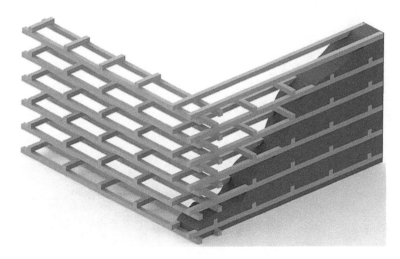

Fig. 2 Greek laced masonry based on the work of Sakellaropoulou [30], without scale (credit A. Tavares)

in the corner while offering the possibility of limited horizontal movement, allowing for energy dissipation during a seismic event, while limiting deformation and cracking.

Masonry structures laced with horizontal timber elements are also observed across the earthquake prone regions between the Eastern Mediterranean and the foothills of the Himalayas [32], a large geographical area where such forms have been developed for several centuries as *taq* (Kashmir) or *hatil* (Turkey) [30].

The main difference between the Greek system and *taq* systems is that the latter rely on the dimensions of the horizontal timber elements, which are larger and assume the function of beams lining the walls. Additionally, the distance between horizontal timber layers is greater, and the layers generally have connections with the beams of the floor.

The tradition of lacing masonry with timber, squared or in the round was common throughout the territory of the Ottoman Empire.

Hughes considers the Turkish *hatil* system related with the *cator* and cribbage systems (Fig. 3) of northern Pakistan [33], despite its difference in the number of timber elements, which can be much less in the *hatil*, the *hatil* also has the same type of timber longitudinal elements embedded on the walls. In the *cator* system Hughes explains that the timber elements have generally a square section of 50–120 mm and horizontal beams are placed on both sides of the wall, at vertical spacing of 0.30–1.30 m. "In better constructed walls the face timbers are tied together through the wall thickness with joined/nailed cross pieces at 1–4 intervals. Where the beam is of insufficient length for the whole length of the wall, two or more pieces are connected with tension resisting scarf joints. Breaks in the integrity of the ring beam may occur at doors and windows" [33, p.3]. The image of Alti Fort Tower in Hunza, Pakistan, presented by Hughes show a more dense use of horizontal timber elements given its

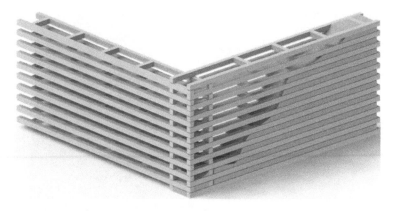

Fig. 3 Scheme based on the work of Hughes R. [33] of an antique defensive tower wall, Pakistani *cator* and cribbage system, without scale (credit A. Tavares)

defensive use. The *cator* system shows also clear similarities with the Greek traditional system as, for example, the system found in Lesvos Island. Both of these systems use horizontal timber elements as beams facing both sides of the width of the wall. In addition to the function of confining the masonry infill, a more regular spacing of the horizontal elements and their larger dimensions makes at the walls corner more robust than for the Greek system.

Sakellaropoulou emphasizes that the placement of horizontal timber elements within masonry bearing walls serves not only to increase the structural resistance of each wall, but also ensuring the continuity of the load transfer between masonry walls [30]. The objective of introducing ductility to the structure, adding cohesion to both sides of the wall and avoiding the disintegration of the in-fill materials is the most important purpose of the horizontal timber lacing. This construction type is also found in earthquake prone area of Peru', where timber logs in the round are used to lace adobe brickwork and in Nepal, where the bracing system confines the masonry panels adjacent to the windows and is connected with pegs to the horizontal structure. Interestingly in the Nepalese system the brick masonry walls only form the outer shell of the building, while the interior is divided by timber frames.

The coupling of timber frame and masonry is the characteristic of a second set of similar solution also found in earthquake prone areas. The already mentioned *Casa Baraccata* (Fig. 4) of the architect Vincenzo Ferraresi published by Vivenzio G., 1783 [15] shows two horizontal ring beams at both sides of the base of the walls at the platform level of the ground floor, which was at least 0.60 m above the ground. Ferraresi's drawings show the same connection arrangement between the two orthogonal walls, which had two horizontal timber beams, as describe for the *hatil* system, however in the Baraccata's drawing, the transversal connecting elements appear as laid on the same plane as the timber beams and dovetailed into them, to form a continuous ring beam. However the major difference is the presence at corner of four vertical posts, braced by the end of the beams of the two

Fig. 4 The lower part of
the system *Casa Baraccata*
(1783) based on the drawings
in the book of Vivenzio G.
[15] (credit A. Tavares)

orthogonal walls. Vivenzio's solution assumes that, for important larger 2–3 storey buildings, the wall material in-fill could be cut stone linked with cramps, connecting all the elements along the horizontal layers (as was used in some Roman constructions, as for instance the masonry blocks of the Coliseum [5, 34]. Moreover the frame panels are regularly braced with diagonal elements which surround the openings. Vivenzio also shows a second solution for one storey building, with post of square cross section, 0.20×0.20 m^2, maintaining a regular distance between them of 3.0 m proximately for one direction and less than 2.0 m for the spacing's of the other direction. The diagonal elements had sections of 0.10×0.10 m^2 combined with horizontal elements with 0.15×0.7 m^2 to form the timber frame.

Although some researchers have emphasised the similarities between the *Baraccata* and the *Pombalino* cage, they are characterised by a very different bracing system and hence they behave quite differently from a seismic point of view. In particular the cross bracing of the *Pombalino* cage ensure a continuous truss action which is not present in the *Baraccata* given the lay-out of the elements. The *Pombalino* cage is characterized by a regular timber-frame structure with standard dimensions, with masonry infill, with specific connections to the floors, roof and exterior walls. The adoption of Saint' Andrews crosses in most of the wall panels increases the stability to the structure, which is an important measure in seismic prone areas. In this book, the *Pombalino* system is discussed in detail in a specific chapter.

The *Casa Baraccata* was introduced in Calabria after the earthquake of 1783. This type of construction system is described by some researchers through the presentation of buildings most of them dating from the end of the nineteenth century and the beginning of the twentieth century (see for example: [35, 36]). However, later solutions present great differences compared to the eighteenth century solutions of Vincenzo Ferraresi, namely in terms of the type of connections used.

According to Mecca [36] these later solutions for the *Baraccata* had a timber frame, with vertical, horizontal and oblique chestnut or oak beams placed at a distance of approximately 1.20 m to create a truss structure. A weave of wickers and reeds was bonded to the main structure with thin chestnut laths and covered with an earth mortar. In some cases adobe filled the structure. For the interior walls, the *incannucciata* technique, a mesh of interwoven canes or branches covered by a clay plaster, was frequently used [36].

In the *Baraccata* solution, "some particular issues as symmetry, reduced heights construction and the use of a timber structure were positive aspects in terms of enhancement of seismic behaviour," as described by Tobriner S. [37, pp. 72]. However, the symmetry was incomplete due to the position of certain interior walls which could induce some torsional effects. In addition, the different volumetric heights of the building could also induce damages in a seismic event.

The construction system *pontelarisma* (Fig. 5) is an ancient solution found on the Lefkada Island (Greece) as described by Sakellaropoulou [30] and Karababa [38] the load-bearing masonry walls of the ground floor were double-leaf walls with a width of approximately 0.5–1.2 m and between 2.5 and 3.0 m in height, that were constructed from local stones (sedimentary rock or limestone). The external leaf is constructed of roughly cut stones, although for the corners of the buildings the masons utilised quoins to ensure bracing between walls. The internal leaf is made of rubble stones, while pieces of bricks or small stones mixed with mortar are used for the in-fill between the two leaves [30]. The common plan dimensions are 4.0–5.0 m along the one axis and 7.0–15.0 m along the other, while the openings are as a rule symmetrically arranged in plan.

(a)   **(b)**

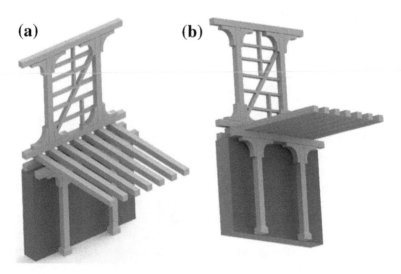

Fig. 5 *Pontelarisma* based on the work of Touliatos [32] and drawings of Sakellaropoulou [30], without scale (credit A. Tavares)

Karababa [38] describes the foundations as solid structure to prevent the differential settlement of the upper superstructure. On the top of the stone masonry walls all around its perimeter is placed a timber beam inside the wall to which is connected the timber frame of the upper floor. Steel ties are used to secure the connection between the masonry wall and the timber beam in regular spacing. Steel ties are also addict to secure timber elements around windows and openings [38]. The floors have timber joists with cross sections of 0.20×0.20 m^2, placed at 0.40 m centres, which are mortised into the sole beams embedded along the perimeter of the stone masonry [38].

The load-bearing timber frame of the ground floor consists of columns and beams that are arranged at the inner perimeter of the stone masonry wall and away from the wall 0.5–0.10 m, which allow the independent movement and deformation of the two systems during a seismic event avoiding pounding effects [38]. The timber columns have a cross-section of 0.15–0.20 m^2 [30] or even between 0.12 and 0.22 m and are built on stone bases secured onto them with steel ties embedded into the stone [38].

On the top of the floor joists are placed the timber frame of the upper floor linked by a horizontal sill beam with a cross section usually between 0.12 and 0.20 m. Each wall is composed by a grid divided and erected at 1.0–2.0 m centres [38]. The exterior posts have a cross-section between 0.12 and 0.22 m. The elbows or "bratsolia" provide stiffness at the corners between posts and beams [30] and are cut from a single piece of olive tree branches [38]. The roof structure is usually a truss system often arranged in more than one dimension to ensure adequate stiffness [38].

The key difference between this solution and those discussed previously is the idea of a double method of support, through the use of an exterior stone wall and the interior timber columns. The main objective of this structure is to guarantee that the upper floor will not collapse in a seismic event because it is supported by an interior timber structure.

The underlying idea of the acceptable collapse of the stone masonry wall is also presented in the *Pombalino* cage. This predicted collapse is restricted to the exterior wall of the ground floor and the identical objective of resistance of the timber structure to seismic loads without collapsing is for the safety of the people inside the building.

Pontelarisma has another link with other types of traditional construction: the vertically aligned double timber beam. Between these two beams, the floor structure was placed, comprised of timber beams with regular spacing that could have the tops outside the exterior vertical alignment of the wall, once again with the purpose of controlling the effects of horizontal loads and allowing restricted structural movement without collapsing. This characteristic is shared by other traditional solutions, including also the insertion of diagonal elements strategically placed in the construction, as can be observed in the *hımış* solution, a Turkish traditional construction system present in the western Anatolia region and Marmara.

The *hımış* (Fig. 6) solution also has a skeleton formed by placing wooden posts vertically and diagonally. The resulting space is filled with in-fill materials such as fired clay bricks, adobe blocks, or stone, which can be easily and economically

Fig. 6 Scheme of a corner of the *hımış* construction system based on the work of Bilge Isik [42] without scale (credit A. Tavares)

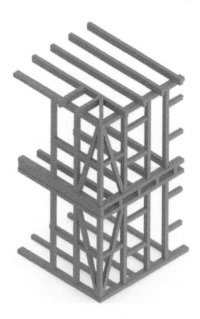

obtained in the region. The use of mud mortar for the in-fill masonry is also widespread [39].

Hımış has rubble stone foundation walls reinforced by timber beams placed on the interior and exterior faces of the wall about 1.0 m apart from one another, a tradition dating back to prehistoric periods in Anatolia [40]. Sahin affirms that the main function of the timber beams was to form a frame around the masonry and to confine and ensure the unity of the wall [40]. The base of the foundation was sunk more than 0.45 m below the ground surface [41].

The ground floor was of stone or adobe masonry and the upper floor has a timber frame construction defined by two main timber beams, a header and footer, which are linked to vertical timber posts cross-section between 0.12×0.12 m^2 and 0.15×0.15 m^2 with at 1.5 m intervals [42]. Between these load-bearing main posts are intermediate vertical elements every 0.60 m and horizontal elements to hold the infill with cross-sections from 0.6×0.12 to 0.6×0.1 m^2 [42]. Diagonal elements with the same cross-section of the main posts connect the foot beam to the main post and have a relevant function on the stability of the whole structure [42].

The infill material used can be fired clay bricks, adobe blocks, or stone, which can easily and economically be obtained in the region. The use of mud mortar for the infill masonry is widespread and the walls are either left exposed or plastered with mud and then whitewashed [39].

In Turkey, mixed construction techniques within single buildings are common and houses were often designed with the laced bearing wall (*hatıl*) construction on the ground floor level and *hımış* used for the upper stories [43].

This review of existing composite structural systems in earthquake prone country helps to identify the elements that are recurring and hence recognised as essential to a successful seismic performance through the ages. It is interesting that they are similar in zone with similar seismic exposure even when it is not immediate to establish a cross fertilisation among these regions in time. These elements can be summarised as the timber lacing (or ladder) composed of two initially round the squared runners along the top of the wall of within its height at regular intervals and connected by transversal elements. The connection of these ladders pertaining to orthogonal walls at the corner was made using simple nails or lapped or scarfed joints. In more prone seismic regions the progressive introduction of timber frames to work in adjacency to the masonry walls or independently from it, showing a full awareness of the use of two separate systems, one to resist gravity loading, the masonry, one to resist seismic loading. Finally the almost complete substitution of the masonry with braced timber frame at the upper storeys, to achieve lighter structures, with natural frequency further from the earthquake content of high amplification frequency. In these cases the masonry is only used as a plinth to isolate the timber frame for the ground and prevent its decay.

5 Main Construction Systems

The construction systems may be grouped in traditional and industrial, according to their conception, considering that each of them has associated specific approaches for questions related to durability and safety. This aspect was already discussed in Sects. 2 and 3. Apart from these two groups, it has to be mentioned that in certain periods composite construction systems were used as consequence of the architectural demands or as a simple evolution of traditional methods introducing new materials. For this reason, in this chapter a division of the construction systems into 3 main groups is proposed, highlighting other aspects besides the structural systems, since this will be discussed in detail in the following chapters of the book.

Construction systems of buildings, such as some of those which are analysed in the subsequent chapters of this book, can be organized into three main groups according to the origin of the materials used, namely if they are raw materials or if they are industrially processed:

- Group A—mostly made of natural raw materials;
- Group B—mostly made of industrial materials;
- Group C—combination mix of natural and industrial materials in different constructions parts or components.

Group A is mostly composed of traditional construction systems such as earth construction solutions (adobe, rammed earth, cob, among others), timber structural systems, composite systems made of masonry and timber (half-timbered, *hımış*, *bağdadi, quincha, taquezal, bahareque, dhajji dewari, pontelarisma, pombalino, baraccata*, among others), stone masonry structural systems.

Group B is defined by the modern engineered systems, such as those involving reinforced concrete (RC) or steel structures, or with industrial brick masonry.

From this group, only the reinforced concrete structures will be discussed in this book (Chap. 9). Buildings belonging to the Modern Architectural Movement or Modernism, which lasted until the 1950s, are examples of the application of reinforced concrete structures.

Group C is defined by traditional construction materials and systems combined with engineered solutions. Many of these solutions were adopted after World War I in residential buildings, particularly until the 1940s. Composite systems were used in some European regions, mostly with reinforced concrete (RC) elements [44, 45] some of them belonging to the first stage of the Modernism architectural movement. Although it can be said that some of the structural recommendations/solutions for seismic regions presented in building codes include the use of composite systems with RC. The use of RC ring beams in load-bearing masonry wall structures of adobe or rammed earth at the level of their foundations and openings can be seen as an example of this situation. The Turkish Building Regulations (1998) actually refers to such procedure by stating that "masonry foundations should be built with reinforced concrete footings and lintels" [41]. The use of reinforced concrete elements in adobe construction was also applied by the Peruvian rural community housing after the 2001 earthquake in Moquegua [46]. The use of ring beams or lintels, or even thin slabs of reinforced concrete in adobe construction was common in the 1930s in Portugal [44, 47], although they were very thin and were not intended to be a measure to improve the seismic behaviour of the structure.

5.1 Group A

Group A includes the most important expression of ancient construction systems which may soon disappear in several regions of the world, such as the timber buildings in Turkey, mentioned in UNESCO reports, and the wide range of earth architecture all over Europe, as mentioned by several researchers involved in the project Terra Europae (2011) [48]. This book includes some chapters about earth construction, due to the growing interest in the preservation of these structures considering their cultural value. The restoration of earthen buildings is being subsidized by governmental rehabilitation schemes or through other institutional financial support such as in Cyprus [49] and in Sardinia, although just in a very limited number of cases. In addition, this activity is also growing in some countries such as Austria, Germany [50], Australia and the USA due to the interest in using such materials for new architectural proposals, which can also be a way to change mentalities and promote the desired protection of this heritage.

There is a wide variety of methods of earth construction and many researchers have noted the existence of earth buildings in almost all continents since the Neolithic age. Their presence in Europe has been recorded, for example, through the remains of the Etruscan civilisation (dating back to the 7th and 6th centuries B.C. [51]. In Germany, earthen buildings had a similar form to wattle and daub buildings from approximately 4000 B.C. [52]; in Aegean areas (Bulgaria, Greece, Cyprus and Malta) and Italy, construction styles changed little since the Neolithic era, a phenomenon also observed in parts of northwest and southeast Europe [53]. In Cyprus, earth constructions were built as early as 9000 B.C. [36]. During the Roman Empire, earth construction techniques such as adobe masonry were used in southeast and east central Europe [36, 50]. The colonisation of the Eastern Mediterranean by the Romans and the arrival of Muslims in some regions are considered to be some of the forces that led to the dissemination of earth construction techniques [54]. Some remaining evidence of this process is rammed earth in France (Poitiers) and military fortifications made of rammed earth in Portugal [55].

Nowadays, there are still earth constructions in many European countries, most of them from the first half of the twentieth century. However, buildings from the eighteenth and nineteenth centuries can be found in Greece, Portugal, Italy, Cyprus, Denmark and many others as noted by the project Terra Europae (2011) [48].

At least twelve main techniques of construction using earth have been identified by Houben and Guillaud [56]. The construction systems identified in this book representing some of the most widespread techniques are rammed earth (monolithic), Fig. 7, adobe (masonry), Figs 8 and 9, and tabique (load-bearing structures), Fig. 10.

The evolution of the earth construction system depended, as in other systems, on the awareness and understanding of ways to improve the system.

In earth construction the foundations present some variations including the use of stone to achieve a more efficient resistance to water effects, such as rain impact and capillary rise in saturated soil. A concern present in ancient treatises which also proposed foundations with stakes and pointed the most durable timber species

Fig. 7 Rammed earth construction, Portugal (credit H. Varum)

Fig. 8 Adobe production, India (credit A. Costa)

Fig. 9 Adobe construction in Portugal (credit A. Tavares)

to use on the most vulnerable situations as silt soils. In regions with less stone available, brick was used sometimes mixed with stone masonry with the same objective. This solution can be seen in the Mediterranean regions as in Greece where the foundation is extended to a height of 0.40–1.20 m [31] or in Portugal [57] and in Cyprus until 1950s [49]. Also it can be seen in Turkey and Macedonia [29, 58]. This characteristic was linked to the desire to improve the building's durability, i.e. to isolate the timber structure and the adobe masonry from direct contact with the ground. This improves not only the resistance to decay due to moisture, but is also very effective in preventing termites attack in hot climates.

Fig. 10 Traditional *tabique* construction in Portugal (credit A. Tavares)

In some regions as in Central region of Portugal or in Istanbul (Turkey), adobe constructions included a ventilation space between the level of the ground floor and the soil level, with holes in the foundation walls to allow air circulation (Figs. 11 and 12). This allows for better conservation of the timber and the permanent drying of the wall base [57]. As mentioned in Alberti's treatise (see Sect. 3), this is a very important issue that can also be observed in old stone masonry constructions. Another example is appointed by Sahin in relation to the base of traditional buildings in Turkey: a timber beam is placed on a stone platform inside the foundation walls and joists are set on this beam. Between the stone platform and

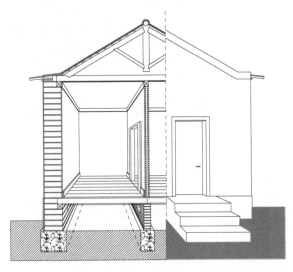

Fig. 11 Scheme of earth (adobe) construction system with ventilation space below (credit A. Tavares)

Fig. 12 Ventilation space
in the base of an adobe
construction (credit
A. Tavares)

the pressed earth there is a minimum of 0.20–0.30 m of air space for ventilation [40]. In the central region of Portugal this ventilation space is usually more than 0.50 m in constructions from the end of the 19th century until the half of the 20th century.

However, other traditional construction systems lack ventilation space and present some problems, namely due to the use of the floor girders placed above timber beams (which are inserted into the walls), a common occurrence in Turkish construction [38]. This hampers the inspection of the structure bellow, including the replacing of timber members that have deteriorated due to the lack of ventilation—one of the causes of structural degradation. The ventilation spaces in the base of the construction thus allow for a better control of fluctuations in the high moisture levels responsible for the degradation of the timber elements less resistant to humidity effects.

Although earth construction systems may be seen to have a number of shapes, the size of the openings remained with limited dimensions. Additionally, the width of the walls has varied in relation to the height of building according to local environmental conditions, i.e., thicker in cold regions and thinner in warmer ones [36].

The insertion of beams into the construction walls is another issue solved by traditional construction in a variety of ways. As for vernacular stone masonry or earth constructions, beams could be placed outside the wall supported by buttress elements connected to the wall; in holes in the wall; resting above one or two timber bond beams; or resting on a timber beam and projecting outside the exterior line of the wall. The kind of insertion used for timber beams has implications for both the durability and the structural behaviour. Again, Alberti's treatise emphasises this subject and recommends that the number of voids in the walls must be as low as possible (see Sect. 3). The ventilation of the top of the beam inside of the wall was the major difficulty and induced, in many cases, its degradation due to higher levels of moisture in that area. In addition, the degradation of the top of the timber beams causes a lack of stability of the floors and also of the connection between opposite walls, which has implications on the structural behaviour. This problem is also observed when the constructions (as earth construction and stone masonry buildings) have a timber ring beam in the top of the exterior wall to connect the structure of the roof. Its degradation due to lack of ventilation, lack of roof maintenance and damages on the roof covering is one of the major problems of this type of construction. Such problem is sometimes responsible for damages on the walls due to the introduction of horizontal loads by the changed action of the roof structure, which makes the upper connection of the walls difficult and can introduce out-of-plane movements particularly relevant in seismic regions.

Concerns related with the construction of the walls and corners in vernacular architecture revealed some links with what is presented in treatises, such as with the use of stone blocks and their imbrication namely in rubble masonry walls. Pilasters introduced in the architectural conception for the corners increase the width of the wall to strengthen that area. However, in some earth constructions, fired bricks were used to strengthen the corners. Again, areas most susceptible to degradation include the arches and vaults, and the areas around openings. Usually the lack of care on the construction of these elements induces damages observed as diagonal cracks. In addition, the distance between openings and between the openings and the corners are important to control diagonal cracks and to achieve a greater structural performance (namely in seismic regions), as seen in the treatises recommendations and in the characteristics of ancient construction systems.

The strengthening of the corner can be observed in Italian rammed earth constructions of Piemonte but also in other regions as Alessandria [36]. Similar procedures can also be found in adobe constructions of the central region of Portugal in the basement of the construction or in rubble masonry.

Other types of earth construction such as wattle and daub can be seen to exhibit similar concerns with the stability of corners, namely by suing diagonal timber elements placed in the corners connecting the timber columns to guarantee the stability of the structure. Diagonal elements are also important in timber constructions since the lack of these elements can totally damage the building, as can be observed in Fig. 13 where the diagonal elements from an abandoned salt warehouse were removed. The damage of this traditional timber building resulted from the movement of the entire construction in one direction which may then lead to its collapse.

Timber buildings were one of the traditional construction systems with higher level of growing standardization and improvements of the method of construction [59, 60] which allowed the prefabrication of solutions from Scandinavian countries, Germany, Austria and England that reached a high level of production in

Fig. 13 Traditional timber constructions—Portugal (credit A. Tavares)

the period between the two World Wars for exportation [61]. The system's main advantages are the simplicity, cheapness, and ease of construction and, obviously, its transportation. Another factor influencing the use of natural materials in the construction is the need for other materials, such as steel for the War equipment.

At the beginning of twentieth century, there was already a particular interest related with the correct position of the timber elements on the building construction as well as to the need of standardised procedures for assigning working stresses to timber. This situation attracted special attention to the debate involving the best kind of tests to obtain the mechanical properties of timber [59]. Military housing, bridges, industrial plants, warehouses, and shipyard facilities used timber wherever possible. Some of these buildings were constructed associated to specific rebuilding programs and were conceived as temporary structures with an anticipated life of no more than 5 years. Nevertheless, at least 36 years later, some still remain in place [59]. The most frequent problem was related to changes in use and occupancy that imposed different loadings than those for which buildings were originally constructed [59]. The same situation can be observed in many regions where timber constructions were the main traditional method applied.

Until recently, most domestic buildings in Turkey were built of wood and most of them were constructed up to the 1960s [62]. As in many parts of the world, the adoption of reinforced concrete almost destroyed traditional construction practices. However, following the earthquakes of 1999, there was a renewed interest in these typologies due to the fact that such construction system had a high seismic performance [63]. Other regions with many timber buildings such as Paramaribo in South American, Suriname, a World Heritage town since 2002, had programs for their maintenance. Nevertheless, the interest of international organizations such as UNESCO, the example of the relevant heritage of timber buildings of Zeyrek (Turkey) shows that the traditional timber buildings are being deteriorated because of man-made reasons including abandonment, fire, wrong choice of wood materials, material fatigue, economic insufficiency, air pollution, lack of laws to protect these structures and municipal indifference, defective workmanship and incorrect attempts of restoration as appointed by SeçKin [64]. The same problems are being observed in many other regions of the world.

Even if this group of traditional construction systems is in a critical situation, the knowledge dissemination of their characteristics is considered still an important action for their protection and for the implementation of proper rehabilitation programs in the future, including the proper capacity evaluation when they are located in seismic prone regions.

5.2 Group B

Reinforced concrete (RC) changed the panoramic of our landscape in an irreversible manner. The beginning of its use was a predictable and desirable fact at the time with strong links with architectural Movements but also with technical demands. A desire that was easily widespread all over the world and for this reason can be assumed as associated to the globalization of a product and a technique. The Modern Movement of architecture is now a cultural value in many regions of the world with a growing interest in its protection, namely in Europe and in USA. The high level of experiences, in terms of space conception and technological improvements to support the new way of living, presented one of the most extraordinary experiences all over the world in almost the same period of time. A contemporary multicultural technology that now shows its vulnerability in seismic regions, because of less caution in its construction methods and, in addition, due to the questionable durability of the materials.

The effects of the Industrial Revolution and the World Wars had a global impact on construction methods. At the beginning of 1915, discussions involving politicians, technicians and the industry were taking place in the United States about the need to establish building codes to safeguard public health and safety. These concerns were also addressed in many other countries in Europe, some of them in an earlier stage in relation to their construction codes [65].

The need for quality control and, especially, for a standardisation of procedures to achieve economical profits and the maintenance of the industry itself imposed significant pressure on construction methods and the selection and creation of materials. The most astonishing new method was the progressive and wide introduction of reinforced concrete. In many countries researchers identify this fact as the one most responsible for the gradual decrease of the use of traditional construction systems. The migration from rural regions to urban centres also contributed to this situation in Europe, as reported by several international researchers, namely the ones involved in the project Terra Europae [48].

The desire for new architectural conceptions due to progressive changes in the lifestyle shattered the confines of traditional construction. Nevertheless, the dissemination of reinforced concrete and Modernist architectural shapes in seismic regions presents today potential problems for several reasons. The most important of them is the safety and the effective control of the construction process in those regions.

5.3 Group C

The architectural movement of *Art Nouveau* was the first signal of a controversial change in the way living spaces were understood, besides the resistance to the impact of the Industrial Revolution. Despite its brief existence in many countries, its contradictory objectives of wide cultural and artistic promotion in relation to the significant complexity of production, including high costs, was an important step towards the introduction of new materials and techniques in construction systems [65]. The following *Art Deco* and Modernism movements were responsible for the widespread dissemination of reinforced concrete. The models and their variations adopted in each country, however, present a similar link which can be assumed to be the most multicultural construction process yet. In some European countries, such as Italy and Portugal, both architectural movements caused a progressive change of the construction that included the use of natural materials and traditional techniques in that period of time [45]. This aspect was also observed after World War II in France and Britain, when faced with shortages of industrial materials and the need to relocate affected populations *en masse*, and led to a brief revival of earthen construction [52]. The same happened with specific rebuilding programs implemented in Germany and Poland to house affected populations after World War II in association with policies of urgent rebuilding.

In Portugal, the introduction of the reinforced concrete was gradual in some regions until the 1940s. It was first used in some elements or short beams on the main façades. Then it became more common in beams and, finally, it was also used in columns and thin slabs despite the maintenance of the loadbearing walls of adobe or stone masonry in the construction. What in some regions of Italy was considered to be the preservation of links with traditional materials and architectural volumetric configurations in order the nationalism, was seen in Portugal as a continuity of the social structure and "natural" acceptation of the new material in old frameworks to "improve" traditional systems [45]. Regionalist proposals mixing tradition and restricted innovation were adopted in Portugal from the beginning of the 20th century. However, this period had also a high experimental purpose in terms of technique and architecture which led to interconnected proposals between these two factors. For this reason, buildings with Modernist aesthetic and traditional materials (Fig. 14) were seen to appear in almost the same

Fig. 14 Timber building, modernism, Costa Nova, Ílhavo, Portugal (credit A. Tavares)

Fig. 15 Old building with reinforced concrete elements (credit A. Tavares)

period as mixed systems with reinforced concrete in old and traditional aesthetic shapes (Fig. 15).

Experimental proposals using natural materials such as earth were built by great architects as Le Corbusier with a rammed earth proposal and Frank Lloyd Wright with an adobe proposal.

It is interesting to see that mixed systems are presently applied and studied even in seismic regions as New Zealand and Peru with specific proposals.

6 Final Comments

The debate involving the interaction between the understandings of the particular character of traditional construction versus the wide range of techniques available, need an exigent and accurate sense on the intervention strategy.

The lack of knowledge about traditional construction systems, the difficulty of controlling their structural performance and the exact characterisation of their state of conservation must not be an obstacle to their preservation. This implies that significant background knowledge encompassing multidisciplinary approaches for interventions in vernacular construction systems is necessary. The compatibility of definitions must be the first step of the design process. These is one of the main reason to continue promoting the knowledge around vernacular architecture and its ancient background knowledge as achieved through written sources as from cultural exchange or practical training along the years and linked with traditions and social or production issues.

There are three key reasons to study traditional construction techniques today:

- Protection and preservation based on research and laboratory tests, strengthening solutions proposals associated to conservation actions, which, however, still present difficulties in terms of the communication of this knowledge to technicians who work in the field;
- Dissemination of the knowledge that supports multicultural values based on an understanding of the links between traditional construction methods (which can lead to an interesting debate on identity);
- Use of proper maintenance measures and compatible interventions in terms of materials, techniques and strategies that can guarantee the future use of buildings with high cultural value.

This complex understanding also involves the actual definition appliance for conservation interventions. The problem rests in the difficulty of accurately characterising an existing structure to implement the correct code standards for new actions as conservation, strengthening or retrofitting. In some cases it assumes a high level of contradiction due to the difficulties in the definition of intervention strategies, which rely on the following: the fact that the materials are already in place; the components were produced through a traditional process based mostly on practice; the building has been subject to numerous undocumented alterations; many traditional construction methods are no longer in use or understood; the lifetime of materials and components produce changes in their mechanical behaviour (some materials may require laboratory tests); and finally, the complexity of cultural value in an intervention. All of these issues create constraints on preliminary evaluations and control of the building characteristics.

Nowadays, in the restoration of vernacular architecture, it is not uncommon unfortunately to find inappropriate solutions to be adopted, such as irreversible structural changes of the timber elements or repairing renderings with cement mortars. As was discussed, the concerns on the compatibility between materials and the definition of adequate components were already presented as main issues in ancient treatises. Specific recommendations were made for aspects related to the structural safety of the constructions, putting the emphasis on the information captured from the observation of ancient buildings contributing for the better establishment of the durability requirements. Particularly, the connections of the construction elements (wall-to-wall, wall-to-floors, wall-to-roof) were issues discussed in several treatises as well in the engineered proposals of the 18th century, considering special rules on seismic prone regions. These concerns, in vernacular construction, can be observed in the foundations, corners, distance between openings (doors and windows), as well as in the choice of the materials and in the definition of the construction geometry.

The comparative study of some historic and vernacular construction systems, from different seismic prone regions, shows the relevance given in the construction to the connections between perpendicular walls to improve the global capacity.

Although the type of traditional construction systems present in some seismic regions, may achieve a considerable level of seismic capacity, other do not have guaranty it. For these structures new strengthening solutions should be proposed and applied. Nevertheless, for structural assessment or for strengthening interventions, it should be encouraged the development of knowledge concerning the evolutionary process of the construction and its relation with the surrounding constructions, in order to achieve less intrusive actions, maintaining the cultural value of the constructions.

It is necessary to spread adequate maintenance know-how for traditional construction and presently also for the buildings from the beginning of the Modern Movement which are the following cultural assumed heritage.

References

1. Schmid, T., Testa, C.: Systems Building: an international survey of methods Architecture, pp. 25–31. Pall Mall Press, London (1969)
2. Warzawski, A.: System building—education and research. construction international research. In: Proceedings of 7 CIB Congress, pp. 113–125, UK (1977)
3. Sebestyén, G.: System in construction. In: Proceedings of 8Q Congress of CIB, Building Research World Wide. Oslo, CIB/Norwegian Building Research Institute, vol. 1b, pp. 773–776 (1980)
4. NYS ITS: NYS Project Management Guidebook, New York State Office, Information Technology Services. http://www.cio.ny.gov/pmmp/guidebook2/SystemConstruction.pdf (2011). Accessed 20 Sept 2012
5. Vitruvius, M.I.: Writing the Body of Architecture. The MIT Press, Cambridge (2003). ISBN 0-262-13415-2
6. Rowland, I. (translation of Vitruvius) Vitruvius Ten Books on Architecture, commentary and illustration by Howe Thomas Noble, Cambridge University Press, Cambridge (1999). ISBN 0-521-55364-4
7. Guo, Q.: Yingzao Fashi: twelfth-century Chinese building manual. Architectural history. J. Soc. Archit. Hist. G. B. **41**, 1–13 (1998)
8. Borsi, F.: Leon Battista Alberti The complete Works, Ed. Faber and Faber/Electa, London (1989). ISBN 0-571-142028
9. Rykwert, J., Leach, N., Tavernor, R.: Leon Battista Alberti On the Art of Building in Ten Books. The Mit Press, Cambridge, (1996). ISBN 0-262-01099-2
10. Niglio, O.: Costruzioni di terra nella Calabria del XVIII secolo, in Scritti sulla terra, a cura di Eugenio Galdieri. EDA, Padova (2009)
11. Guidoboni E. Pirro Ligorio Libri di diversi terremoti. De Luca Editori D'Arte. Libri delle Antichità, Archivio di Santo di Torino. vol. 28. Cod.Ja.II.15. Roma, Italy (2005)
12. Milizia, F.: Principj di architettura civile, Bassano, Tipografia Giuseppe e Figli (1813)

13. Bélidor, B F.: La science des ingenieurs dans la conduit des travaux de fortification et d'architecture civile, A La Hate, Paris (1754)
14. França, J-A.: A reconstrução de Lisboa e a arquitectura pombalina. 3ª Ed. Biblioteca Verde. Instituto de Cultura e Língua Portuguesa, vol. 12. Ministério da Educação, Lisbon (1989)
15. Vivenzio, G.: Istoria de' tremuoti avvenuti nella Provincia della Calabria ulteriore e nella città di Messina nell'anno 1783, vol. I. Stamperia Regale, Naples, Italy (1787)
16. Barucci, C.: La casa antisismica: prototipi e brevetti: materiali per una storia delle tecniche e del cantiere. Gangemi Editore (1990)
17. Grimaldi, A.: La Cassa Sacra ovvero la sopressione delle manimorte in Calabria nel secolo XVIII, Stamperia dell'Iride, Archivio di Stato di Catanzaro, Napoli, Italy (1863)
18. Tobriner, S.: La Casa Baraccata: Earthquake-resistant construction in 18th century Calabria. J. Soc. Archit. Hist. **42**(2), 131–138 (1983)
19. Diderot, D., D'Alembert, J-R.: L'Enciclopédie ou Dictionnaire raisonné Des Sciences des Arts et des Métiers, Vol IV, Tomas XVIII-XXVIII, Suite, Readex Microprint Corporation, New York, USA (1969)
20. Grunthal, G. (ed).: European Macroseismic Scale 1998. Cahiers du Centre Européen de Géodynamique et des Séismologie, 15 Luxembourg (1998)
21. D'Ayala, D., Bostenaru Dan, M., Akut A.: Vulnerable dwelling typologies in European Countries affected by recent earthquakes. 13th World Conference on Earthquake Engineering, paper n 823, Vancouver, Canada (2004)
22. D'Ayala, D., Spence, R., Oliveira, C., Pomonis, A.: Earthquake Loss Estimation for Europe's Historic Town Centres. Earthq. Spectr. Earthq. Loss Estim. **13**(4), 773–793 (1997)
23. Payne, A.: The architectural treatise in the Italian Renaissance. University Press, Cambridge (1999). ISBN 0-521-62266-2
24. Aires, C.: Manuel da Maia e os engenheiros militares portugueses no terramoto de 1755, inclui dissertações sobre a Renovação da cidade de Lisboa por Manoel da Maya Engenheiro-Mor do Reino. Imprensa Nacional, Lisbon (1910)
25. D'Ayala, D., Yeomans, D.: Assessing the seismic vulnerability of late Ottoman buildings in Istanbul. Structural Analysis of Historical Construction, Ed. Modena, C., Lourenço, PB., Roca, P., Padova, Italy (2004). 2004-01-01
26. Boethius, A.: Etruscan and early roman architecture, revised by Roger Ling and Tom Rasmussen, Founding Ed.: Nikolaus Pevsner, The Pelican History of Art, (1978). ISBN: 0140561.447
27. Akrotiri, Palyvou C.: Thera: building techniques and morphology in late Cycladic archietcture. P.H.F. Thesis N.T.U.A, Athens (1988)
28. Touliatos, P.: Wood reinforced masonry techniques, pathology, interventions and the principles of the restoration, Restoration of old and modern wooden buildings, EU Raphael programme, Ed. pp.74–82. University of Oulu, Finland, (2000). ISBN 951-42-5658-1
29. Karydis, N.: Eresos: The house, the structure, the settlement. Ed. Papasotiriou, p. 89. Athens, Greece (2003)
30. Sakellaropoulou, M.: Seismic behaviour of vernacular Greek architecture, the development of the earthquake-resistant structural systems: the case of Lefkada Island. Master dissertation of Science in the Conservation of Historic Buildings, pp. 23–57. University of Bath, Bath (2009)
31. Bei, G.: Earthen architecture in Greece. Terra Europae, Earthen Architecture in the European Union, Ed. ETS, p. 126. Pisa, Italy (2011), ISBN 978-88-467-2957-6
32. Touliatos, P.: Seismic Behaviour of vernacular structures. Research report. pp. 13–31. National Technical University of Athens. Athens (2004)
33. Hughes, R.: Cator and Cribbage Construction of Northern Pakistan. In: Proceedings of the UNESCO-ICOMOS International Conference on the Seismic Performance of Traditional Buildings, Istanbul, Turkey 16-18 November 2000; http://www.icomos.org/iiwc/seismic/Hughes-C.pdf (2000). Accessed 19 Sept 2012
34. Croci, G., D'Ayala, D., Conforto, M.L.: Studies to evaluate the origin of cracks and failures in the history of Colosseum in Rome. Nueva Grafica SAL, Canaries (1992)

35. Bianco, A., Cartella, V., Guastella, S., Suraci, A., Surac, G., Tuzza, S., Uccellini, E.: La "Casa Baraccata", Ed. GB Editoria, Rome, Italy (2010). ISBN 978-88-95064-41-3

36. Mecca, S., Dipasquale, L.: Earthen architecture in South-eastern Europe. Terra Europae, Earthen Architecture in the European Union, Ed. ETS, pp. 77–81. Pisa, Italy (2011). ISBN 978-88-467-2957-6

37. Tobriner, S.: Response of traditional wooden japanese construction, Seismological and Engineering aspects of the 1995 Hyogoken – Nanbu (Kobe) Earthquake. Berkeley: Earthquake Engineering Research Center, cap 5 pp. 61– 80. Report n UCB/EERC – 95/10 (1995)

38. Karababa, F.: Local seismic construction practices as means to vulnerability reduction and sustainable development, a case-study in Lefkada Island Greece. Ph.D. thesis, University of Cambridge, Cambridge (2007)

39. Diren, D., Aydin, D: Traditional Houses and Earthquake. http://www.icomos.org/iiwc/seismic/Diren.pdf (2000). Accessed 19 Sept 2012

40. Sahin, N.: A study on Conservation and Rehabilitation problems of historic timber houses in Ankara. Unpublished Ph.D. Thesis, METU, Ankara, Turkey (1995)

41. Akan, A.: Some observations on the seismic behaviour of traditional timber structures in Turkey, Master Thesis (2004)

42. Isik, B.: Seismic Rehabilitation study in Turkey for existing Earthen constructions, Proceedings of the Getty Seismic Adobe Project 2006 Colloquium, pp. 93–100. Los Angeles, USA (2006)

43. Lagenbach, R.: Survivors in the Midst of Devastation, Traditional Timber and Masonry Construction in Seismic Areas. http://www.conservationtech.com (2003). Accessed 19 Sept 2012

44. Tavares, A., Costa, A., Varum, H.: Study of common pathologies in composite adobe reinforced concrete constructions. J. Perform. Constr. Facil. **26**(4), 389–401 (2012). doi 10.1061/(ASCE)CF.1943-5509.0000200 (2012)

45. Tavares, A., Costa, A., Varum H.: The evolutionary process of building heritage influenced by the architecture/engineering borderline decisions. In: ICSA 2013 Conference, Guimarães, Portugal, paper accepted in December 2012 (2013)

46. PREDES Organization – Centro de Estudios y Prevención de Desastres. (2001) Proyect o de Reconstrucción de vivendas y canales de riego en zonas rurales de Sánchez Cerro, Moquegua, http://www.predes.org.pe/tquincha.htm (2001). Accessed 24 Jan 2010

47. Tavares, A., Costa, A., Varum, H.: Adobe and Modernism in Ílhavo, Portugal. J. Archit. Heritage Conserv. Anal. Restor. **6**(5), 525–541 (2012). doi:10.1080/15583058.2011.590267

48. Terra Europae: Earthen Architecture in the European Union, Ed. ETS. Pisa, Italy (2011). ISBN 978-88-467-2957-6

49. Llamps, R., Ioannow, I., Castrillo, M., Theodosiou, A.: Earth construction in Cyprus,Terra Europae, Earthen Architecture in the European Union, Ed. ETS, p. 97. Pisa, Italy (2011). ISBN 978-88-467-2957-6

50. Vegas, F., Mileto, C., Cristini, V.: Earthen architecture in East Central Europe: Czech Republic, Slovakia, Austria, Slovenia, Hungary and Romain. Terra Europae, Earthen Architecture in the European Union, Ed. ETS, p. 68. Pisa, Italy (2011) ISBN 978-88-467-2957-6

51. Dipasquale, L., Mecca, S.: Earthen architecture in Italy. Terra Europae, Earthen Architecture in the European Union, Ed. ETS, p. 137. Pisa, Italy (2011) ISBN 978-88-467-2957-6

52. Guérin, R., Schroeder, H., Jorchel, S., Kelm, T.: Earthen architecture in central Europe: Germany and Poland. Terra Europae, Earthen Architecture in the European Union, Ed. ETS, p. 61. Pisa, Italy (2011). ISBN 978-88-467-2957-6

53. Chabenat, M., Cook, L., O'Reily, B.: Earthen architecture in Northwestern Europe: Ireland, United Kingdom and Northern France. Terra Europae, Earthen Architecture in the European Union, Ed. ETS, p. 49–53. Italy (2011). ISBN 978-88-467-2957-6

54. Ribeiro, O.: Geografia e Civilização. Ed. Livros Horizonte, p. 39. Lisbon, Portugal (1969)

55. Correia, M., Merten, J., Vegas, F., Mileto, C., Cristini, V.: Earthen architecture in Southwestern Europe: Portugal, Spain and Southern France. Terra Europae, Earthen Architecture in the European Union, Ed. ETS, p. 71–75. Pisa, Italy (2011). ISBN 978-88- 467-2957-6

56. Houben, H., Guillaud, H.: Earth Construction: a Comprehensive Guide. Intermediate Technology Publications, London (1994)
57. Tavares, A., Costa, A., Varum, H.: Transition period of the contemporary constructive system in the Modern Movement. In: 9th Ibero-American Seminar on Architecture and Earth constructions, 6th National Earth Architecture Seminar, University of Coimbra, Portugal (2010)
58. Zeren, M T., Karaman, O Y.: Analysis of construction system and damage assessment of traditional Turquish house—Case study of timber framed Kula houses. In: SHATIS' International Conference on Structural Health Assessment of Timber Structures, Lisbon, Portugal (2010)
59. Tuomic R.L. (1979) Moody R.C.: Historical considerations in evaluating timber structures, General Technical Report FPL21, Forest Products Laboratory, Forest Service, U.S. Department of Agriculture, Madison, Wisconsin, USA
60. Lewis, N.B.: Building conservation and Craftsmanship of Traditional Timber Structures. ISS Institute/DEST Fellowship Report, Australia (2006)
61. Beusekom, JW van. (2006) Building in wood in the Netherlands – From early medieval times until World War II, problems and possibilities of preservation. ICOMOS, Netherdlands/ Foundation National Contact Monuments, Proceedings of the 15th International Symposium of the IIWC, Istanbul, Turkey
62. Doğangün, A., Tuluk, Ö.Ġ., Livaoğlur, R., Acar, R.: Traditional wooden buildings and their damages during earthquakes in Turkey. Eng. Fail. Anal. **13**(6), 981–996 (2005)
63. ICOMOS: Preliminary notice. In: ICOMOS International Wood Committee Symposium in Turkey. pp.18–23, Istanbul, Turkey Sept (2006)
64. SeçKin, N.: (2012) An assessment on traditional timber structures in Suleymaniye and Zeyrek districts of Historical Peninsula. ITU A|Z. **9**(1), 56–69 (2012)
65. Tavares, A., Costa, A., Varum, H.: Analysis of the structural behaviour of Art Nouveau heritage (1900–1919) in Ílhavo, Portugal. In: Proceedings of the XIth International Conference on the Study and Conservation of Earthen Architecture, Terra 2012, Lima, Peru (2012)

Structural Behaviour and Retrofitting of Adobe Masonry Buildings

Humberto Varum, Nicola Tarque, Dora Silveira, Guido Camata,
Bruno Lobo, Marcial Blondet, António Figueiredo,
Muhammad Masood Rafi, Cristina Oliveira and Aníbal Costa

Abstract Earth is one of the most widely used building materials in the World. Different types of adobe dwellings are made to assure protection and wellbeing of the population according to the diverse zones needs. Therefore, it is important

H. Varum (✉) · D. Silveira · B. Lobo · A. Figueiredo · C. Oliveira · A. Costa
Civil Engineering Department, Aveiro University, 3810-193, Aveiro, Portugal
e-mail: hvarum@ua.pt

D. Silveira
e-mail: dora.silveira@ua.pt

B. Lobo
e-mail: bmlobo@ua.pt

A. Figueiredo
e-mail: ajfigueiredo@ua.pt

C. Oliveira
e-mail: cferreiraoliveira@ua.pt

A. Costa
e-mail: agc@ua.pt

N. Tarque · M. Blondet
Civil Engineering Section, Engineering Department, PUCP, 1801,
San Miguel, Lima, Peru
e-mail: sntarque@pucp.edu.pe

M. Blondet
e-mail: mblondet@pucp.pe

G. Camata
University "G. D'Annunzio", 65129 Pescara, Italy
e-mail: camata@unich.it

M. M. Rafi
Department of Earthquake Civil Engineering Department,
NED University of Engineering and Technology, Karachi 75270, Pakistan
e-mail: rafi-m@neduet.edu.pk

A. Costa et al. (eds.), *Structural Rehabilitation of Old Buildings*,
Building Pathology and Rehabilitation 2, DOI: 10.1007/978-3-642-39686-1_2,
© Springer-Verlag Berlin Heidelberg 2014

to study the structural behaviour of the adobe masonry constructions, analysing their seismic vulnerability, which may help in preventing social, cultural and economic losses. In the present chapter, an explanation of the seismic behaviour of adobe buildings, a summary of recent research outputs from experimental tests conducted on adobe masonry components and from numerical modelling of full-scale representative adobe constructions are reported. In addition, different rehabilitation and strengthening solutions are presented and results from the testing of retrofitted adobe constructions and components are discussed.

Keywords Adobe • Masonry • Seismic vulnerability • Mechanical characterization • Numerical modelling • Retrofitting solutions

1 Introduction

Adobe derives from the Arabic word *atob*, which literally means sun-dried brick, being one of the oldest and most widely used natural building materials, especially in developing countries (Latin America, Middle East, north and south of Africa, etc.), many of which are also characterized by moderate to high seismic hazard. The use of sun-dried blocks dates back to approximately 8000 B.C., and until the end of the last century it was estimated that around 30 % of the World's population lived in earth-made constructions [1]. Adobe construction presents some attractive characteristics, such as low cost, local availability, the possibility to be self/owner-made with unskilled labour (hence the term "non-engineered constructions"), good thermal insulation and acoustic properties [2]. Adobe buildings present high seismic vulnerability due to the low tensile strength and fragile behaviour of the material, which constitute an undesirable combination of mechanical properties. Earthen structures are massive and thus attract large inertia forces during earthquakes; on the other hand, these structures are weak and cannot resist large forces. Additionally, this type of construction has a brittle behaviour and may collapse without warning [3]. The seismic capacity of an adobe house depends on the mechanical properties of the materials (blocks and joints), on the global structural system (structural geometry, connections, etc.), on building foundations, and also on the quality of the construction and maintenance [4]. Each time an earthquake occurs in a region with abundant earth-construction, enormous human, social and economic losses are recorded, as has been the case in El Salvador (2001), Iran (2003), Peru (1970, 1996, 2001 and 2007), Pakistan (2005), and China (2008 and 2009).

2 Adobe Constructions in the World

As previously said, 30 % of the World's population lives in earth-made constructions [1], with approximately 50 % of them located in developing countries. Adobe constructions are very common in some of the World's most

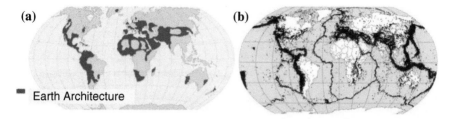

Fig. 1 Map of earthen constructions around the World and distribution of earthquake epicentres. **a** Distribution of earthen constructions [5]. **b** Earthquake epicentres (1963–1998) [6]

hazard-prone regions, such as Latin America, Africa, the Indian subcontinent and other parts of Asia, Middle East and Southern Europe [5]. Figure 1 compares the distribution of earthen constructions with the seismic hazard distribution in the World.

2.1 Typology of Adobe Dwellings in Southern Europe

A wide range of earth buildings can be found in Europe, particularly in Southern regions, as Portugal, Italy and Spain. But, in countries as Germany, England and France it can be also found a significant number of earth buildings. In France, for example, 15 % of the population lives in earth buildings, constructed with different techniques like adobe and rammed earth (*'pisé'*) 0.

Italy, being one of the countries with the tradition in adobe construction, has a very pronounced historic and cultural earth built heritage. In old houses in urban areas, the load-bearing walls at ground floor are made mainly of stone masonry, in the intermediate floors walls are made of stone and/or adobe masonry, and in the upper floor walls are made of adobe masonry. In certain regions, the rural houses have their foundations made of stone and the walls of rammed earth or adobe masonry [7].

In Portugal, earth construction dates back to several hundreds of years ago, located mainly in the centre and south coastline [8–10]. These buildings are examples of the vernacular architecture, part of the Portuguese built heritage. In the district of Aveiro, a particular type of adobe was produced and used in construction. In this region, adobe blocks were stabilized with lime, to improve its mechanical strength and durability. In this region, the use of adobe in the construction was widely applied until mid of the twentieth century. After, the reinforced concrete as being used as the main building material/solution. The typical adobe houses in urban areas are two or three storey buildings (Fig. 2), sometimes covered with high valued tiling. In the city of Aveiro, it can be found several examples of buildings influenced Art-Nouveau architectural style [11, 12]. The majority of the constructions in rural areas are simpler and with only one storey, as can be seen in Fig. 3.

Fig. 2 Historical constructions in urban areas of Aveiro city

Fig. 3 Typical rural house in Aveiro surroundings

2.2 Other Typologies of Adobe Dwellings in the World

In South America, adobe is mainly used by low-income families. Adobe houses are predominant in rural areas, though they also exist in urban areas generally with a better quality of construction.

The typology of adobe dwellings is similar in most countries, with a rectangular plan, single door entry and small lateral windows. The walls are made with adobe blocks connected with mud mortar. The stucco is made with mud and sometimes mixing mud and gypsum. The foundation, if present, is made of medium to large stones joined with mud or coarse mortar. The roof supporting structure is made of wood joists resting directly on the walls or supported inside indentations on top of the walls. The type of roof covering depends principally on the family incomes and on the location of the constructions, and could be made of corrugated zinc sheets or clay tiles [13], this last preferable at the Peruvian highland.

In Peru, the percentage of the country's population living in earth dwellings has gone down from 54 to 43 % in the last 15 years [14] with most of the concentration in urban and coastline areas. Figure 4 shows some typical Peruvian adobe dwellings located at the countryside. But also in Lima, capital of Peru, some Colonial houses made of adobe can be found (Fig. 4c). These houses are typically characterized by a first floor made with adobe and clay bricks, and the second and third floors made with *quincha* (wooden frame with infill of mud and cane). Due to fragilities presented by adobe construction, some South American countries have been restricting its construction.

In Asia and Middle East (e.g.: Pakistan, Iran), see Fig. 5, adobe is a very traditional construction material. Luxurious adobe residences are constructed by wealthy families, and modest adobe houses by poor families.

In Iran, walls from rural constructions are usually built with adobe mixed with mud, stone, wood or bricks and concrete blocks [15]. In more than 4 million rural houses, at least 26 % have adobe and mud walls. Adobe dwellings in Iran are mainly characterized by the type of roof. The most common types are the vault

Fig. 4 Typical Peruvian adobe houses. **a** Adobe house located at the peruvian coast. **b** Adobe house in peruvian highland. **c** Colonial adobe houses of two and three storeys in Lima, Peru

Fig. 5 Adobe houses in Pakistan. **a** Adobe house in Sindh. **b** Adobe house in Khuzdar (*Courtesy* Prof Sarosh Lodi)

and dome roofs. Vault roofs are built following a semi-cylindrical shape, with two plates or semi-spherical caps at their ends, being locally called '*tharby*'. When the vault covers only part of the roof it is called '*kalil*'. Other type of roofs are the quadripartite arched and crescent-arched roofs, in addition to the typical flat roof formed by wooden beams covered with branches of trees and mud.

In the northern regions of Africa, like in Morocco, the typologies of earth buildings vary from place to place. In Morocco, the Atlas chain divides the type of adobe dwellings. In the Drâa region, the constructions are built in fortified villages known as '*ksur*'. In these villages, the external walls are made from a technique called '*pisé*' (rammed earth), which allows the construction of very high walls, depending on the wall width. In this type of construction, adobe is used in the construction of columns and decorative elements both inside the patios and at the top of buildings [16].

3 Fragilities of Adobe Constructions

As stated before, adobe has been used in construction since ancient times due to cultural, climatic and economic reasons. Adobe structures can be in fact durable, but they present fragilities that have to be accounted for [9, 17].

Particular care should be taken with adobe constructions in high rainfall or humid areas, due to the susceptibility to of the materials to water and humidity. To avoid rising damp from the soil, an adequate solution at the foundation should be considered. Adobe masonry walls may also suffer erosion due to the wind actions. In order to protect adobe houses from the rain and winds, the covering of the exterior adobe masonry walls should be restored regularly to prevent the development of more severe damages, cracks and crumbling.

But, one of the most important fragility of adobe houses is related to their limited capacity to resist to earthquake demands, presenting poor behaviour for moderate to strong ground shakings, as observed in previous earthquakes, with important human losses and structural damages associated to adobe constructions.

As previously referred, the capacity of an adobe construction to resist earthquakes depends on the individual adobe block and mortar characteristics, on the mechanical behaviour and characteristics of the masonry system (considering blocks and joints), dimensions of the adobe wall (especially its thickness), building location and building geometry, as well as on the quality of construction and maintenance [4]. The extent of damage to an adobe structure depends on several factors such as: the severity of the ground motion; the geometry of the structure; the overall integrity of the adobe masonry; the existence and effectiveness of seismic retrofit measures; and the structure conservation state when the earthquake strikes.

The seismic vulnerability of adobe buildings is mainly associated to the perverse combination of the mechanical properties of their materials (low tensile strength and rupture in a brittle manner) with high density of their walls. As a consequence, every significant earthquake that has occurred in regions where earthen construction is common has produced life losses and considerable material damage (see examples in Fig. 6).

From the damage survey carried after the Peruvian earthquake of 2007 (Mw 7.9, 510 fatalities), it was concluded that the most common failure observed in non-reinforced earthen buildings, especially in those with slender walls, was the overturning of the façade walls and their collapse onto the street [18]. This happened because the effectiveness of the wall connection at the intersection between the façade wall and the perpendicular walls was too low to withstand the earthquake demands. The walls collapsed as follows: first vertical cracks occur at the wall's corners, originating damage in the adobe blocks in that area (Fig. 7). This triggered the walls to disconnect until finally the façade wall overturns. This is the most common collapse mechanism of adobe walls under earthquake actions. Observations made after the Peruvian earthquake have shown that the magnitude of damage suffered by the buildings was directly related to quality of the

Fig. 6 Destruction of adobe houses due to earthquakes. **a** El Salvad 2004 [4]. **b** Pisco, Peru 2007

Fig. 7 Vertical crack at the
corner of an adobe house,
Pisco earthquake, Peru, 2007

Fig. 8 Collapse of adobe houses during the Pisco earthquake in 2007, Peru. **a** Roof supported on
the facade walls. **b** Roof supported by transversal walls

connection between the roof's wooden joists and the top of the façade walls. When
the roof joists were supported by the façade wall, its collapse affects the roof support conditions, and finally the roof to collapse as well (Fig. 8a). If, on the other
hand, the joists were supported by the walls that were perpendicular to the façade
wall, the roof didn't fall apart (Fig. 8b).

Lateral seismic forces acting within the plane of the walls generate shear forces
that produce diagonal cracks, which usually—but not always—follow stepped
patterns along the mortar joints. The diagonal cracks often start at the corners
of openings, such as doors and windows, due to the stress concentration at these
locations (Fig. 9). If the seismic movement continues after the adobe walls have
cracked, the wall breaks into separate pieces, which may collapse independently in
an out-of-plane mode.

Dowling [4] makes a brief description of the common damage patterns of
adobe dwellings based on a damage survey carried out after El Salvador earthquake (Mw 7.7, 825 fatalities) in 2001, where more than 200,000 adobe houses
were severely damaged or collapsed.

Fig. 9 Typical X-cracks on adobe walls due to in-plane actions

As exemplified in Fig. 10, the more common damage patterns can be summarized as follows:

- Vertical cracking at corner angles associated to large relative displacement between orthogonal walls. This type of failure is very common because demands are largest at the wall–wall interface. Cracking occurs when the material strength is exceeded in either shear or tension;
- Vertical cracking and overturning of upper part of wall panel. Bending about the vertical axis causes a splitting-crushing cycle generating vertical cracks in the upper part of the wall;
- Overturning of wall panel due to vertical cracks at the wall intersections. Here the wall foundation interface behaves as a pin connection, which has little strength to overturning when an out-of-plane force is applied. This type of failure has been seen in long walls without other lateral restraints along the wall;
- Inclined cracking in walls due to large in-plane demands, which generates maximum tensile stresses in directions of about 45° relatively to the horizontal;
- Dislocation of corner. Initial failure is due to vertical corner cracking induced by shear or tearing stresses. The lack of connection at wall corners allows greater out-of-plane displacement of the wall panels, which generates a pounding impact with the orthogonal wall. The top of the wall is subjected to larger displacements, which tends to cause larger pounding, thus inducing greater stresses that may lead to failure;
- Horizontal cracking in upper section of wall panels, and displacement and deformation of the roof structure;
- Falling and slipping of roof tiles.

Webster and Tolles [19] conducted a damage survey on 20 historic and 9 older adobe buildings in California after the Northridge earthquake (Mw 6.7, 60 fatalities), in 1994. They concluded that ground shaking levels between 0.1 and 0.2 g PGA are necessary to initiate damage in well-maintained, but otherwise unreinforced, adobes. This study confirms that the most typical failure mechanism is due to out-of-plane flexural damage (Fig. 11). These cracks initiate as vertical cracks at the intersection of perpendicular walls, extending vertically or

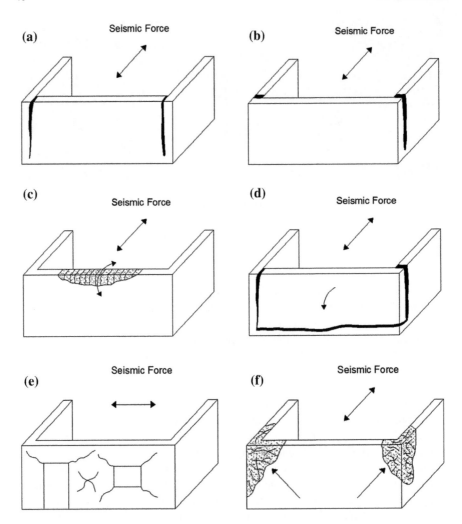

Fig. 10 Typical damage and failure mechanisms on adobe masonry constructions [4]. **a** Vertical corner angle cracking due to shear forces. **b** Vertical corner angle cracking due to to out-of-plane forces. **c** Diagonal cracking in wall due to in-plane shear forces. **d** Global overturning of wall panel. **e** Diagonal cracking in wall due to in-plane shear forces. **f** Sequence leading to corner dislocation

diagonally and running horizontally along the base between the transverse walls. Then, the wall rocks out-of-plane, back and forth, rotating round the horizontal cracks at the base. The gable-wall collapse is more specific for historic buildings. For long walls, the separation of these walls with the perpendicular ones results in out-of-plane moving of the wall. Diagonal cracks (X-shape) result from shear forces in the plane of the wall, and these cracks are not particularly serious unless the relative displacement across them becomes large. When the building is located at the corner of a building aggregate, some diagonal cracks appear at

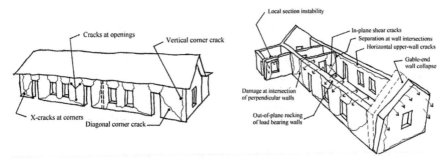

Fig. 11 Typical damages observed in adobe buildings after the Northridge earthquake [19]

exterior walls since they form wedges that can easily move sideways and downward as the building shakes, and also vertical cracks at the intersection of walls due to the out-of-plane actions.

4 Mechanical Characterization of Adobe Units and Masonry Panels

The properties of the adobe material are quite scattered and their values principally depend on the type of soil used for the fabrication of the blocks. In addition, the property values change if some binder or additive is included in the composition of bricks or mortar. Results of tests performed at two universities are described as follows showing the variation that can occur in terms of the material properties and mechanical behaviour of the adobe masonry.

4.1 Tests on Adobe Walls

4.1.1 Tests Performed at Aveiro University

A series of tests on 10 adobe masonry walls was developed at the Civil Engineering Department of Aveiro University. The adobe walls presented dimensions of $1.26 \times 1.26 \times 0.29$ m^3 and were built using adobe blocks from a demolition in Aveiro region and lime mortar formulated in the laboratory with a composition similar to the traditionally used [9].

Tests were performed according to the recommendations of ASTM E519 [20] and EN 1052-1 [21]. Five walls were tested in diagonal compression and the other 5 walls were tested in compression perpendicular to the horizontal joints. Tests were performed in a closed reaction frame using a servo-hydraulic actuator with a maximum load capacity of 300 kN to impose the displacements on the walls. The deformations of the walls were recorded during the test with a set of displacement potentiometers.

The adobe blocks used presented mean dimensions of $46 \times 32 \times 12$ cm³, specific weight of approximately 15 kN/m³, mean compressive strength of 0.56 MPa and mean tensile strength of 0.13 MPa. A detailed analysis and discussion on the mechanical properties of the adobe blocks from the region of Aveiro can be found in [22–24].

Mean shear strength of 0.026 MPa and mean modulus of rigidity (shear modulus) of 40 MPa were obtained in the diagonal compression tests. The results in terms of stress versus deformation measured in both directions (vertical and horizontal) during the test are presented in Fig. 12.

The other five walls were tested in compression perpendicular to the horizontal joints. The scheme of the test and the distribution of instrumentation are presented in Fig. 13. From the tests, a mean compressive strength of 0.33 MPa and a mean modulus of elasticity of 664 MPa were obtained.

4.1.2 Tests Performed at the Catholic University of Peru

For the evaluation of the tensile strength, 10 square wallets of $0.6 \times 0.6 \times 0.2$ m³ were built using $0.2 \times 0.40 \times 0.08$ m³ adobe bricks, which imply 6 layers of 1 ½ adobe bricks [25]. The load was applied at two opposite corners of each wallet. Instrumentation to measure the diagonal deformations was applied in each adobe panel, and was used to compute the shear modulus. For a second group of tests [26], 7 panels were built and tested. More precise equipment for load application and to read deformations was used. From all the tests, the mean maximum shear strength stress was 0.026 MPa and the mean modulus of rigidity was 39.8 MPa, similar to the values reported by Aveiro University.

For the evaluation of the compressive strength of adobe prisms, a total of 120 samples were built by Blondet and Vargas [25] and Vargas and Ottazzi [26].

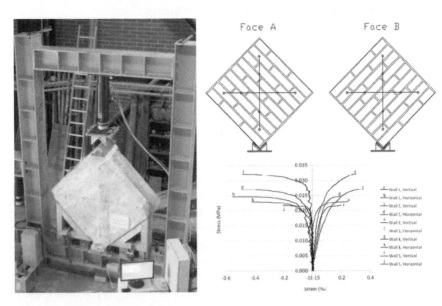

Fig. 12 Diagonal compression tests

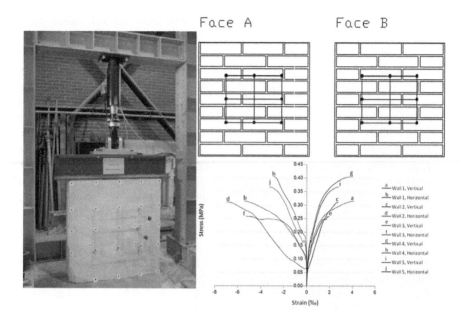

Fig. 13 Perpendicular compression tests

Specimens presented different slenderness ratios (thickness:height): 1:1, 1:1.5, 1:2, 1:3, 1:4 and 1:5. The adobe bricks had dimensions of $0.20 \times 0.40 \times 0.08$ m^3, and were laid on top of each other with mortar in between: 89 specimens were built with mud mortar and 31 with a combination of cement, gypsum and mud mortar.

Only the tests on adobe prisms built with mud mortar are reported here. The irregularity of the top surface of each prism was corrected by adding a cement/ sand mortar. Two steel plates of $0.20 \times 0.40 \times 0.02$ m^3 were placed at both ends of each pile and were loaded axially. The axial load was applied perpendicularly to the joints with 2.45 kN increments until failure of the specimen. The test was force controlled. The axial deformation was measured in each prism tested. In all cases, the observed failure was brittle, and cracks did not follow a common pattern. As a preliminary conclusion it was established that the compression strength for prisms of slenderness 1:4 is between 0.80 and 1.20 MPa, depending on the specimen's age. The modulus of elasticity computed from full adobe wall tests was 170 MPa.

5 Tests on Full-scale Structures

5.1 Double-T wall Tested at Catholic University of Peru

Blondet et al. [27] carried out a displacement controlled cyclic test (pseudo static push–pull) on a typical adobe wall at the Catholic University of Peru. With the first test it was intended to analyse the cyclic response of the wall and the damage

pattern evolution caused by in-plane forces. The wall presented a double-T shape
in-plan view (Fig. 14a), and the main longitudinal wall (with a central window opening) was 3.06 m long, 1.93 m high and 0.30 m thick. The structure also
included two 2.48 m long transverse walls that were intended to: (a) simulate the
influence of the connection between transversal walls found in typical buildings;
(b) avoid rocking due to in-plane actions. The specimen was built on a reinforced
concrete continuous foundation beam. A reinforced concrete beam was built at the
top of the adobe wall to provide gravity loads corresponding to a roof composed
of wooden beams, canes, straw, mud and corrugated zinc sheet. The top beam also
ensures a more uniform distribution of the horizontal forces applied to the wall.
The window lintel was made of wood. The horizontal load was applied in a series
of increasing load cycles. Loads and displacements were applied slowly in order
to avoid dynamic effects. Each displacement cycle was repeated twice. During the
test, the cracks started at the windows' corners and advanced diagonally up to the
top and down to the base of the wall. During reversal loads, the cracks generated the
typical X-shape crack due to in-plane forces. Figure 14b shows the crack patterns
that settled in the adobe wall during the tests. Considering, as control displacement,
the lateral displacement at the base of the top beam, the cracks observed after two
consecutive cycles for the displacement levels of 1, 2, 5 and 10 mm are marked in

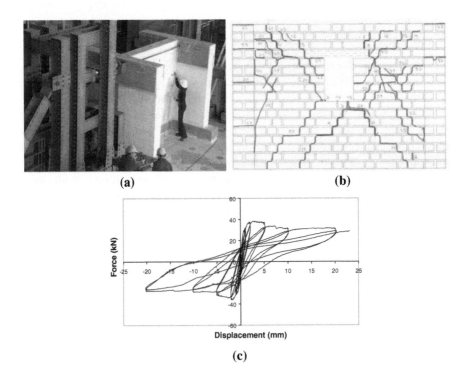

Fig. 14 Cyclic test carried out on a double-T wall [27]. **a** Tested wall. **b** Damage pattern evolution during the cyclic test. **c** Cyclic test carried out on a double-T wall [27]

Fig. 14b. After an imposed displacement of 10 mm at the top of the wall, it was observed sliding of the adobe wall panels, as can be interpreted from Fig. 14c.

5.2 Dynamic Tests on Adobe Modules

A dynamic test analyses the response and the damage pattern evolution of adobe masonry structures when subjected to seismic actions. The unidirectional dynamic test was performed on an adobe module built on a reinforced concrete ring beam to simplify the anchorage of the specimen to the unidirectional shaking-table [3]. With this module it was intended to represent a typical Peruvian adobe construction located at the coast. The total weight (module + foundation) was approximately 135 kN. The weight of the concrete beam was 30 kN. The adobe bricks and the mud mortar used for the construction of the module had a soil/coarse sand/straw volume proportion of 5/1/1 and 3/1/1, respectively. The module consisted of four walls 3.21 m long, identified as right, left, front and rear wall. The thickness of the walls was 0.25 m, except for the right wall which had a thickness of 0.28 m because it was plastered with mud stucco (Fig. 15).

The adobe module was subjected to three levels of unidirectional displacement signals, which were scaled to present maximum displacements ate the base

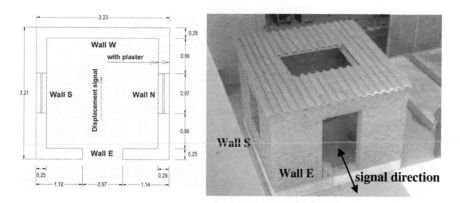

Fig. 15 Adobe module tested on PUCP shaking-table. **a** Plan dimentions. **b** 3D view

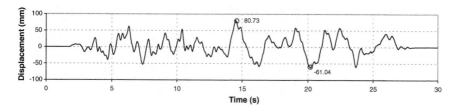

Fig. 16 Displacement input used in Phase 2 and scaled to present maximum amplitude of 80 mm

of 30 (Phase 1), 80 (Phase 2, Fig. 16) and 120 mm (Phase 3), to comparatively represent the effects of a frequent, moderate and severe earthquake. The input signals were scaled from an acceleration record of the Peruvian earthquake occurred in 1970.

At the end of Phase 1 and during Phase 2 typical vertical cracks appeared at the walls intersections causing the separation of the walls (Fig. 17), as typically occurs during moderate ground motions. Subsequently, X-shape cracks appeared at the longitudinal walls, and cracks appeared at the transverse walls due to horizontal and vertical bending. The anchorage of the steel nails that connected the wooden beams to the walls was lost during the movement, and as a result the roof was supported by the walls just through its own weight and friction. Major damage was observed at the end of Phase 2 and total collapse was observed during Phase 3.

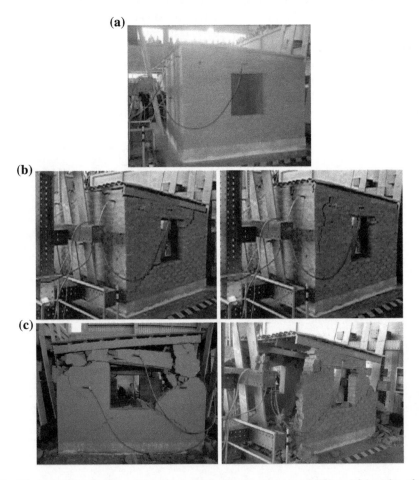

Fig. 17 Views of the adobe module during and after the dynamic test [3]. **a** Adobe odue after Phase 1. **b** Snapshots of the adobe module during Phase 2. **c** Adobe modue after Phase 3

Wall W, which was itself broken more or less into 3 big blocks (typical of walls supported only by three sides), presented a rocking behaviour due to out-of-plane actions. During Phase 3, with a maximum displacement at the base of 130 mm, the walls perpendicular to the movement (walls W and E) fell down at the beginning of the input signal, while the parallel walls S and N were completely cracked (Fig. 17). As the roof was supported by the lateral walls, it did not collapse. The formation of vertical cracks caused the separation of walls, allowing them to move independently.

5.3 Double-T wall Tested at Aveiro University

With the objective of conducting a thorough evaluation of the performance of adobe structures with and without seismic retrofit, an experimental study with a double-T wall was carried out at Aveiro University. A full-scale adobe wall was built in the Civil Engineering Laboratory using adobe blocks from a demolition in Aveiro region. A series of pseudo-static cyclic in-plane tests was carried out on the wall in order to evaluate and characterize the existing adobe construction in the region.

With the intention of considering the influence of adjacent walls, the wall was built in the shape of double-T in-plan view (Fig. 18a). The real-scale adobe wall presented the following dimensions: height of 3.07 m, length of 3.5 m and mean thickness of 0.32 m. The adobe blocks used in the construction of the wall presented mean dimensions of $24 \times 44 \times 12$ cm^3, specific weight (approximately) 15 kN/m^3, mean compressive strength of 0.42 MPa, and mean tensile of 0.14 MPa.

A vertical uniform load was added at the top of the wall through an equivalent mass of 20 kN to simulate the common dead and live loads on typical adobe constructions. A cyclic horizontal demand of increasing amplitude was applied 2.5 m above the base of the wall, until failure.

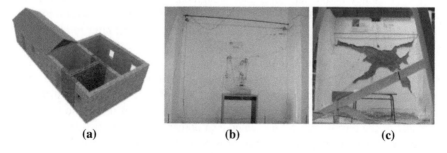

| (a) | (b) | (c) |

Fig. 18 Cyclic test on the double-T wall. **a** 3D View of the idealized adobe wall. **b** Final damage pattern of the original wall. **c** Final damage pattern of the strengthened wall

From the cyclic tests, the maximum lateral force obtained was 58.14 kN, with a corresponding shear strength capacity of 57.28 kPa and a maximum drift of 0.61 %.

After the first cycle, the wall's strength registered a strong decrease, with an important stiffness reduction. The failure mode was fragile, as expected for adobe constructions. An important factor for the decrease in the strength of a masonry wall is the strength capacity of the bond between the mortar and the adobe blocks. The initial development of cracks is mainly in the diagonal direction (Fig. 18b) [28, 29].

After the first experimental test series, the wall was repaired by pressure injection of a hydraulic lime gum into the cracks. Afterwards, the original plaster was removed, and a synthetic mesh was applied on the surface of the wall. The mesh was fixed to the wall with PVC angle pins and angle profiles, using highly resistant nylon thread on all concave vertices of the wall [28].

After repairing and retrofitting, the strengthened wall was tested using the same test procedure described for the original (non-strengthened) wall.

In the second test, the maximum shear strength of the wall was approximately 70.69 kPa with a corresponding force of 71.75 kN. The shear strength obtained for 1 % drift was approximately 45 kPa (70 % of its maximum shear strength) and the maximum imposed deformation was 1.6 % with a corresponding displacement of 45 mm. During the first cycles the wall response was almost linear, even though some small cracking occurred (Fig. 18b).

In order to evaluate the efficiency of the retrofit solution, a stress-drift plot was built with the results obtained in the two tests conducted on the original and retrofitted wall. With this plot it is possible to compare the wall responses before and after strengthening.

After repair and strengthening, the stiffness of the wall improved becoming very close to the stiffness of the original wall. The maximum resistant shear capacity of the wall increased 23.43 % after retrofit and the maximum deformation tripled.

The fragility of the wall post peak force decreased, and the ductility and energy deformation capacity increased. In consecutive cycles, a lower degradation of strength was observed.

The efficiency of the repairing and strengthening measures conducted on the wall was also evaluated by the observation of the values of the natural frequency of the wall before testing, in the original state of the wall and after reinforcement (Fig. 19b) [30].

By analysing the values obtained for the first frequency of the wall before and after retrofitting, it is clear that the rehabilitation process restored the original stiffness corresponding to the undamaged wall. The first frequency displayed in the graph corresponds to the wall before the beginning of the cyclic tests. The response of the retrofitted wall presented a smoother decrease of stiffness and consequently of natural frequency. Hence, it is possible to conclude that the repairing and strengthening measures performed are beneficial to the behaviour of the wall when subjected to horizontal displacements.

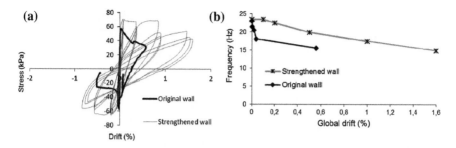

Fig. 19 Comparative results obtained in the test performed on the original and strengthened wall. **a** Comparative response of the original and strengthened adobe wall (stress versus drift). **b** First natural frequency evolution of the original and strengthened adobe wall

These significant improvements suggest that this solution can be used efficiently by construction and rehabilitation companies in the preservation and strengthening of existing adobe structures.

6 Numerical Modelling: Actual Knowledge and Needs for Research

6.1 Introduction

Masonry is a composite material made of bricks and mortar joints, each of the constituents with its own material properties. The level of accuracy in the numerical models strongly depends on the knowledge of the material properties (e.g. constitutive laws, isotropic or orthotropic behaviour, etc.), on the type of analyses conducted (e.g. linear, nonlinear), on the model used (e.g. shell elements, brick elements), and on the solution scheme adopted (e.g. implicit or explicit). Since adobe material is almost brittle, elastic analysis can give only information on the first cracking zones and not on the cracking process and cracking developing. In order to properly describe the seismic behaviour it is necessary to conduct nonlinear analyses.

The description of the nonlinear behaviour of adobe masonry is more complex than in the case of other materials (e.g. reinforced concrete, steel). The non-homogeneous nature and variability of the material, the lack of information on the constituent material properties and the numerical convergence problems due to brittle behaviour are challenges that need to be overcome to properly analyse this material.

In addition, the cracking pattern observed in adobe walls subjected to horizontal loading is quite complex and difficult to predict. Generally, the mortar is weaker and softer than the bricks and therefore cracking tends to follow the mortar joints (Fig. 20a). However, sometimes failure of masonry may involve crushing and tensile fracturing of masonry units [31], in particular in adobe walls where the mortar has the same material properties as the bricks.

Fig. 20 Typical crack patterns in adobe masonry walls. **a** Stair crack shape (concentrated on mortar). **b** Vertical crack (fracture of bricks)

Detailed modelling of the masonry components, describing constituents separately, or modelling masonry as an equivalent and homogenous material are two possible options. Simplification by considering a homogeneous isotropic material for adobe is acceptable since adobe bricks and mortar are made of the same material, raw earth, and with the same binder when it is used.

6.2 Numerical Methods for Nonlinear Analysis of Adobe Structures

Numerical analysis of unreinforced masonry structures (URM) can be performed using different methods, such as: limit analysis, finite element method, discrete element method, amongst others [32–35]. Simplified approaches consist of idealizing the structure through an equivalent frame where each wall is discretized by a set of masonry panels where the nonlinear response is concentrated at the pier and spandrels [36–40]. In all cases the nonlinear information of the adobe material is important to describe properly the material behaviour. Each of the mentioned methods has advantages and disadvantages, and the analyst should adapt any of these methods according to his experience and expertise, computational facility available and data information. In the following sections some relevant methods are described.

6.3 Finite Elements Method

Following the finite element method, the analysis of masonry structures (e.g. clay brick, adobe, stone, etc.) can be classified according to the level of accuracy [33], Fig. 21:

- Detailed micro-modelling: Bricks and mortar joints are represented by continuum elements, where the unit-mortar interface is represented by discontinuous elements [41–44]. Any analysis with this level of refinement is computationally

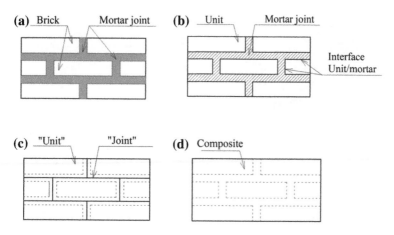

Fig. 21 Modelling strategies for masonry structures within a finite element approach [33]. **a** Masonry sample. **b** Detailed micro-modelling. **c** Simplified micro-modelling. **d** Macro-modelling

intensive and requires a well-documented representation of the properties (elastic and inelastic) of the constituents;

- Simplified micro-modelling: Bricks are represented by continuum elements, where the behaviour of the mortar joints and unit-mortar interface is lumped in discontinuous elements [34, 45–47]. This approach can be compared with the discrete element method, originally proposed by Cundall [48] in the area of rock mechanics, where a special procedure is used for contact detection and contact force evaluation [46];
- Macro-modelling or continuum mechanics finite element: Bricks, mortar and unit-mortar interface are smeared out in the continuum and masonry is treated as a homogeneous isotropic/orthotropic material. This methodology is relatively less time consuming than the previous ones, but still complex because of the brittle material behaviour.

The first two approaches are computationally intensive for the analysis of large masonry structures, but they accurately describe the behaviour of adobe and are an important research tool in comparison with the costly and often time-consuming laboratory testing. The third approach is faster than the previous ones and, in the case of adobe structures, does not significantly reduce the accuracy of the results. In the case of macro modelling, the selection of the nonlinear model used to represent the soil behaviour is very important to achieve accurate results (e.g. Mohr–Coulomb model, Drucker-Prager model, Concrete Damage Plasticity model, Smeared Cracking model, amongst others).

6.4 Discrete Elements Method

In the discrete element method the masonry structure is represented by an assembly of blocks with special nonlinear behaviour at their boundaries (e.g. mortar joints). The walls are modelled in a micro-scale level. This methodology

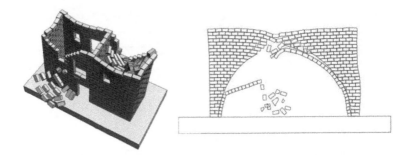

Fig. 22 Modelling of an unreinforced masonry structure (*left*) and a vaulted wall (*right*) within a discrete element approach [49]

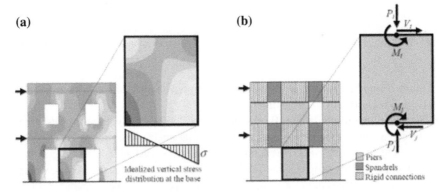

Fig. 23 Modelling strategies for masonry structures based on macro-elements [36]. **a** Modelling with FEM. **b** Modelling with the equivalent approach

allows the representation of large movements and complete block separation with results in changes in the structural geometry and connectivity as seen in Fig. 22. The algorithm recognizes new contact zones between the bodies (blocks, units, particles) as the analysis progresses [35]. As it is usual in a micro-scale model, the computational effort is demanding for the analysis of real adobe masonry structures. Normally, an explicit solution procedure is selected for the non-linear analyses.

6.5 Approaches Based on Macro-models

In the equivalent approach the masonry walls are modelled in a macro-scale level. This is a simplified method which can be used to evaluate the global strength of the building. Each masonry wall is represented by pier elements, spandrel beam elements and joint rigid elements (one-dimensional beam-column elements), as seen in Fig. 23. The piers are the principal vertical and horizontal seismic resistant elements; while the spandrels couple the piers in case of seismic loading.

The effective height of the pier should be able to represent well the non-linear behaviour of the panel. Nonlinear force-drift relationships represent the damage in the masonry panels due to flexion, shear and sliding. This equivalent approach is more accurate when the masonry walls are well connected by horizontal floors (not necessarily rigid floors) and the openings have a regular vertical distribution [40]. It is assumed that the global building response is more influenced by the in-plane response than the out-of-plane response of the walls. The out-of-plane is evaluated locally or it can be indirectly measured by the drift at roof level.

Each of the previous methodologies uses algorithm to compute the equilibrium at each step of the analysis. Those are called solution schemes and are divided into implicit and explicit procedures as it is explained in the next section.

6.6 Solution Schemes for Nonlinear Material Behaviour: Implicit and Explicit

One of the important issues in solving the nonlinear response of structures is the type of solution schemes used to indirectly evaluate the equilibrium of the system and the accuracy of the results. The solution of the problem is found by checking the convergence or the non-divergence of the structures under the application of incremental loading (force or displacement). There are two schemes: implicit and explicit.

6.6.1 Implicit

An implicit analysis is an iterative procedure used for checking the equilibrium in terms of internal and external forces of the system at each time step. This analysis implies the solution of a group of non-linear equations from time t to time $t + dt$ based on information of $t + dt$. For example, if $Y(t)$ is the current system state and $Y(t + dt)$ is the state at the later time, so the Eq. (1) should be solved to find $Y(t + dt)$.

$$G\left(Y\left(t\right), Y\left(t + dt\right)\right) = 0 \tag{1}$$

Amongst the different solution procedures used in the implicit finite element solvers, the Newton–Raphson solution is the faster intended for solving non-linear problems under force control. The convergence is measured by ensuring that the difference between external and internal forces, displacement increment and displacement correction are sufficiently small.

6.6.2 Explicit

Explicit solution was originally created to analyse high-speed dynamic events and models with fast material degradation (such as almost-brittle materials), which may cause convergence problems when analysed with implicit procedures. The explicit method solves the state of a finite element model at time $t + dt$ exclusively based on information at time t (Eq. 2). It implies no iterative procedure and no evaluation of the tangent stiffness matrix, which results in advantages comparing with the implicit procedure.

$$Y (t|dt) = F (Y (t)) \tag{2}$$

Here, the movement equations are integrated using the central difference integration rule, which is conditionally stable (this means the necessity of a small time increment). The stability limit for this integration rule is normally smaller than the Newton–Raphson procedure. The explicit stability limit is the time that an elastic wave spends to cross the smallest shell element dimension in the model.

6.6.3 Particular Aspects for the Modelling of Adobe Structures

In the case of adopting implicit solutions when analysing complex problems of brittle materials, such as adobe masonry, the algorithms to evaluate the nonlinear dynamic response may face convergence problems. In such cases, an explicit solution procedure could be used. The explicit modelling strategies can provide a robust method for large nonlinear problems. However, the accuracy of the model must be controlled, for example, by the crack pattern evolution and by the quantification of the energy (e.g. internal energy, external energy, etc.).

Normally, a non-linear analysis within an explicit scheme needs more running time that an analysis within the implicit one. This is because the time step for the explicit method should be small to avoid divergence in the results.

In addition, when dealing with quasi-static problems, like in pushover analyses, with an explicit scheme, the inertial forces should be reduced as much as possible. This procedure is not straight forward and some assumptions, like mass scaling, should be applied to the model. In this case (pseudo-static analyses), an implicit scheme could be more convenient for obtaining preliminary results of the numerical models [50].

7 Seismic Vulnerability Assessment of Adobe Constructions

7.1 Earthquake Loss Estimation

Earthquake loss estimation is considered one of the important components of disaster management programmes. In Chap. 1 this topic is addressed in more detail and in this section are just briefly introduced the general concepts and the results of recent research on earthquake loss estimation of adobe constructions. Loss estimation models can be used by the experts in the insurance industry, emergency planning and seismic code drafting committees [51]. The studies aimed at determining losses are based on earthquake risk, which is a product of hazard and vulnerability. Hazard refers to the probability of earthquake occurrence at a place within a specified time period, whereas seismic vulnerability refers to the potential damage of elements at risk. The elements at risk include buildings, infrastructures, people, services, processes, organisations, etc. [52]. For building infrastructure, vulnerability is expressed as expected damages to structures during ground shaking. The loss estimation studies can be carried out using either scenario studies or probabilistic analysis.

7.2 Seismic Hazard Analysis

Hazard is described in terms of a ground motion parameter such as peak ground displacement (PGD), peak ground acceleration (PGA), spectral displacement (Sd) and earthquake macroseismic intensity. The selection of a suitable parameter is dependent on the type of vulnerability analysis [53]. There are two approaches for carrying out a seismic hazard analysis: deterministic seismic hazard analysis (DSHA) and probabilistic seismic hazard analysis (PSHA). The size and location of the earthquake is assumed to remain unchanged in DSHA and the hazard evaluation is based on a particular seismic scenario. On the other hand, PSHA allows considering uncertainties in the size, location and rate of occurrence of earthquakes and variation in the ground motion due to these factors. Nevertheless, identification and characterisation of potential earthquake sources, which are capable of producing significant ground motion at a site, are common elements in both approaches. The characterisation of a source includes its location, geometry and earthquake potential. The determination of earthquake size on macroseismic intensity scale is based on human observations which are made during an earthquake regarding the damage of natural and built environment [54]. Different macroseismic scales are employed, such as, Modified Mercalli Intensity (MMI), European Macroseismic Scale (EMS-98) [55], parameter scale of seismic intensity (PSI scale) [56], etc. These provide a qualitative assessment of the effects of earthquakes on the building taxonomy, at a particular location.

7.3 Seismic Vulnerability Analysis

Vulnerability assessment methods provide a relationship between the intensity of ground shaking (hazard) and expected building damages in terms of mean damage grade (μ_D). This relation is termed as vulnerability curve. The results obtained from vulnerability curves can be extended to develop building fragility curves, which provide an estimate of conditional probability of exceeding a damage state of a building, or portfolio of buildings, under a given level of earthquake loading. In probabilistic terms, fragility is a cumulative density function (CDF) which represents the vulnerability of a building or building stock to failure [57].

Vulnerability analysis can be carried out with the help of building inventory for the area of interest. Different vulnerability assessment methods have been proposed in the published technical literature. These can be divided into four categories: empirical methods, judgement-based methods, mechanical methods and hybrid techniques. Of these, the first two are qualitative whereas the last two are quantitative methods. A brief description of these is given in the following paragraphs:

- Empirical Methods: These methods are based on the data of observed damage which is collected from the post-earthquake field surveys. These are the oldest seismic vulnerability assessment methods which were developed as a function of macroseismic intensities [58]. Different empirical methods include damage probability matrix (DPM) [59], vulnerability index [60], vulnerability curves [61], continuous vulnerability functions [62], etc.
- Judgement-based Methods: These methods employ the information provided by the earthquake engineering experts based on their judgement. The experts provide estimates of probability of damage likely to be experienced by different structure types at several ground shaking intensities. A judgement-based method was first employed by the Applied Technology Council (ATC) for California and a summary of the method is available in ATC-13 [63].
- Mechanical Methods: Nonlinear numerical analyses of computer models of the buildings are carried out for different intensity earthquakes. The information on damage distribution obtained from the analyses is statistically analysed to develop vulnerability curves. These methods are employed when the available earthquake damage data is insufficient.
- Hybrid Techniques: Different researchers, e.g. [64–66], have employed hybrid methods of seismic vulnerability assessment. The development of these is a result of deficiencies in the aforementioned methods in carrying out damage assessment, such as incomplete damage statistics from surveys, bias in the opinion of experts, limitations of computer models, etc. The development of fragility curves in hybrid methods is based on the combination of several damage prediction methods.

The empirical and judgement-based methods are suitable for a set of buildings whereas mechanical and hybrid methods are employed for individual building analysis.

7.4 Vulnerability Assessment of Adobe Buildings

Various approaches have been employed by researchers in the vulnerability assessment of adobe structures. Owing to the nonstandard nature of adobe materials and differences in construction practices these studies are region specific and results present significant variability.

Giovinazzi and Lagomarsino [67] developed a method for carrying out a macroseismic vulnerability assessment of building infrastructures in a given area. The seismic hazard for this method can be described using EMS-98 [55] or any other macroseismic intensity scale. In addition, the method can be employed either with the field survey data or with statistically obtained data which varies in origin and quality as compared to Europe. The method is based on determining vulnerability index (V) and ductility index (Q). The former is a measure of the ability of a building/building stock to resist lateral seismic loading. The higher is the value of V, the lower the building resistance and vice versa. The ductility index describes the ductility of a building/building stock and controls the rate of increase in the damage with earthquake demand level. The distribution of building damage is represented using beta distribution. For a continuous variable x, which ranges between a and b, the shape of distribution is controlled by the beta-distribution parameters designated as t and r. The values of a and b are taken as 0 and 6, respectively. The mean value of x (μ_x) can be related to μ_D through a third degree polynomial (Eq. 3).

$$\mu_x = 0.042\mu_D^3 - 0.315\mu_D^2 + 1.725\mu_D \tag{3}$$

The distribution parameters t and r are correlated with μ_D, as given in Eq. (4):

$$r = t(0.007\mu_D^3 - 0.0525\mu_D^2 + 0.287\mu_D \tag{4}$$

An analytical expression was obtained by [67], based on probability and fuzzy set theory, that relates μ_D with V and Q (Eq. 4).

$$\mu_D = \left[1 + tanh\left(\frac{I + 6.25V - 13.1}{Q}\right)\right] \tag{5}$$

A good agreement between μ_D calculated from Eq. (5), and the observed damage data was found. The values of V, Q and t came out to be 0.84, 2.3 and 6, respectively, for the European adobe buildings. Figure 24 illustrates vulnerability curves for adobe buildings in Europe. It is noted in Fig. 24 that the building vulnerability increases rapidly at lower intensity levels, as compared to higher levels. Figure 25 presents probability of exceedance of damage to adobe buildings versus macroseismic intensity. These are developed using the method described above for damage grades given by EMS-98. These damage grades are: D1 = slight damage; D2 = moderate damage; D3 = heavy damage; D4 = very heavy damage and D5 = total collapse. It is noted in Fig. 25 that a small percentage of buildings face the risk of collapse up to an intensity of VIII. This percentage rises above 90 % at an intensity of XII.

Demircioglu [68] employed a macroseismic approach for carrying out seismic risk studies of building typologies in Turkey. Adobe buildings were part of the building typologies and were included in this analysis. Damage states for the buildings were selected as defined by EMS-98 [55]. The vulnerability assessment methods proposed by Giovinazzi and Lagomarsino [67] and Modified KOERI [69] were employed. The latter is based on the vulnerability relationships suggested by Coburn and Spence [56]. The results from both the employed models

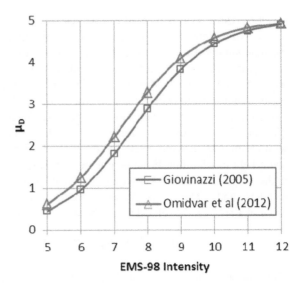

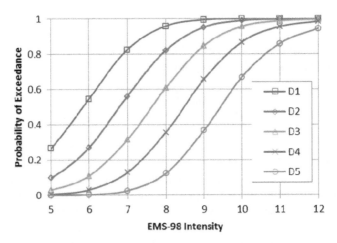

Fig. 25 Fragility curves for adobe buildings in Europe [53]

were compared with the observed earthquake damage data and a good agreement
between them was found. Table 1 shows the results of V, Q and t, as obtained from
the two models for adobe buildings in Turkey.

Omidvar et al. [70] conducted vulnerability analysis for Iranian building typologies which included adobe construction. DPM of all the building types were
developed based on the observed damage data from the Bam earthquake in 2003.
The macroseismic method, as suggested by Giovinazzi [67], was employed. The
results indicated that adobe building damage was initiated at an earthquake intensity of VIII on EMS-98 intensity scale. The vulnerability index (V) for adobe

Table 1 Comparison of parameters for vulnerability assessment models [68]

Adobe building	Modified KOERI method [69]			Giovinazzi and Lagomarsino method [67]		
	V	Q	t	V	Q	t
Low rise	0.6	2.3	4	0.84	2.3	6
Mid rise	0.6	2.3	4	0.84	2.3	6
High rise	0.6	2.3	4	0.84	2.3	6

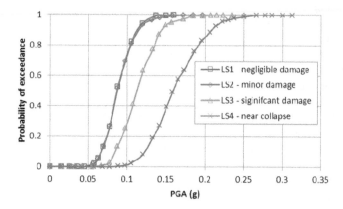

Fig. 26 Fragility curves for in-plane adobe behaviour [72]

buildings in Iran came out to be 0.9. The results co-related well with those pre-sented by JICA [71] for adobe buildings in Tehran on the basis of Manjil earth-quake in 1990.

Tarque [72] carried out a seismic risk analysis of adobe buildings in Cusco (Peru) which considered both the in-plane and out-of-plane adobe wall behaviour. Displacement response spectra of Cusco for different earthquake return periods were obtained using PSHA. These were employed to determine the demand for adobe structures. The capacities of the buildings were calculated for different limit states (LS) in terms of displacement capacity and period of vibration (mechani-cal displacement based procedures). The probabilities of failure for the employed return periods were calculated by comparing the capacity with the demand. Fragility curves were plotted for conditional seismic risk analysis. These relate failure probability in each limit state, as a function PGA (and its associated return period) for both in-plane and out-of-plane behaviour. For the unconditional analy-sis, all the ground motions with their return periods were considered for a specified time window, up to 100 years. The results of fragility curves indicated that 77 % of adobe buildings will have in-plane failure and 75 % buildings will develop wide cracks at the wall junction due to out-of-plane wall bending at a PGA of 0.18 g which is recommended by the Adobe Peruvian Code for Cusco region. Figure 26 illustrates fragility curves for in-plane behaviour at different limit states.

8 Strengthening and Retrofitting Solutions

Walls are the fundamental structural elements in earthen buildings. Earthquakes cause the sudden formation of cracks in the adobe walls at the beginning of any ground motion. Adequate seismic reinforcement solutions are needed to assure the safety of adobe construction by controlling the displacements of cracked walls. Furthermore, due to the fact that the large majority of adobe dwellings are located in developing countries, the implementation of low-cost seismic strengthening solutions using widely available materials is desirable.

Several studies to achieve this goal have been conducted [2, 27]. The main objectives of the developed strengthening or reinforcement schemes are to assure a proper connection between construction elements and to reach overall stability.

One simple and effective method for structural rehabilitation of adobe structures in general and also valuable for seismic retrofit is the injection of grouts into the cracks existing on the constructions [73]. Other traditional techniques used to repair cracks in adobe constructions can be very disturbing and intrusive when compared to grout injection. This could be, however, a non-reversible technique, which can originate durability and compatibility problems if non-suitable materials are chosen to compose the grout, particularly for earthen structures [74]. Earthen grouts could be good enough to get a restitution of the low tensile strength of earthen construction.

The improvement of the mechanical behaviour requires a fluid grout with very good penetrability and bonding properties, while durability requires the development of a microstructure as close as possible to the microstructure of the existing materials. Currently, a design methodology for grout injection of earthen constructions is under study [75], which could represent an important step forward in the repair of these structures. However, mechanical injection techniques are not yet totally developed.

Regarding seismic retrofit, grout injection must be complemented with other techniques that could increase ductility to dissipate seismic energy. From several possible solutions, the following are mentioned: cane or timber internal reinforcements, cane external reinforcement, reinforced concrete as internal reinforcement and synthetic mesh strengthening systems.

8.1 Cane or Timber Internal Reinforcements

This type of reinforcement consists of placing an internal grid, with vertical and horizontal elements, able to bond efficiently with the structure, improving its seismic performance (see example in Fig. 27). The vertical elements should be conveniently anchored to the foundation and to a ring beam on top of the walls. The spacing of the vertical or horizontal elements should be such to provide an efficient connection to the structure. Bamboo canes or eucalypt dry timber is

Fig. 27 Internal cane mesh
reinforcement [77]

recommended for these reinforcements [76, 77]. It should be noted that this type
of reinforcement can only be done in new constructions.

The placement of the horizontal layers should be carefully carried out, as
these can become weak points, which, under seismic forces, can cause horizontal cracks. For adobe structures, in order to provide an effective bonding, mortar
thickness between two rows of adobe blocks, with reinforcement in between, can
become larger than desirable [78]. Laboratory tests proved that high thickness
mortars correspond to lower wall masonry strength.

Full-scale shaking table tests were conducted with adobe houses using this kind
of reinforcement, demonstrating a good response to save lives [79]. The model
reinforced with an internal cane mesh suffered significant damage, but did not collapse. A major restraint in using this strengthening solution is the fact that cane or
adequate timber is not available in all seismic regions.

8.2 Cane External Reinforcement

For repair or seismic retrofit of existing structures, an external reinforcement using
a grid of canes and ropes can be a good solution. Canes are placed vertically and
externally to the wall, on both sides, inside and outside. Ropes are then positioned
horizontally tying the vertical canes along the walls and involving the structure.
Different rows of horizontal ropes are placed along the height of the wall with a
spacing of 30–40 cm. In order to connect the two grids—outside and inside grids—
and thus confine the earthen structure, small extension lines are placed connecting
the two grids, crossing the wall from one side to another through holes, made at
each 30–40 cm. This reinforcement grid can then be covered with plaster for adequate finishing, providing at the same time more confinement to the earth structure.

Fig. 28 External cane-rope
mesh reinforcement [80]

Figure 28 shows an example of this type of reinforcement applied to a real-scale model tested at the PUCP, where only part of the structure was covered with plaster.

The main limitation of this type of reinforcement is the fact that a great quantity of cane is required. As cane is not available in all regions, industrial material must be studied and tested.

8.3 Reinforced Concrete Elements as Reinforcement

This technique consists of building first the adobe walls with gaps in the corners, or connections with other walls to be filled by concrete. Steel bars are then placed and the concrete is poured in order to form a confined system with columns and collar beam. This solution is rather expensive, conducting to a high stiffness system with low ductility [78]. Furthermore, important collapses in earthen construction with reinforced concrete elements were reported, implying that this can be an inadequate reinforcement solution, though more studies on the subject are required. Figure 29 shows examples of collapses after reinforcement actions using concrete: Tarapacá Cathedral, Iquique earthquake, 2005, Chile, and San Luis de Cañete Church, Pisco earthquake, 2007, Peru. In Tarapacá Cathedral, the bending of the reinforced concrete beam destroyed completely the main adobe wall with 1.30 m of thickness. In San Luis de Cañete Church, the reinforced concrete frames changed the global behaviour of the structure, creating discontinuities, and the adobe walls overturned during the earthquake.

8.4 Synthetic Mesh Strengthening Solutions

Reinforcement solutions with synthetic meshes (geogrids) involving the walls have been studied and tested, proving its applicability, simplicity and efficiency. Figure 30 shows examples of application. In Figueiredo et al. [28] and Oliveira

Fig. 29 Examples of collapses after reinforcement using concrete. **a** Tarapacá Cathedral, Chile, 2005 (courtesy: Chesta J.). **b** San Luis de Cañete Church, Peru, 2007 [81]

Fig. 30 Synthetic meshes used in adobe structures strengthening. **a** Double-T adobe wall [82]. **b** Adobe module [83]

et al. [82], is described a test performed in a real-scale adobe wall with and without reinforcement and the results of both tests were compared. The solution for filling the wall cracks (injection of hydraulic lime grout) combined with the strengthening solution (synthetic mesh incorporated in the plaster) proved to be very effective. Figure 31 shows the comparison between the results obtained for the original wall and the strengthened wall. The tests on the retrofitted wall demonstrated that the lateral strength increased slightly, and the ductility and the energy dissipation capacity improved significantly. The wall was able to recover its initial stiffness.

In Blondet et al. [3], several similar full-scale adobe housing models with different density and types of synthetic mesh were tested in a shaking-table. The results showed that the damage decreased as the amount of synthetic mesh placed involving the walls increased. In Vargas et al. [81], the use of geogrid in adobe constructions is extensively explained, with comprehensible details on how to cut and place the grid with the objective of improving seismic performance.

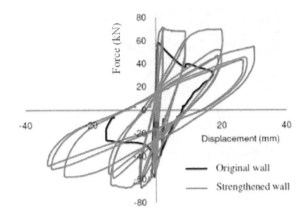

Fig. 31 Horizontal force versus lateral displacement for the original and strengthened walls [82]

The use of synthetic mesh bands involving the adobe walls and covered with mortar is also possible. The mesh is placed in horizontal and vertical strips, following a layout similar to that of beams and columns. This solution is able to provide additional strength to the structure, though the failure mode observed was brittle and dangerous. The use of cement also makes this an expensive solution.

9 Final Comments

The adobe bricks are basically composed of farm soil mixed with other materials, like straw or lime, depending on the soil characteristics. The bricks are sun-dried and the building construction process passed from generation to generation, in a learning by doing process. In comparison to other building materials, as reinforced concrete, the adobe is costless, for this reason it is massively used by families with low-incomes, without paying adequate attention on the structural safety and reinforcement issues [84]. In many regions of the World, vernacular buildings have thickness of around 0.30 m, which may influences on the stability of walls due to their pronounced slenderness.

Adobe is classified as a brittle material with a fragile behaviour, in particular when subjected to horizontal forces, like those induced by earthquakes. Adobe constructions are heavy, brittle and during earthquakes may attract large inertia forces that lead to the collapse of the adobe constructions. Survey of damage and experimental test results on adobe buildings subjected to earthquake demands reveals that one of the most common failure mechanisms in these constructions is the overturning on the walls out-of-plane. For this reason, a simple improvement of the diaphragm (roof) may stabilize and promote the "box behaviour" of the construction. In recent years, some solutions have been proposed and studied to improve the seismic behaviour of adobe constructions, and some low-cost and easy to apply solutions have been implemented in adobe constructions as in Pisco (Peru).

The results of tests and numerical models developed and calibrated allow for a better understanding of the behaviour of adobes constructions under different loading conditions, which may help in designing economical and sustainable rehabilitation and strengthening solutions for the existing constructions.

The rehabilitation and strengthening solutions studied by different research groups showed to be efficient, as observed in their test results. It was concluded that these techniques are affordable and easy to apply techniques to improve the seismic behaviour of adobe masonry constructions.

The information given in this chapter pretends to summarize the actual knowledge and research outputs on the structural behaviour of adobe buildings. It is clear the need for further research in this field, and particularly for other types of constructions, as monuments, to which the strengthening should provide seismic capacity, but using non-intrusive solutions.

References

1. Houben, H., Guillard, H.: Earth Construction: a Comprehensive Guide. Practical Action, London (1994)
2. Memari, A.M., Kauffman, A.: Review of existing seismic retrofit methodologies for adobe dwellings and introduction of a new concept. In: Proceedings of SismoAdobe2005, Pontificia Universidad Católica del Perú, 15, Lima, Peru. (2005)
3. Blondet, M., Vargas, J., Velásquez, J., Tarque, N.: Experimental study of synthetic mesh reinforcement of historical adobe buildings. In: Lourenço, P.B., Roca, P., Modena, C., Agrawal. S. (eds) Proceedings of Structural Analysis of Historical Constructions, pp. 1–8, New Delhi, India (2006)
4. Dowling, D.: Adobe housing in El Salvador: Earthquake performance and seismic improvement. In: Rose, I., Bommer, J.J., López, D.L., Carr, M.J., Major, J.J. (eds) Geological Society of America Special Papers, 281–300 (2004)
5. De Sensi, B.: Terracruda, la diffusione dell'architettura di terra. www.terracruda.com/architet turadiffusione.htm (2003)
6. Lowman, P.D., Montgomery, B.C.: Preliminary determination of epicenters of 358,214 events between 1963 and 1998, (2008). http://denali.gsfc.nasa.gov/dtam/seismic/. Accessed in June 2011
7. Fratini, F., Pecchioni, E., Rovero, L., Tonietti, U.: The earth in the architecture of the historical centre of Lamezia Terme (Italy): Characterization for restoration. Constr. Build. Mater. **53**, 509–516 (2011)
8. Fernandes, M., Portugal, M.V.: Atlântico versus Portugal Mediterrâneo: Tipologias arquitectónicas em terra. In: International Conference TerraBrasil 2006, Ouro Preto, Minas Gerais, Brazil (2006)
9. Varum, H., Figueiredo, A., Silveira, D., Martins, T., Costa, A.: Outputs from the research developed at the University of Aveiro regarding the mechanical characterization of existing adobe constructions in Portugal—Informes de la Construcción, doi: 10.3989/ic.10.016, July–Sept 2011, **63**(523), 127–142 (2011)
10. Tavares, A., Costa, A., Varum, H.: Adobe and modernism in Ílhavo, Portugal. Int. J. Architect. Heritage, Taylor & Francis, ISSN 1558–3058, Paper reference ID UARC-2011-0357.R1, **6**(5), 525–541, doi: 10.1080/15583058.2011.590267 (2012)
11. Sandrolini, F., Franzoni, E., Varum, H., Niezabitowska, E.: Materials and technologies in Art Nouveau architecture: Façade decoration cases in Italy, Portugal and Poland for a consistent

restoration—Informes de la Construcción, Instituto de Ciencias de la Construcción Eduardo Torroja (IETcc), ISSN 0020-0883, doi 10.3989/ic.10.053, **63**(524), 5–11 (2011)

12. Dell'Acqua, A.C., Franzoni, E., Sandrolini, F., Varum, H.: Materials and techniques of Art Nouveau architecture in Italy and Portugal: A first insight for an European route to consistent restoration—International Journal for Restoration of Buildings and Monuments, Aedificatio Verlag Publishers, ISSN 1864–7251 (print-version); ISSN 1864-7022 (online-version), **15**(2), 129–143 (2009)

13. Blondet, M., Villa-García, G., Brzev S.: Earthquake-resistant construction of adobe buildings: In: Greene, M. (ed) A Tutorial, Report, EERI/IAEE world Housing Enciclopedy, Okland, California, USA (2003)

14. INEI (National Institution for Statistics and Informatic): Census 2007: XI de Población y VI de Vivienda" Report. National Institute of Statistics and Informatics, Lima (2008)

15. Mousavi, S.E., Khosravifar, A., Bakhshi, A., Taheri, A., Bozorgnia, Y.: Structural typology of traditional houses in iran based on their seismic behaviour.In: Proceedings of 8th U.S. National Conference on Earthquake Engineering, (2006)

16. Baglioni, E., Fratini, F., Rovero, L.: The material utilised in the earthen buildings sited in the Drâa Valley (Morocco): Mineral and mechanical characteristics. 6ATP, 9SIACOT, (2010)

17. Ruano, A., Costa, A.G., Varum, H.: Study of the common pathologies in composite adobe and reinforced concrete constructions. ASCE's J. Perform. Constr. Facil. Am. Soc. Civil Eng. doi 10.1061/(ASCE)CF.1943-5509.0000200, **26**(4), 389–401 (2012)

18. Blondet, M., Vargas, J., Tarque, N.: Observed behaviour of earthen structures during the Pisco earthquake (Peru). In: Proceedings of 14th World Conference on Earthquake Engineering, Beijing, China (2008)

19. Webster, F., Tolles. L.: Earthquake damage to historic and older adobe buildings during the 1994 Northridge, California Earthquake. In: Proceedings of 12th World Conference on Earthquake Engineering, Auckland, New Zealand (2000)

20. ASTM E519/E519 M–10: Standard Test Method for Diagonal Tension (Shear) in Masonry Assemblages. ASTM International, West Conshohocken (2010)

21. EN 1052–1: Methods of test for masonry—Part 1: Determination of compressive strength. European Committee for Standardization (CEN), Brussels (1998)

22. Martins, T., Varum, H.: Adobe's Mechanical Characterization in Ancient Constructions: The Case of Aveiro's Region—Materials Science Forum, Trans Tech Publications, Switzerland, ISSN 0255-5476, vol. 514–516, pp. 1571–1575 (2006)

23. Silveira, D., Varum, H., Costa, A.: Influence of the testing procedures in the mechanical characterization of adobe bricks. Constr. Build. Mater. J. doi 10.1016/j.conbuildmat.2012.11.058, **40**, 719–728 (2013)

24. Silveira, D., Varum, H., Costa, A., Martins, T., Pereira, H., Almeida, J.: Mechanical properties of adobe bricks in ancient constructions—Construction & Building Materials, Elsevier, Manuscript reference CONBUILDMAT-D-11-00604, doi 10.1016/j.conbuildmat.2011.08.046, **28**, 36–44 (2012)

25. Blondet, M., Vargas, J.: Investigación sobre vivienda rural. Report, Division of Civil Engineering, Pontificia Universidad Católica del Perú, Lima, Peru (1978)

26. Vargas, J., Ottazzi, G.: Investigaciones en adobe. Report, Division of Civil Engineering, Pontificia Universidad Católica del Perú, Lima, (1981)

27. Blondet, M., Madueño, I., Torrealva, D., Villa-García, G., Ginocchio, F.: Using industrial materials for the construction of safe adobe houses in seismic areas. In: Proceedings of Earth Build 2005 Conference, Sydney, Australia, (2005)

28. Figueiredo, A., Varum, H., Costa, A., Silveira, D., Oliveira, C.: Seismic retrofitting solution of an adobe masonry wall. Mater. Struct. RILEM, ISSN 1359-5997, doi 10.1617/s11527-012-9895-1, **46**(1–2), 203–219 (2013)

29. Tareco, H., Grangeia, C., Varum, H., Senos-Matias, M.: A high resolution GPR experiment to characterize the internal structure of a damaged adobe wall. EAGE First Break **27**(8), 79–84 (2009)

30. Antunes, P., Lima, H., Varum, H., André, P.: Optical fiber sensors for static and dynamic health monitoring of civil engineering infrastructures abode wall case study. Measurement **45**, 1695–1705 (2012)
31. Stavridis, A., Shing, P.B.: Finite Element Modeling of Nonlinear Behavior of Masonry-Infilled RC Frames. J. Struct. Eng. ASCE 2010, **136**(3), 285–296 (2010)
32. Kappos, A.J., Penelis, G.G., Drakopoulos, C.G.: Evaluation of simplified models for lateral load analysis of unreinforced masonry buildings. J. Struct. Eng. **128**(7), 890 (2002)
33. Lourenço, P.B.: Computational strategies for masonry structures. Ph.D. Thesis, Delft University, Delft, The Netherlands (1996)
34. Page, A.W.: Finite element model for masonry. J. Struct. Eng. **104**(8), 1267–1285 (1978)
35. Roca, P., Cervera, M., Gariup, G., Pela, L.: Structural analysis of masonry historical constructions. Classical and advanced approaches. Arch. Comput. Methods Eng **17**(3), 299–325 (2010)
36. Calderini, C., Cattari, S., Lagomarsino, S.: In plane seismic response of unreinforced masonry walls: comparison between detailed and equivalent frame models. In: Papadrakakis, M., Lagaros, N.D., Fragiadakis, M. (eds) ECCOMAS Thematic Conference on Computational Methods in Structural Dynamics and Earthquake Engineering, Rhodes, Greece, (2009)
37. Gambarotta, L., Lagomarsino, S.: Damage models for the seismic response of brick masonry shear walls. Part I: The mortar joint and its applications. Earthquake Eng. Struct. Dynamics **26**(4), 423–439 (1997)
38. Gambarotta, L., Lagomarsino, S.: Damage models for the seismic response of brick masonry shear walls. Part II: The continuum model and its applications. Earthquake Eng. Struct. Dyn. **26**(4), 441–462 (1997)
39. Lagomarsino, S., Galasco, A., Penna A.: Non-linear macro-element dynamic analysis of masonry buildings. In: Proceedings of ECCOMAS Thematic Conference on Computational Methods in Structural Dynamics and Earthquake Engineering, Rethymno, Crete, Greece, (2007)
40. Magenes, G., Della, F.A.: Simplified non-linear seismic analysis of masonry buildings. In: Proceedings of Fifth International Masonry Conference, British Masonry Society, London, England, (1998)
41. Ali, S.S., Page, A.W.: Finite element model for masonry subjected to concentrated loads. J. Struct. Eng. **114**(8), 1761 (1987)
42. Cao, Z., Watanabe, H.: Earthquake response predication and retrofitting techniques of adobe structures. In: Proceedings of 13th World Conference on Earthquake Engineering, Vancouver, Canada, (2004)
43. Furukawa, A., Ohta, Y.: Failure process of masonry buildings during earthquake and associated casualty risk evaluation. Nat. Hazards, **49**, 25–51 (2009)
44. Rots, J.G.: Numerical simulation of cracking in structural masonry. Heron **36**(2), 49–63 (1991)
45. Arya, S.K., Hegemier, G.A.: On nonlinear response prediction of concrete masonry assemblies. In: Proceedings of North American Masonry Conference, Boulder, Colorado, USA, pp 19.1–19.24 (1978)
46. Lotfi, H.R., Shing, P.B.: Interface Model Applied to fracture of masonry Structures. ASCE **120**(1), 63–80 (1994)
47. Lourenço, P.B., Rots, J.G.: Multisurface interface model for analysis of masonry structures. J. Eng. Mech. **123**(7), 660 (1997)
48. Cundall, P.A.: A computer model for simulating progressive large scale movements in blocky rock systems. In: Proceedings of Symposium on Rock Fracture (ISRM), Nancy, France, (1971)
49. Alexandris, A., Protopapa, E., Psycharis, I.: Collapse mechanisms of masonry buildings derived by distinct element method. In: Proceedings of 13th World Conference on Earthquake Engineering, (2004)
50. Tarque, N.: Numerical modelling of the seismic behaviour of adobe buildings. PhD thesis. Universitá degli Studi di Pavia, Istituto Universitario di Studi Superiori. Pavia, Italy, (2011)

51. Bommer, J.J., Scherbaum, F., Bungum, H., Cotton, F., Sabetta, F., Abrahamson, N.A.: On the use of logic trees for ground-motion prediction equations in seismic hazard analysis. Bull. Seismol. Soc. Am. **95**(2), 377–389 (2005)
52. Foerster, E., Krien, Y., Dandoulaki, M., Priest, S., Tapsell, S., Delmonaco, G., Margottini, C., Bonadonna, C.: Methodologies to assess vulnerability of structural systems, pp. 1–139. Seventh Framework Programme, European Commission (2009)
53. Giovinazzi, S.: The Vulnerability Assessment and the Damage Scenario in Seismic Risk Analysis. PhD Thesis, University of Florence and Technical University of Braunschweig, (2005)
54. Kramer, S.: Geotechnical Earthquake Engineering. Prentice Hall International Series, Ohio (1996)
55. Grünthal, G.: European Macroseicmic Scale (EMS-98), Cahiers du Centre Européen de Géodynamique et de Séismologie 15. Centre Européen de Géodynamique et de Séismologie, Luxembourg (1998)
56. Coburn, A., Spence, R.: Earthquake Protection. Wiley, USA (2002)
57. Jozefiak, S.: Fragility Curves for Simple Retrofitted Structures. CM-4 Consequence, (2005)
58. Calvi, G.M., Pinho, R., Magenes, G., Bommer, J.J., Restrepo-Velez, L.F., Crowley, H.: Development of seismic vulnerability assessment methodologies over the past 30 years. J. Earthquake Technol. paper no 472, **43**(3), 75–104 (2006)
59. Whitman, R.V., Reed, J.W., Hong, S.T.: Earthquake damage probability matrices. In: Procedings of 5th European Conference on Earthquake Engineering, Rome, 25–31, (1973)
60. Benedetti, D., Sulla, Petrini V.: Vulnerabilità di Edifici in Muratura: Proposta di un Metodo di Valutazione. L'industria delle Costruzioni **149**(1), 66–74 (1984)
61. Lagomarsino, S., Giovinazzi, S.: Macroseismic and mechanical models for the vulnerability and damage assessment of current buildings. Bull. Earthq. Eng. **4**, 445–463 (2006)
62. Spence, R., Coburn, A.W., Pomonis, A.: correlation of ground motion with building damage: the definition of a new damage-based seismic intensity scale. In: Proceedings of the Tenth World Conference on Earthquake Engineering, Madrid, Spain, vol. 1, pp. 551–556 (1992)
63. Applied Technology Council (ATC): Earthquake damage evaluation data for California, Applied Technology Council, ATC-13. Redwood, California (1985)
64. Barbat, A.H., Moya, F.Y., Canas, J.A.: Damage scenarios simulation for seismic risk assessment in urban zones. Earthquake Spectra **12**(3), 371–394 (1996)
65. Kappos, A.J., Panagopoulos, G., Panagiotopoulos, C., Penelis, G.: A hybrid method for the vulnerability assessment of R/C and URM buildings. Bull. Earthquake Eng. **4**, 391–413 (2006)
66. Vicente, R., Parodi, S., Lagomarsino, S., Varum, H., Silva, J., Mendes, A.R.: Seismic vulnerability and risk assessment: Case study of the historic city centre of Coimbra, Portugal. Bulletin of Earthquake Engineering, Springer, Manuscript Ref. BEEE325R1, doi 10.1007/s10518-010-9233-3, vol. 9, 1067–1096 (2011)
67. Giovinazzi, S., Lagomarsino, S.: Fuzzy-Random Approach for a Seismic Vulnerability Model. In: Proceedings of ICOSSAR, Rome, Italy (2005)
68. Demircioglu, M.B.: Earthquake Hazard and Risk Assessment for Turkey. PhD Thesis. Bogaziçi University, Turkey, (2010)
69. DEE-KOERI.: Earthquake Risk Assessment for the Istanbul Metropolitan Area. Report prepared by Department of Earthquake Engineering-Kandilli Observatory and Earthquake Research Institute, Bogazici University Press, Istanbul (2003)
70. Omidvar, B., Gatmini, B., Derakhshan, S.: Experimental vulnerability curves for the residential buildings of Iran. J. Nat. Hazards **60**(2), 345–365 (2012)
71. JICA.: The study on seismic microzoning of the greater Tehran area in the Islamic Republic of Iran, Final report (2000)
72. Tarque, N., Crowley, H., Varum, H., Pinho, R.: Displacement-based fragility curves for seismic assessment of adobe buildings in Cusco Peru. Earthquake Spectra J. **28**(2), 759–794 (2012)
73. Silva. R.A., Schueremans. L., Oliveira. D.V.: Grouting as a repair/strengthening solution for earth constructions. In: Proceedings of the 1st WTA International PhD Symposium, WTA publications, Leuven, pp. 517–535 (2009)

74. The Getty Conservation Institute, Pontificia Universidad Católica del Perú: Interdisciplinary Experts Meeting on Grouting Repairs for Large-scale Structural Cracks in Historic Earthen Buildings in Seismic Areas. Peru (2007)
75. Vargas-Neumann, J., Blondet, M., Ginocchio, F., Morales, K., Iwaki, C.: Uso de grouts de barro líquido para reparar fisuras estructurales en muros históricos de adobe. V Congresso de Tierra en Cuenca de Campos, Valladolid (2008)
76. NTE E.080: Norma Técnica de Edificación. "Adobe Peruvian Code". In: Spanish, MTC/SENCICO. Peru (2000)
77. Blondet, M., Vargas, J., Tarque, N., Iwaki, C.: Seismic resistant earthen construction: the contemporary experience at the Pontificia Universidad Católica del Perú. J. Informes de la Construcción 63(523), 41–50 (2011)
78. Minke, G.: Manual de construcción para viviendas antisísmicas de tierra. Universidad de Kassel, Alemania (2001)
79. Ottazzi, G., Yep, J., Blondet, M., Villa-Garcia, G., Ginocchio, J.: Shaking table tests of improved adobe masonry houses. In: Proceedings of Ninth World Conference on Earthquake Engineering. Japan (1988)
80. Torrealva, D., Acero, J.: Reinforcing adobe buildings with exterior compatible mesh. In SismoAdobe 2005: Architecture, Construction and Conservation of Earthen Buildings in Seismic Areas, Lima, Pontificia Universidad Católica del Perú, Lima, Peru, May 16-19 2005 [CD], ed. Marcial Blondet (2005) http://www.pucp.edu.pe/eventos/ SismoAdobe2005
81. Vargas-Neumann, J., Torrealva, D., Blondet, M.: Construcción de casas saludables y sismor-resistentes de adobe reforzado con geomallas. Fondo Editorial, PUCP (2007)
82. Oliveira, C., Varum, H., Figueiredo, A., Silveira, D., Costa, A.: Experimental tests for seismic assessment and strengthening of adobe structures. 14th ECEE (2010)
83. Blondet, M., Aguillar, R.: Seismic Protection of Earthen Buildings. Conferencia Internacional en Ingeniería Sísmica, Peru (2007)
84. North, G.: Waitakere City Council's Sustainable Home Guidelines—Earth building; Waitakere City Council. http://www.waitakere.govt.nz/abtcit/ec/bldsus/pdf/materials/earthbu ilding.pdf (2008)

Conservation and New Construction Solutions in Rammed Earth

Rui A. Silva, Paul Jaquin, Daniel V. Oliveira, Tiago F. Miranda, Luc Schueremans and Nuno Cristelo

Abstract The conservation and rehabilitation of several sites of cultural heritage and of the large housing stock built from rammed earth requires adopting intervention techniques that aim at their repair or strengthening. The present work discusses the main causes of the decay of rammed earth constructions. The intervention techniques used to repair cracks and lost volumes of material are also discussed. Regarding the strengthening of rammed earth walls, the discussion is focused on the techniques that improve the out-of-plane behaviour. Special attention is given to the injection of mud grouts for crack repair in rammed earth walls, including the presentation of the most recent developments on the topic, namely

R. A. Silva (✉) · D. V. Oliveira
ISISE – Institute for Sustainability and Innovation in Structural Engineering,
Department of Civil Engineering, University of Minho, Guimarães 4800-058, Portugal
e-mail: ruisilva@civil.uminho.pt

D. V. Oliveira
e-mail: danvco@civil.uminho.pt

P. Jaquin
Integral Engineering Design, Bath, UK
e-mail: pauljaquin@gmail.com

T. F. Miranda
C-TAC – Territory, Environment and Construction Research Centre
Department of Civil Engineering, University of Minho, Guimarães 4800-058, Portugal
e-mail: tmiranda@civil.uminho.pt

L. Schueremans
Department of Civil Engineering, Catholic University of Leuven/Frisomat,
Leuven, Belgium
e-mail: luc.schueremans@frisomat.be

N. Cristelo
CEC – Construction Studies Centre Engineerings Department,
Trás-os-Montes e Alto Douro University, Vila Real 5000-801, Portugal
e-mail: ncristel@utad.pt

A. Costa et al. (eds.), *Structural Rehabilitation of Old Buildings*,
Building Pathology and Rehabilitation 2, DOI: 10.1007/978-3-642-39686-1_3,
© Springer-Verlag Berlin Heidelberg 2014

regarding their fresh-state rheology, hardened-state strength and adhesion. Finally, the use of the rammed earth is discussed as a modern building solution. In addition, several typical techniques for improving rammed earth constructions are discussed, aiming at adequate those to modern demands. In addition, the alkaline activation of fly ash is presented and discussed as a novel improvement technique.

Keywords Rammed earth • Damage • Repair • Conservation • Strengthening • Grout injection • Alkaline activation • Fly ash

1 Introduction

Nowadays, building with earth is still considered to be one of the most popular solutions for shelter and housing. In fact, thirty per cent of the World's population, i.e. nearly 1500 million people live in a house built with raw earth [1]. In general, earth constructions have great presence in developing countries, where economical and technical limitations often make this type of construction the only feasible alternative. On the other hand, in developed countries the practice of building with raw earth has fallen into disuse over the past century, as a consequence of the technological development and extensive use of modern building materials (mainly concrete sand steel). Despite that, there is a large housing stock built from earth, widely distributed around the world comprising many monuments and buildings of acknowledged historic, cultural and architectural value [2, 3].

The earth construction concept includes several building techniques that have different constructive features [1], which depend mostly on local limitations related to the properties of the available soil and other resources, and thus this type of constructions is often associated with vernacular architecture. From all the traditional earth construction techniques, rammed earth, adobe masonry, wattle-and-daub are the most commonly found around the World, while CEB (compressed earth blocks) masonry is the most common from the modern techniques [4].

Rammed earth, also known as *"taipa"*, *"taipa de pilão"*, *"tapial"*, *"pise de terre"*, *"pisé"* or *"stampflehm"*, consists in compacting layers of moist earth inside a removable formwork, therefore building monolithic walls (Fig. 1). In fact, the formwork constitutes a key feature, which differentiates this technique from the others. The conception and design of the formwork have been an evolving process, which resulted in several configurations. In order to grant the quality of the rammed earth wall, this element must present adequate strength and stability to support the dynamic loads of the compaction process, and adequate stiffness to mitigate the consequent deformations [4].

Another important feature that requires tight control is the weight of the formwork, which must be sufficiently low in order to make it easily handled by the workers (in the assembling and disassembling operations). Typically, the formwork is supported directly on the wall and is dislocated as the rammed earth blocks that form the wall are built, and whereby this is called crawling formwork [5].

Fig. 1 Construction of a new rammed earth house in Odemira, Portugal

The construction is carried out by courses (like masonry), in which the formwork runs horizontally around the entire building perimeter and then is moved vertically to build the next course. Alternatively, a formwork externally supported can also be used to perform rammed earth, thus implying the assembly of a scaffolding structure (see Fig. 2a). In Spain, there are some pre-Muslim rammed earth sites reported, where this type of formwork was used, which is identified by the absence of putlogs holes (which usually result from the removal or deterioration of timber elements used to support a crawling formwork, and get embedded in the wall when the block is compacted) [5].

The elements of traditional formworks are made of timber, but not exclusively, as ropes and tying needles, for example, can be used for holding the shutters in place (Fig. 2b). Nowadays, the trend is to use the same metallic shutters used in concrete technology to constitute the formwork; a crawling formwork is composed by two metallic shutters or a self-supporting formwork is constituted by assembling the shutters in such way that they cover the entire wall.

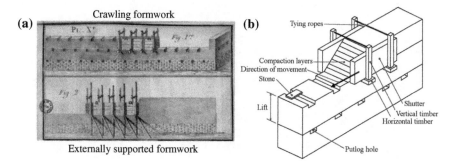

Fig. 2 Rammed earth formworks: **a** Crawling and externally supported formwork [6]; **b** elements of a traditional crawling formwork

The compaction of rammed earth blocks is traditionally carried out by manual means, resorting to rammers that in general are made of timber (Fig. 3a). Nowadays, the compaction process has been simplified by the introduction of mechanical apparatus such as pneumatic (Fig. 3b) and vibratory rammers, which comparatively reduce substantially the labour and time consumed in the construction. The dimensions of the rammed earth blocks are very variable from country to country, from region to region or even within the same region; for example in Alentejo (Portugal) the length of rammed earth blocks from typical dwellings may vary from 1.4 to 2.5 m, the height from 0.4 to 0.6 m and the thickness from 0.4 to 0.6 m [7].

The conservation of rammed earth constructions according to modern standards and principles requires the development of repair, consolidation and strengthening solutions that are compatible, but at the same time effective. For example, cement based materials have good mechanical properties, but their use as a repair material should be avoided, since their properties are rather different from those of rammed earth. In extreme situations, such differences can result in severe damage to the construction. Currently, the trend is to use earth based materials that grant the requirements of the construction demanding repair measures. Some of these repair solutions will be further discussed in detail, as well as solutions for structural strengthening.

Nowadays, rammed earth construction is regarded as a sustainable building solution for housing, but in most countries rammed earth is considered as a non-standard building material, which can be an obstacle to its utilization. On the other hand, there are some countries that have developed documents to regulate rammed

Fig. 3 Rammers used to compact rammed earth: **a** manual rammer; **b** pneumatic rammer (courtesy of Bly Windstorm from Earthdwell)

earth construction [2], but the demanded properties often limit the use of local soils. In most cases, this is surpassed by chemical stabilization of the soil, using binders such as cement and lime, which results in solutions less attractive from economical and sustainable points of view.

2 Rammed Earth in the World

From the history of all earth construction techniques, rammed earth is relatively recent [8], but its origin is not consensual. According to Houben and Guillaud [1] this technique was first developed in its "true" form in China during the Three Kingdoms period (221–581 AD). On the other hand, Jaquin et al. [3] argue that the technique had two independent origin focuses: in China and around the Mediterranean basin. It first appeared in China, where Jest et al. [9] claim that remnants of rammed earth walls and houses found in Qinghai, Tsaidam (between Tibet and Central Asia), date from the Muomhong period (2000–500 BC). The Phoenicians used rammed earth in their settlements in Iberian Peninsula and in northern Africa from around 800 BC, and a type of rammed earth vernacular construction has existed in Iberian Peninsula since that time [3]. Rammed earth technique also developed in northern Africa, particularly as a military building material. The spread of Islam from northern Africa to Iberian Peninsula from 711 AD brought rammed earth as a quick construction technique. As a consequence, the rammed earth technique is documented in several Arabic documents from the 8th century AD in military constructions of settlements and in alcazabas such as that of Badajoz [8].

In the sixteenth century AD, the rammed earth technique was used in South America by the Portuguese and Spanish settlers and later on (eighteenth and nineteenth centuries) in North America and Australia by European settlers. It was also used in the construction of many European settlements in South America, such as Sao Paulo, where there is a rich earth construction heritage. The publication of construction manuals by Cointeraux in 1791 [6] marked and stimulated rammed earth construction in Europe, which was re-introduced as a fireproof alternative solution to the typical timber constructions of that period. Then, with the invention of Portland cement in the nineteenth century, rammed earth construction waned. However, this technique helped solving the housing problem in Germany generated after the end of World War II [8]. More recently, there has been a growing interest on earth construction (including rammed earth) which led, for example, to the creation of the CRATerre group by University of Grenoble, one of the most important international centres concerning earth construction. Most recently, rammed earth has been championed in Australia, New Zealand and North America, where recommendations and standards to regulate earth construction have been developed. Rammed earth is still used as a vernacular construction technique in parts of Nepal and Bhutan, and is being developed in many other parts of the world.

This short overview on the history of rammed earth shows that this technique is worldwide spread. Figure 4 shows that earth construction (where rammed earth is

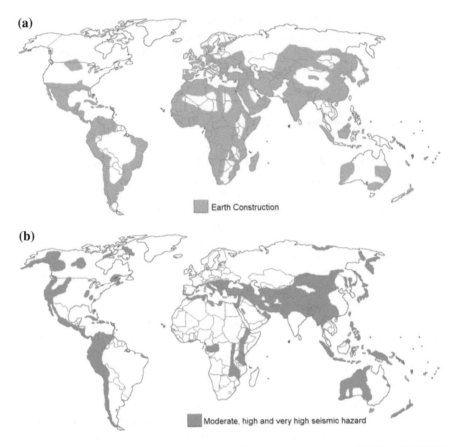

Fig. 4 World's geographical distribution: **a** earth construction (source: CRATerre-ENSAG); **b** zones of moderate to very high seismic hazard (based on De Sensi [10])

included) is present in all five inhabited continents. Moreover, it is shown that the geographical distribution of earth construction is almost coincident with zones of moderate to very high seismic hazard.

3 Damage and Rehabilitation

3.1 Damaging Agents and Pathologies

Like other building materials, rammed earth is also subjected to different types of degradation, but in this case the degradation rate is in general faster, especially if compared to modern materials [11]. In general, the earth constructions decay may be attributed to: (i) material deficiencies; (ii) foundation problems; (iii) structural defects; (iv) thermal movements; (v) water; (vi) biological activity; (vii) wind; (viii) natural disasters.

Material deficiencies are normally related to the texture of the soil used in the construction or to the composition of the earth mixture. If the soil has low clay content and presents excessive percentage of stones and gravel, it may result in rammed earth walls with low compressive strength and water resistance. On the other hand, if the clay content of the soil is excessive, it may result into excessive cracking due to shrinkage [11]. The addition of natural materials, such as straw and manure, to the earth mixture is a common procedure in earth construction. Straw has beneficial effects in terms of tensile strength, but at long-term and if adverse conditions are observed, it will decompose, leaving undesired voids that have negative impact on the mechanical properties.

Commonly to other types of structures, foundation deficiencies may result in terrible consequences to the structural integrity, such as severe cracking or even collapse (Fig. 5a). These problems are related to the capacity of the foundation soil to bear the loads transmitted by the massive rammed earth walls. Moreover, the poor tensile strength of rammed earth walls does not allow them to absorb differential settlements caused by seasonal variations of the water table or by the presence of different foundation soils along the length of the wall.

Structural defects may also result into severe damage to rammed earth construction. These may be caused by incorrect or inexistent design or due to construction errors. For example, timber A-frame trusses are often used as support structure of the roof of rammed earth constructions, but if they are incorrectly designed they may transmit horizontal thrust that cannot be absorbed by the walls, resulting into leaning and cracking of the walls [12].

Thermal movements are often ignored as decay agent, since it is assumed that the inherent softness and pliability of earthen structures may render them immune to such problems, but in general this is not true [11]. These movements

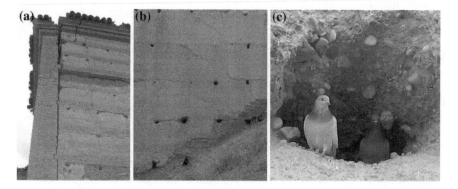

Fig. 5 Damage in rammed earth constructions: **a** corner crack caused by settlements of the foundations or by horizontal thrust of the roof; **b** basal erosion caused by rainfall; **c** damage caused by nesting pigeons

normally result in vertical cracks, found through the walls length (spaced at regular intervals) and at walls junctions, which debilitate the monolithic behaviour and stiffness.

Water is considered to be one of the worst enemies of rammed earth constructions. The water-related problems start immediately after finishing the construction, due to the inherent drying shrinkage, which typically results in cracking of the material, reducing its strength and exposing it to other decay agents. The shrinkage behaviour of earthen materials is function of the water content of the earth mixture upon compaction, i.e. the higher the water content, the higher the shrinkage. However, there is a maximum limit to this volume decrease, set at the shrinkage limit of the soil [4, 12].

On the other hand, the strength of rammed earth walls is strongly affected by its moisture content; the higher the moisture content, the lower the strength. Therefore rammed earth walls must be protected against the direct influence of rainwater after being built. Additionally, the direct impact of rainwater may also have other prejudicial long-term effects, such as the occurrence of basal erosion that reduces the bearing section of the wall (Fig. 5b). These problems are usually mitigated by protecting walls with roof overhangs, external renderings and masonry plinths [12]. This last element serves also to avoid the direct contact of rammed earth with the ground to prevent the occurrence of rising damp. It should be noted that renderings with low water vapour permeability (e.g.: from cement mortar) are harmful for rammed earth constructions, since the interior moisture accumulates inside the walls. In fact, this is judged to be the underlying cause of some recent earth constructions collapses [12].

The high moisture content in an earth construction, may also lead to the appearance of biological activity. The development and penetration of plant roots results in cracking, due to the tensile stress caused by their expansion. The voids, resulting from shrinkage cracks or from decomposition of the straw, attract animals searching for food or shelter (Fig. 5c), extending even more the damage, by drilling tunnels and feed from the incorporated organic matter (e.g. straw) [11]. Small animals, like insects, are more damaging than the larger (e.g. rodents), since huge colonies may develop extending the damage to larger areas.

Wind has mainly an erosive action over earth constructions. However, it can also contribute to other decay agents, like shrinkage or rainwater impact.

Natural disasters, such as earthquakes and floods, can inflict severe damage and even lead to the collapse of rammed earth constructions. Nevertheless, earthquakes are the natural disasters with the highest catastrophic effects, thus being of special interest among researchers dealing with this kind of constructions. In general, the seismic performance of earth constructions is very deficient when compared with contemporary structures, due to their low strength and high deadweight. Deficient constructive details also greatly contribute to the poor seismic behaviour of earth constructions. Such deficiencies are typically related to the lack of connection between the elements composing the earthen structures. Therefore, a strong earthquake may lead them to collapse or may inflict severe structural damage, by originating harsh cracks and reducing the overall structural stiffness [13].

3.2 Structural Intervention Techniques

The rammed earth building stock is large, worldwide spread and includes both ordinary and monumental buildings. The conservation of this stock requires the adoption of intervention measures that must follow specific principles, but this is a topic almost absent from technical documents on conservation of historical buildings. However, it is generally accepted that the intervention measures must allow the structure to function in a soft mode, and that excessive stiffening of a structure can lead to further damage. This was the result of a technical meeting after the TERRA 2000 conference, where it was accepted that not only guidelines for clear identification of structural and seismic cracks are necessary, but also that improved methods for the repair of such cracks are required [14].

The purpose of a structural intervention on a rammed earth construction can be at repairing or/and at improving its structural capacity. In the first case the intention is to recover the original structural behaviour partially or totally, which is in agreement with the principles referred to previously. The second case intends to improve the structural behaviour significantly, and thus such interventions are not so prone to follow these principles.

There is little documentation reporting and describing structural interventions specifically for rammed earth constructions. Generally, such documents propose intervention solutions that are adapted from other earth construction techniques, similar to rammed earth (e.g. cob), which result from an extensive practical experience on such constructions. In the following paragraphs the most significant intervention techniques on rammed earth will be discussed, including a novel technique based on the injection of mud grouts.

3.2.1 Refill Lost Volumes

A repair intervention on rammed earth walls can have two main goals: to refill lost volumes of material and to connect cracked parts. The absence of great volumes of material is a common pathology of rammed earth constructions, which can be attributed to several factors such as: partial collapse caused by an earthquake or another external factor, presence of biological activity, basal erosion caused by rising damp and erosion caused by rainwater impact. The obvious repair solution for this problem is to replace the missing material with new material, and for that there are some techniques that serve this purpose.

A common technique used to repair basal erosion consists in using a mix of soil (previously studied) similar to that of the rammed earth construction, which is compacted in place (Fig. 6a). First, the missing section is regularized and the loose material is removed. Then, a single-sided formwork partially covering the damaged section is assembled, and the earth mixture is compacted from the uncovered zone in horizontal layers. One of the main problems with this technique is associated to the difficulty in compacting adequately the last layers due to the limited

Fig. 6 Repair of the basal erosion of Paderne's Castle [16]: **a** by compacting new rammed earth material in place; **b** by projecting earth at high speed

vertical space within the formwork. This situation is normally overcome by leaning the last layers, which allows compacting them from outside the wall surface. Another problem comes from the shrinkage of the new rammed earth material upon drying, which is responsible for difficulties in bonding the repair material to the original rammed earth. Therefore, steel reinforcements and plastics meshes are often included. An alternative to this procedure, which also mitigates the shrinkage problem, consists in prefabricate rammed earth blocks that are put in place and then bonded with earth mortar. Researchers also reported cases where bricks, adobes and even concrete were used as replacing materials [11, 15].

Projected earth (Fig. 6b) is a recent intervention technique used to repair basal erosion, which was applied successfully in the conservation of Paderne's Castle (Algarve, Portugal) [17] and in the Alhambra of Granada (Spain) [18]. This technique consists in projecting a mixture of moist earth at high speed against the damaged area, whereas the impact energy is supposed to grant a similar compaction to that of the original rammed earth. This technique has some aesthetical limitation, since the rammed earth layers are not recreated, and instead a much more homogenous surface texture is obtained. The shrinkage is another problem of this technique, since the projection of the earth is only adequately executed by the wet method [18], which requires high moisture content of the earth mixture in order to provide adequate flowability for the projection apparatus. It should be noted that this technique is recent and thus further investigation must be carried out.

3.2.2 Reconnect Cracked Parts

The presence of structural cracks is a common pathology of rammed earth, and whose cause is often attributed to settlements of the foundation, to concentrated loads, to horizontal thrusts applied by the roofs or to earthquakes. The repair of such cracks is essential to recover the original monolithic behaviour of the rammed earth walls. Moreover, cracks are weak points for water infiltration, which are prone to cause further decay to the construction. Several techniques are used to repair these cracks, but whose efficiency greatly varies from case to case. The most basic method used to repair structural cracks consists in simply filling it with earth mortar. The mortar, with a composition similar to that of the rammed earth, is introduced from the faces, and wherever the crack is extremely thinner the wall needs to be cut back. In terms of re-establishing the connection of the crack, this method is inefficient, but if the problem is strictly the water infiltration then it constitutes an adequate solution. The filling of the crack by grout injection is a good alternative to this method as it will be further discussed.

Stitching the crack is another solution to reconnect cracked parts, and consists in creating a mechanical connection between the two sides of the crack ("stapling" the crack), see Fig. 7. According to Keefe [12], this is carried out by cutting horizontal chases in the wall with a determined spacing along the crack, which are then filled with mud bricks and earth mortar prepared with a soil similar to that of the original rammed earth.

Fig. 7 Repair of cracks by stitching [5]

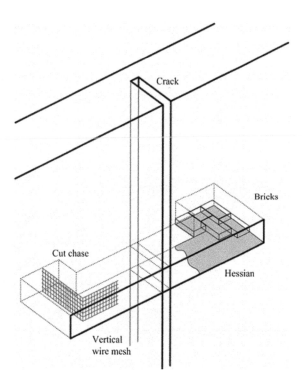

3.2.3 Strengthening

The introduction of stone masonry buttresses (see Fig. 8a) is probably one of the most basic and ancient solutions for out-of-plane strengthening of rammed earth walls. This solution serves both as amendment for the seismic performance and as solution to solve the problem of leaning walls caused by horizontal thrusts transmitted by traditional timber roofs. The introduction of steel tie bars to connect opposed facades is another method to improve the out-of-plane behaviour of rammed earth walls, see Fig. 8b. The placement of external beams tied by wires on opposed walls is an alternative to the previous technique. This constitutes a more global strengthening, as the forces from the beams are transmitted along the walls. On the other hand, the forces in the tie bars are transmitted to the walls in localized points by anchorage plates [19], whose dimensions are designed accordingly.

The connection of perpendicular walls is generally weak, since very often cracks are observed at this type of connection, which debilitate the structural behaviour, specially its seismic performance. The placement of grouted anchors connecting perpendicular walls is a valid technique for solving this problem, but Pearson [21] argues that their use in earth walls is limited because of the low shear strength of earthen materials, and they should not be relied upon to stitch major structural cracks.

As it is generally accepted, the tensile strength of earthen materials is very low and constitutes a key factor contributing to the high seismic vulnerability of earth constructions. The improving of the tensile strength of rammed earth is possible through the fixing of timber elements on the facades of the building, which is then rendered with a compatible earth mortar. As alternative to timber elements, metallic meshes or geo-meshes can be used, see Fig. 9. From an aesthetical point of view, this solution should only be used where the rammed earth facades are covered by a rendering.

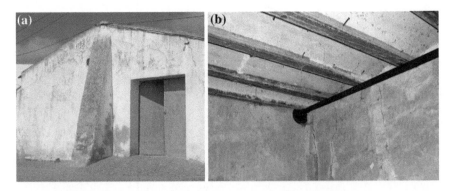

Fig. 8 Strengthening of rammed earth walls by means of: **a** buttressing [20]; **b** tie bars [5]

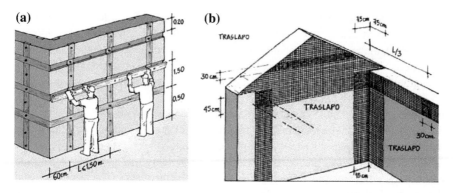

Fig. 9 Strengthening by fixation: **a** timber elements [22]; **b** metallic meshes [23]

4 Repair by Injection of Mud-grouts

4.1 Mud Grouts Background

The repair of structural cracks is essential for re-establishing the structural integrity of an earth construction. However, in rehabilitation interventions on earth constructions, cracks are often not truly repaired. Most of the times, they are "cosmetically" hidden by covering with a plaster or by filling them with mortar. These procedures do not grant an appropriate continuity to earth constructions, and in some situations they cause more problems than they solve [12]. Grout injection may constitute a more feasible, efficient and economic repair solution. Nevertheless, grouts compatible with earthen materials must be employed, which are not similar to those used for consolidating historic masonry (based on fired brick or stone units), including the recently developed binary and ternary grouts [24]. The obvious trend is to adopt grouts incorporating earth, also called mud grouts [11], but the knowledge on these grouts is still limited.

Currently, there are several published works on grouts for the consolidation of historical masonry. However, there are only a few cases where mud grouts are studied or applied for repairing earth constructions [25]. Roselund [26] describes a grouting solution applied to the restoration and strengthening of the Pio Pico mansion in Whittier, California, which is built in adobe. The damage to this mansion was mainly in the form of cracks in the adobe walls, as a consequence of the 1987 California earthquake. The cracks were repaired by injecting a modified mud grout, whose composition consisted of earth, silica sand, fly ash and hydrated lime. The design of this mud grout (whose hardening relies not only on clay but also on another binder) was mainly focused on obtaining a material with adequate consistency, acceptable shrinkage, and with hardness, strength and abrasion resistance similar to those of the original adobes. The results of the preliminary study of the intervention project showed that the tested unmodified mud grouts presented excessive shrinkage, while the modified ones presented low shrinkage, justifying the preference for the latter type of grout. After this intervention, the Pio Pico mansion was struck by the 1994 Northridge earthquake. The resulting minor to moderate damage, when compared to the damage in other buildings in a similar condition, showed that grout injection in combination with other consolidation/ strengthening measures is effective in preventing serious additional damage [27].

Jäger and Fuchs [28] also used grout injection for consolidating the remaining adobe walls of the Sistani House at Bam Citadel in Iran, severely damaged during the 2003 Bam earthquake. A modified mud grout composed of clay powder, lime and wallpaper paste was employed for this purpose. The decision on the grout composition was preceded by a composition study that included testing the addition of other materials (such as cement and water glass). The shrinkage and the mechanical properties, namely the compressive, flexural and splitting tensile strengths, were the properties tested in this study.

On the other hand, Vargas et al. [29] defend the employment of unmodified mud grouts (mud grout whose hardening relies only on the clay fraction when it dries out) rather than modified ones. This point of view is supported by an extended set of splitting tests carried out on specimens consisting of adobe sandwiches bonded by a layer of mud grout. Several compositions were tested, including unmodified and mud grouts modified by the addition of different percentages of cement, lime or gypsum. Their results showed that, in general, the unmodified mud grouts have better adhesion capacity. In addition, the results of diagonal compression tests performed on adobe masonry wallets repaired by injection of an unmodified mud grout showed that it is possible to recover the initial strength of the damaged walls. Furthermore, the addition of binders, such as cement or hydraulic lime, greatly increases the Young's modulus of a mud grout, which may be from one to two orders of magnitude higher than that of earthen materials [30, 31]. Despite the advantages of modified mud grouts regarding shrinkage and resistance to water, its employment must be carefully evaluated, since excessive stiffness constitutes an important drawback with respect to satisfying mechanical compatibility with earthen materials.

The role of the construction and its interaction with hardened grout is another key point that is highlighted by Vargas et al. [29], while other documents mainly focused on the material properties of the grout.

4.2 Grout Injection Approach in Earth Construction

As pointed out by Silva et al. [32], the design of a mud grout is a complex process. For instance, it must consider the construction demands, namely recovering the structural behaviour and granting durability, see Fig. 10, in a similar approach to that used in the conservation of historical masonry [24]. Then, the composition is defined such that the properties of the mud grout meet the demands. The complexity of the design lies in the interdependence between each of the grout properties, which becomes evident when the composition is adjusted. Therefore, it is essential

Fig. 10 Design methodology
of a mud grout [32]

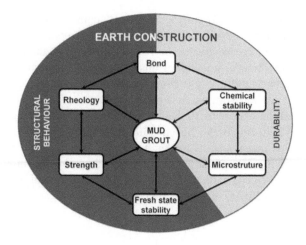

to understand how the composition of a mud grout affects its properties in order to design it effectively. In the subsequent paragraphs, the different material properties and their interdependencies are outlined in detail, namely rheology, strength, bond, chemical stability, stability against sedimentation of the fresh mix and the microstructure compatibility.

4.2.1 Rheology

The mud grout must hold a rheological behaviour that allows it to enable complete filling of cracks or voids laying in the earth constructions. This is essential to assure the continuity of the repaired earthen walls and to re-establish the monolithic behaviour of the earthen structure. Thereby, all the factors influencing the rheology, such as the texture (particles size, distribution and shape), the interaction between particles (dispersion or flocculation), the quantity of solids in the grout, the mixing procedure and the action of super-plasticizers or dispersants, must be carefully accounted for during the design of the mud grout. Moreover, the capacity of the dry earthen materials in quickly absorbing high amounts of water requires that the grout presents a great water retention capacity during its injection. This is essential in order to maintain adequate fluidity and penetrability during the full injection process.

4.2.2 Strength

The grout strength should correspond to the level demanded by the structure and be compatible with the original material. The use of grouts much stronger than the original earthen material should be avoided. Also, stiff grouts can cause problems of mechanical compatibility, since the hardened grout is hardly able to follow the displacements occurring in the earth construction, resulting in damage to the intervention. Therefore, the addition of stabilizers has to be carefully considered, since these increase greatly the Young's modulus of the earthen materials. For example, the Young's modulus of unstabilized adobe from an old construction can range between 80 and 500 N/mm^2 [33], while a mud grout prepared with 40 % binder (50 % of hydraulic lime and 50 % of cement) and 60 % earth (mixture of 30 % of kaolin clay with 70 % of sand) can present a Young's modulus ranging between 10340 and 10590 N/mm^2 [30]. Also, not using a binder in a mud grout composition means that its hardening relies exclusively in the drying of the material. The drying of the mud grout is processed essentially through the water absorption by the original earthen material. The incorporation of hydraulic binders gives the mud grout a more independent behaviour, when regarding its hardening.

4.2.3 Bond

An efficient mud grout design requires that enough bonding develops between the hardened grout and the original earthen material. This is essential for granting continuity throughout the earthen structure, both for structural and durability reasons. The swelling/shrinkage nature of earth constitutes a major drawback. During the drying of a mud grout, its shrinkage can lead to cracking and consequently the required bond cannot be established. This is a problem that can also be found in reparation works using earthen materials [12]. In order to overcome the shrinkage problem of mud grouts, several options can be considered. The first one consists in adopting selected clays with low shrinkage ratio in the composition of the grout, such as kaolinite clays. The texture of the mud grout is another parameter which can be intervened, by decreasing the clay content of the grout and by correcting the particles size distribution with addition of "unshrinkable" fine material (such as fly ash, silica fume, calcium carbonate powder, quartz powder, among others). Another possibility consists in decreasing the water content. However, this solution has significant consequences over the rheology, which can only be counteracted by using dispersants. Using stabilizers is also an alternative to solve the shrinkage problem, but other problems can arise from this decision, namely those concerning the compatibility between materials.

4.2.4 Chemical Stability

A mud grout also needs to present chemical stability over time. The salts content has to be limited in order to avoid efflorescence and crypto-efflorescence problems. Moreover, the grout must have adequate resistance against aggressive compounds existing in the original earthen materials. For example, if a mud grout contains Portland cement, there is a possibility that the formation of expansive products will occur, since the presence of sulphates is very common in earthen materials.

4.2.5 Stability Against Sedimentation of the Fresh-State

While the mud grout is in its fresh-state, the solid particles in suspension must not suffer sedimentation. For that, it needs to present limited bleeding, no segregation and adequate water retention. These technical aspects are essential in order to assure that the mud grout maintains its fluidity and penetrability during the injection and remains homogeneous after hardening. For example, using a soil with large dimension particles in the composition of the mud grout is an obvious solution to help solving the shrinkage problem, but such option can constitute a major drawback since these larger and heavier particles are prone to sedimentation.

4.2.6 Microstructure Compatibility

Obtaining a hardened mud grout microstructure compatible with that of the original earthen materials is essential for fulfilling the durability requirement. The water vapour transport occurring within the original material should not be disrupted by the hardened mud grout. For example, the injection of a grout with low porosity (possible case of mud grouts modified by cement addition) can constitute a barrier that condensates the water vapour at its interface with the earthen material. This may be harmful for the construction, depending on the level of the intervention. The condensed water can leech the material around the grout disturbing the already created bond, and damage the intervention. In those cases where a big grout barrier is created, the condensed water can lead to the weakening of the earthen materials, that can be responsible for a possible collapse at long-term. Therefore, the incorporation of materials such as cement has to be carefully evaluated, since it reduces the porosity of the mud grout. The thermal properties of the hardened grout must also be closer to the ones of the original earth materials. This is even more important in monolithic earth constructions (for example rammed earth), where the grout has to be able to follow the thermal displacements of the earthen material.

4.3 New Developments on Unmodified Mud Grouts

Regarding the development of unmodified mud grouts, the authors have been researching the topic and present here an overview of the research findings validated by an extensive laboratorial investigation.

A mud grout composition study has been carried out, which assessed the influence of the mud grout composition on its fresh and hardened properties, namely rheological behaviour, strength and adhesion. By reproducibility reasons, the solid phase of the tested mixes was composed by commercial materials with controlled production quality, namely kaolin powder (Wienerberger, Kaolin RR40) and limestone powder (Carmeuse, Calcitec 2001 S). The kaolin represents the clay fraction

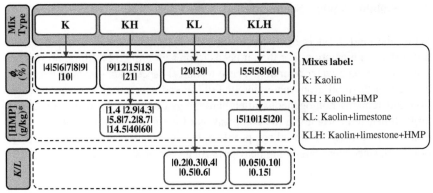

* weight (g) of HMP / weight (kg) of kaolin

Fig. 11 Matrix of the tested mixes

of the mud grout, while the limestone powder represents the silt fraction. Sodium hexametaphosphate (HMP) was also used as admixture for fluidity improvement.

Several mixes with different proportions of kaolin, limestone powder and HMP were tested. The variables of the study were the volumetric solid fraction (ϕ_v), the amount of HMP added as function of the kaolin content ([HMP]) and the ratio between the weight of kaolin and limestone powder (K/L). These variables are combined in the matrix given in Fig. 11, where the mixes are grouped according to the composition: kaolin (K); kaolin and HMP (KH); kaolin and limestone powder (KL); and kaolin, limestone powder and HMP (KLH). The same mixing procedure was followed for all mixes, see [34] for details. The fresh-state rheological behaviour of the mixes was assessed by means of Marsh funnel test (ASTM C 939 [35]) and by determining their flow curves using a Viskomat PC mixer-type rheometer [34].

The strength of the hardened KLH mixes was assessed through flexural and compression tests, according to EN 1015-11 [36]. The specimens were tested after achieving the equilibrium moisture content (constant mass of the specimens) at room temperature of 20 °C and relative humidity of 65 %.

The adhesion developed between earthen materials and three selected mixes (Table 1) was tested on 18 earthen beams with dimensions $160 \times 40 \times 40$ mm^3, see Fig. 12a, prepared with three types of soil typically used in the construction of rammed earth houses in Alentejo (Portugal). The soils were sieved to remove

Table 1 Composition of the mud grouts used for assessing the adhesion

Mud grout	[HMP] (g/kg)	K/L	ϕ_v (%)	W/S (wt.)	% Clay	% Silt	% Sand
MG_55	20	0.15	55	0.30	16	58	26
MG_58	20	0.15	58	0.27	16	58	26
MG_60	20	0.15	60	0.25	16	58	26

Table 2 Properties of the soils used in the bond test after being sieved

Soil	% Clay	% Silt	% Sand	PL	LL
S1	19	47	34	28	44
S2	15	37	48	17	32
S3	18	29	53	20	35

Clay < 0.002 mm/*silt* ≥ 0.002 mm and < 0.060 mm/*sand* ≥ 0.060 mm and < 2.0 mm (fractions usually defined for earth construction)
PL Atterberg's plastic limit/*LL* Atterberg's liquid limit

Fig. 12 Preparation of the earthen beams: **a** for adhesion tests; **b** injection of the broken earthen beams after 3-point bending tests

particles larger than 2 mm, which resulted in the properties given in Table 2. The beams were tested under bending, being repaired afterwards by injecting selected KLH mixes (2 beams of each soil per mix) into the crack between the two parts of each beam, see Fig. 12b. After 15 days, the repaired beams were tested again under bending.

4.3.1 Fresh-state Rheology

The results of the flow time tests of the K mixes (Fig. 13a) showed that the higher the ϕ_v, the higher the measured flow time. Moreover, the flow through the Marsh's funnel is not possible after a critical solid fraction (ϕ_{vcr}) is reached, which is approximately 9 to 10 %. Higher solid percentages are not adequate to prepare a mud grout. This behaviour is a consequence of the colloidal behaviour of the kaolinite particles in suspension, which interact with each other under Brownian

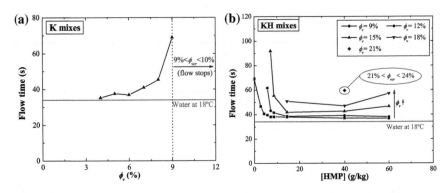

Fig. 13 Flow time measurements: **a** K mixes; **b** KH mixes

and/or hydrodynamic motion, generating two possible particle states: deflocculated or flocculated [37]. The interaction between kaolinite particles is governed by Derjaguin-Landau-Verwey-Overbeek (DLVO) forces, namely electrostatic forces (repulsion between like electric double layers and attraction between unlike charged surfaces) and van der Waals attractive forces. If the attractive forces are favoured, the clay particles tend to flocculate, forming an internal structure (house-of-cards or scaffold structure) that opposes the flow.

The aforementioned internal structure is disturbed by the addition of HMP in the KH mixes, where it is responsible for the following deflocculating mechanisms: (i) increase in the overall negative surface charge by the adsorption of anionic HMP polymeric chains onto the kaolinite surface, especially at the edges of the kaolinite particles [38, 39]; (ii) stabilisation caused by the steric hindrance effect of the adsorbed HMP chains [40]; (iii) complexing of the dissolved alkaline-earth cations and replacing them by lower valence Na^+ cations, which increases the thickness of the electric double layers [38]. These mechanisms allow further increase in ϕ_{vcr} (between 21 and 24 %), where the higher the [HMP], the higher the fluidity (Fig. 13b). However, at high [HMP] the fluidity decreases due to a concentration of linear polyphosphate chains that is above a critical value and promotes the association of kaolinite particles instead of their repulsion [40].

Figure 14 shows that the shear stress required to shear the KH mixes decreases with increasing [HMP]. The addition of increasing amounts of HMP reduces substantially both Bingham's parameters of the KH mixes (parameters obtained by linear fitting of the descending branch of the flow curves). However, the yield stress seems to have the greatest contribution to the flowing resistance, and furthermore this parameter is the main responsible for the failure registered with some Marsh funnel tests.

The results of the Marsh's funnel tests for the KL mixes (Fig. 15a) show that the increase in clay content (increase in K/L) increases the flow time. Moreover, it is shown that substituting the clay content with a silt size material (limestone powder), significantly increases ϕ_{vcr} when comparing to the K and KH mixes.

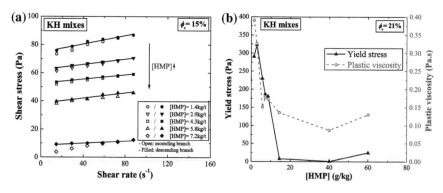

Fig. 14 Flow curves: **a** KH mixes with $\phi_v = 15\ \%$ (with Bingham's model fitted to the descending branch); **b** Bingham's parameters of KH mixes with $\phi_v = 21\ \%$

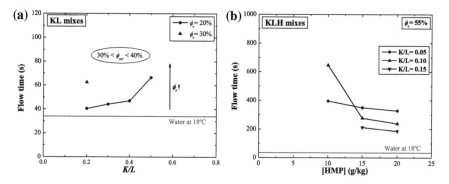

Fig. 15 Flow time measurements: **a** KL mixes; **b** KLH mixes with $\phi_v = 55$

The KLH mixes showed the combined effect of the addition of HMP and the incorporation of limestone powder (Fig. 15b) by further increasing ϕ_{vcr}, since flowing suspensions with ϕ_v between 55 and 60 % were obtained. However, the flow times increased substantially when compared with those of the K, KH and KL mixes. The addition of increasing amounts of HMP results in the reduction of the flow resistance of the KLH mixes (Fig. 16a), but it seems to have greater impact on the reduction of the yield stress than it has on the plastic viscosity (Fig. 16b). In fact, the addition of high quantities of HMP brings the yield stress to values close to zero, which is an important finding regarding the success of a grouting intervention [41]. This importance relies in the fact that if the injection pressure of a mud grout inside a crack is not sufficient to keep the shear stress at the head higher than the yield stress, the flow stops, which does not allow the complete filling of the crack by the grout.

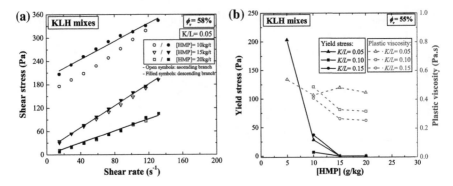

Fig. 16 Flow curves: **a** KLH mixes with $\phi_v = 55$ % and K/L $= 0.05$; **b** Bingham's parameters of KLH mixes with $\phi_v = 55$ %

4.3.2 Strength

The flexural strength of the KLH mixes with a *K/L* ratio of 0.15 is presented as function of ϕ_v in Fig. 17a. It appears that there is no relation between these parameters, with the exception of the mixes with an [HMP] of 20 g/kg. On the other hand, the compressive strength of the mixes seems to be favoured by an increasing ϕ_v, see Fig. 17b. In regard to the effect of clay content, increasing *K/L* promoted positive development of both strength parameters (Fig. 18). Moreover, the mixes with a low *K/L* developed flexural and compressive strengths that are quite satisfactory when compared with those of earthen materials [34].

4.3.3 Adhesion

The results of the three-point bending tests performed on the repaired earthen beams, as well as the repair efficiency of the selected mud grouts, are presented

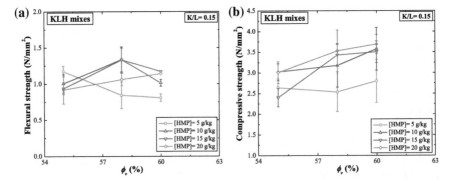

Fig. 17 Strength of the KLH mixes with *K/L* $= 0.15$ (both average values and scatter are outlined): **a** flexural strength; **b** compressive strength

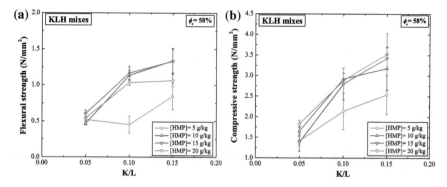

Fig. 18 Strength of the KLH mixes with $\phi_v = 58$ % (both average values and scatter are outlined): **a** flexural strength; **b** compressive strength

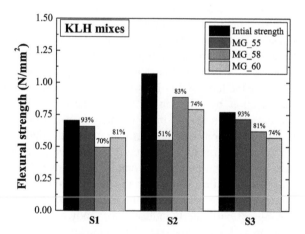

Fig. 19 Adhesion capacity of the selected mud grouts for the three soil types

in Fig. 19. The grouts failed at re-establishing completely the original strength of the earthen beams. However, the repaired beams developed a flexural strength of at least 0.5 N/mm^2, which is above the minimum flexural strength for adobes required by the New Mexico [42] and New Zealand [43] standards (0.35 and 0.25 N/mm^2, respectively). The volumetric solid fraction ϕ_v of the selected grouts does not seem to interfere significantly with the repair efficiency, which probably means that the water content of the mud grout may be further increased in order to favour its injectability properties, but accounting for the possible occurrence of excessive drying shrinkage.

5 New Horizons for Rammed Earth

5.1 Rammed Earth as a Modern Building Solution

Rammed earth is in general considered to be a non-standard material, since its use in the construction industry does not follow any industrialized processes [44]; rammed earth walls are entirely built on site, which makes their properties extremely dependent on the characteristics of the available soil and workmanship. In addition, there are only a few countries that have codes/standards for earth construction (e.g. [43, 45, 46]), which together with the limited knowledge on the technique and resulting material, discourage the option for this building solution, especially in countries where those documents are absent. However, the recent environmental concerns in the building industry have been recalling rammed earth construction as a modern building solution, mostly due to its recognized sustainability and interior comfort benefits [4, 47].

On the other hand, the mechanical and durability requirements for rammed earth demand the use of soil with adequate properties [48]. Such requirements, within the framework of modern building codes, can be excessively severe to the traditional unstabilized rammed earth, especially in what concerns to local hazards such as earthquakes. Therefore, finding a soil at the construction site with adequate properties to meet these modern requirements is hardly possible. However, this limitation can be overcome by transporting an adequate soil to the construction site, by improving the local soil or by improving the construction process.

Chemical stabilization by addition of binders, such as bitumen, lime and cement is a recurrent solution used to improve the properties of the soil. The stabilization by addition of bitumen aims at improving mainly the durability of rammed earth, namely in what concerns to the resistance against water (erosion and swelling). The bitumen is added in an aqueous emulsion and mixed with earth. Upon drying, the bitumen forms a thin strong film that holds together the particles of the soil, which also improves the cohesion of the material and thus its strength. This stabilization technique is particularly indicated for sandy or sandy-gravel soils [1].

Lime stabilization is historically related to rammed earth construction, since millenary fortresses from the Iberian Peninsula still exist nowadays due to the durability and mechanical strength promoted by this stabilization technique. The cementitious compounds resulting from pozzolanic reactions between the lime and the clay fraction in the soil are the main responsible for this improvement. Thus, this technique is indicated for clayey soils.

The addition of cement is currently the preferred stabilization technique for new rammed earth constructions, since it significantly improves the strength of almost all soil types. The stabilization effect is provided by the formation of a cementious matrix that binds the particles of the soil. This matrix results from the hydration reactions of the cement and from the pozzolanic reactions with the clay fraction.

Rammed earth constructions are considered to have a high seismic vulnerability, like most earth construction techniques, and the main reason is the poor out-of-plane and in-plane behaviour of the walls. As a consequence, there are several proposed solutions/techniques that aim at improving the behaviour of rammed earth walls through the building process. The most basic solution consists in building walls thicker than usual, but this result in a significantly higher cost.

The walls from rammed earth constructions can be seen as masonry made of macro-blocks, which are defined by horizontal and vertical joints existing between them. These joints result from the building process and have an important role in shrinkage control. However, the bond (shear-bond and adhesion) between blocks is debilitated by cracking resulting from this phenomenon, which has negative consequences on the in-plane shear behaviour of the walls. However, the building process can be adapted to mitigate this problem. The "*pisé*" technique is a French variation of rammed earth, where a layer of lime mortar between the blocks is included. The lime mortar cures for several weeks, during which it remains plastic and allows shrinkage movement between blocks without cracking [4].

The inclusion of hard materials, such as stones, bricks and tiles, in the horizontal joints is frequently observed in traditional rammed earth dwelling from the Iberian Peninsula (Fig. 20a). This is thought to improve the shear-bond behaviour between blocks and thus it can be adapted for new constructions. In a set of surveys carried out in Columbia, Lacouture et al. [22] found several cases of reinforcement of the vertical joints. These reinforcements consisted of timber elements that were embedded in the rammed earth to connect two contiguous blocks (Fig. 20b). The same principle can be applied in new rammed earth constructions to improve their structural behaviour.

The introduction of vertical reinforcements during the compaction of the walls, such as bamboo canes, can also result in the improvement of their shear and out-of-plane behaviours [49]. However, the presence of the reinforcements renders difficult the compaction process by making it more time consuming and by compromising the density of the material around the reinforcements.

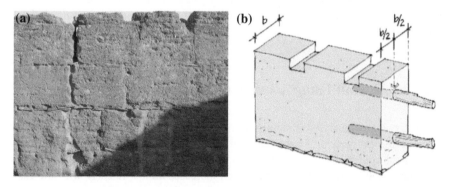

Fig. 20 Strengthening of rammed earth joints: **a** inclusion of schist stones in the horizontal joints (abandoned dwelling); **b** reinforcement of the vertical joint by timber elements (based on [22])

Providing reinforced concrete frames for the rammed earth [50] is currently a usual solution used in southern Portugal to obtain a construction with improved seismic behaviour. The concrete frames assume the structural function of the building, while the rammed earth serves as infill. The uncertainties about the non-standard feature of rammed earth and about the respective design are thus eliminated, and the reinforced concrete structure is designed according to the current codes.

As it was shown, there are several solutions to improve the structural behaviour of rammed earth constructions. Nevertheless, a promising future for rammed earth construction in the global building industry requires the development of knowledge on this building technique and the consequent development and establishment of specific design codes. For example, the recent publishing of the book Rammed Earth: Design and Construction Guidelines [51] led to an increase in rammed earth construction in the UK.

There have been recent advances in the modelling of unstabilised rammed earth samples, by using theories borrowed from unsaturated soil mechanics. These suggest that in an unstabilised soil, the soil particles are held together by small bridges of water acting across the pores between the particles. These bridges are held in place by surface tension and the capillary action due to the relative humidity of the pore air. There are thousands of such bridges holding all the particles together, and these act to provide additional strength and stiffness to an unsaturated soil when compared to a completely saturated or dry soil [52]. Using such a model helps to explain the behaviour of unstabilised rammed earth in the presence of water, why the strength reduces at increasing water contents, and why collapse eventually occurs when a section of rammed earth becomes saturated [53].

The relationship between stabilised rammed earth and water is much more complex, with initially free water being used in the hydration of the cementing products, and further water acting across pores as for an unstabilised sample. In this case, it is difficult to directly infer the behaviour of the rammed earth because of the complex interactions between the water and the cement. There are many aspects of this research are still being investigated, such as the nature of the pore network in rammed earth, and the development of the micro-structure of stabilised rammed earth.

5.2 Stabilization Using Alkaline Activativated Fly Ash

As referred previously, soil stabilisation by addition of cement is often used to improve the properties of rammed earth. According to Reddy and Kumar [54], the cement content has the most important contribution for the embodied energy of cement stabilized rammed earth, which increases almost linearly with increasing cement content. Lax [55] demonstrated that for a specific case, the embodied energy of 8 % cement stabilized rammed earth and that of unstabilized rammed earth is 1.84 times higher. The difference can be even higher if traditional compaction methods are used instead of those mechanical and if no transportation of soil is required, as was assumed in that case. It appears that the impact of cement stabilization on the performance of rammed earth construction is substantial; however, this procedure has a negative impact in the sustainability of this building technique.

Soil stabilization is often required for safety and durability reasons. However, preserving the sustainability of rammed earth requires adopting solutions that include materials with low embodied energy. Adding natural materials, such as cow-dung, sawdust and straw, is a possibility, but the problem lies on their low effectiveness and reliability when compared with addition of cement or lime.

A possibility for reducing the impact of the stabilization process on sustainability of rammed earth consists in using industrial wastes, such as fly ash. Fly ashes result from the combustion of coal in power stations. This thermal treatment makes it a material prone to be used in a technique termed alkaline activation. The alkaline activation of raw materials such as fly ash enables the formation of what is called a geopolymeric binder. This binder can then be mixed with the soil and upon hardening it forms a matrix that involves and binds the particles, forming a soil-binder interface that usually delivers strength levels higher than the soil alone. In general terms, alkaline activation consists in a reaction between alumina-silicate materials and alkali or alkali earth substances, namely: ROH, $R(OH)_2$, R_2CO_3, R_2S, Na_2SO_4, $CaSO_4 \cdot 2H_2O$, $R_2 \cdot (n)SiO_2$, in which R represents an alkaline ion like sodium (Na^+) or potassium (K^+), or an alkaline earth ion like calcium (Ca^{2+}). It can be described as a polycondensation process, in which the silica (SiO_2) and alumina (AlO_4) tetrahedra interconnect and share the oxygen ions. The resulting polymeric structure of Al–O–Si bonds is the main structure of the hardened geopolymer matrix, which is very similar at a molecular level to natural rocks, sharing their stiffness, durability and strength.

The authors have been doing some research on the stabilization of granitic residual soils using alkaline activated fly ash. These soils, typically found in northern Portugal, are formed from the weathering of granite rock. They are characterized by a well graded grain size distribution and very low plasticity indexes, which usually classifies them as silty sands (SM) and clayey sands (SC). In fact, the clay content of these soils is typically very low, making them unsuitable for unstabilized rammed earth construction, or even for stabilization with lime. Escobar [56] tested three granitic residual soils from northern Portugal, and verified that the dry compressive strength of rammed earth specimens (compacted with the standard Proctor test density) barely exceeded 0.4 N/mm². This is a very low value for unstabilized rammed earth construction, and thus building with these soils is only feasible if they are chemically stabilized.

Following the work of Escobar [56], one of the assessed soils (soil S1) was stabilized using alkaline activation of fly ash, using a broad set of compositions, and the unconfined compressive strength as the control parameter (see [57] for details). The activator was composed by sodium silicate and sodium hydroxide and the fly ash was class F (contains less than 20 % CaO and has pozzolanic properties). The effect of several variables was analysed, such as: soil maximum particle size, liquid:solid ratio, activator concentration and Na_2O:ash ratio. A further analysis was performed in order to quantify the effects of additives like hydrated lime, sodium chloride and concrete super-plasticisers.

The main findings of this composition study were that the compressive strength is greatly increased by the geopolymer binder, even in the case of the lowest fly ash content (about 15 % in wt.); the compressive strength varies between 3 and 23 N/mm^2, for curing periods between 1 and 7 days at 60 °C. The application of this technique to the construction of rammed earth walls shows significant potential. Nevertheless, further research is required for optimizing the mixture composition in accordance with the technical and sustainability requirements of rammed earth. This will result possibly in lower binder quantities, since the strength values obtained were higher than needed for this type of structure. Authors are currently assessing the eco-efficiency of this stabilization technique and comparing it with that from soil–cement stabilization.

6 Conclusion and Final Remarks

The rammed earth building stock is large and includes several heritage places and monumental architecture, which are of great interest to preserve. There are several factors contributing to the decay of rammed earth construction; however, the most common pathologies found are cracking and loss of material. Repairing such damage is crucial in order to preserve these constructions. It should be required that the repairing materials should be similar to the original material as much as possible, which requires using earthen materials in such interventions. To refill lost volumes of material, there are several techniques that can be used, such as reconstruction by ramming earth in the damaged area or by projecting it. On the other hand, cracks can be repaired by filling or stitching.

The strengthening of rammed earth constructions is another form of preservation. The majority of the intervention techniques discussed here aim at improving the out-of-plane behaviour of rammed earth walls, which include buttressing, tie rods, grouted anchors and rendering meshes. Finally, the approach of injection of mud grouts for repairing cracks in rammed earth walls was discussed in detail.

It was shown that the clay fraction has great influence on the rheological behaviour of an unmodified mud grout. The clay particles present in the mud grout tend to flocculate to form an internal house-of-cards or scaffold structure, which opposes the flow. The addition of a deflocculant promotes the repulsion between clay particles and promotes their dispersion in the liquid phase, which weakens and disrupts the aforementioned structure. This improves the fluidity of the grouts and allows increasing the solid fraction without compromising this last property. Moreover, adding increasing amounts of silt size materials (such as limestone powder), for decreasing the clay content of the mix, also allows a further increase of the solid fraction. Thus, designing a mud grout with adequate ϕ_v (to avoid excessive drying shrinkage) demands accounting for the previous two effects.

Regarding the strength of unmodified mud grouts, it was shown that the higher the clay content, the higher the flexural and compressive strength. Nevertheless, the maximum clay content of a mud grout must be limited, since excessive clay content has a negative impact on its rheological behaviour. A compromise should be found between these properties.

The adhesion capacity of selected mud grouts was also assessed. The mud grouts recovered a reasonable part of the initial strength of the specimens (between 51 and 93 %), but not completely. However, in practical terms, this shows the great potential of the injection of mud grouts in the conservation of the rammed earth heritage.

Despite the recognized sustainability of rammed earth construction, its future in the building industry requires adopting improvement measures in order to adequate it to modern demands. Both the material and the building process can be improved for this purpose, which includes solutions such as chemical stabilization, reinforcement of the joints between blocks, introduction of vertical reinforcements (bamboo canes) and embedment of reinforced concrete structure.

Finally, it was presented and discussed an alternative chemical stabilization technique, which consists in using alkaline activated fly ash. This is a technique in development, but preliminary results on granitic residual soils showed its great capacity in improving the compressive strength. Further developments are required to optimize the mixture composition in accordance with technical and sustainability requirements.

References

1. Houben, H., Guillaud, H.: Earth Construction: A Comprehensive Guide. CRATerre—EAG. Intermediate Technology Publication, London (2008)
2. Delgado, M., Guerrero, I.: The selection of soils for unstabilised earth building: A normative review. Constr. Build. Mater. 21(2), 237–351 (2007)
3. Jaquin, P.A., Augarde, C.E., Gerrard, C.M.: Chronological description of the spatial development of rammed earth techniques. Int. J. Architect. Herit. 2(4), 377–400 (2008)
4. Minke, G.: Building with Earth: Design and Technology of a Sustainable Architecture. Birkhäuser, Basel (2006)
5. Jaquin, P.A.: Analysis of historic rammed earth construction. Ph.D. thesis, Durham University, Durham, United Kingdom (2008)
6. Cointeraux, F.: Traite dês Constructions Rurales et e Leur Disposition. Paris, France (1791)
7. Correia, M.: Rammed Earth in Alentejo. Argumentum, Lisbon (2007)
8. Fernandes, M.: A Taipa no Mundo. Seminário de Construção e Recuperação de Edifícios em Taipa, Almodôvar (2008)
9. Jest, C., Chayet, A., Sanday, J.: Earth used for building in the Himalayas, the Karakoram and Central Asia—Recent Research and Future trends. 6th International Conference on the Conservation of Earthen Architecture, Las Cruces, New Mexico, Getty Conservation Institute, Los Angeles (1990)
10. De Sensi, B.: Terracruda, La Diffusione Dell'architettura Di Terra—Soil, Dissemination of Earth Architecture (2003)
11. Warren, J.: Conservation of Earth Structures. Butterworth Heinemann, Bath (1999)
12. Keefe, L.: Earth Building: Methods and Materials, Repair and Conservation. Taylor & Francis, London (2005)

13. Tolles, E., Kimbro, E., Ginell, W.: Planning and Engineering Guidelines for the Seismic Retrofitting of Historic Adobe Structures. The Getty Conservation Institute, Los Angeles (2002)

14. Houben, H., Avrami, E.: Summary report, Project Terra research meeting. Plymouth University, Torquay, England (2000)

15. Costa, J.P., Cóias, V., Pifano, A.: Avantages de 1 aterre projetée dans la conservation structurelle du patrimoine en terre. In: Séminare: Le Patrimoine Architectural d'Origine Portugaise au Maroc: Apports à sa Conservation, Rabat, Marrocos (2008)

16. Mileto, C., Vegas, F., López, J.M.: Criteria and intervention techniques in rammed earth structures. The restoration of Bofilla tower at Bétera. Informes de la Construcción **63**, 81–96 (2011)

17. Cóias, V., Costa, J.P.: Terra Projectada: Um Novo Método de Reabilitação de Construções em Taipa. In: Achenza, M., Correia, M., Cadinu, M., Serra, A. (eds.) Houses and Cities Built with Earth: Conservation, Significance and Urban Quality, pp. 59–61. Argumentum, Lisboa (2006)

18. García, R.M.: Construcciones de Tierra. El Tapial. Nuevo Sistema para Construcción y Restauración Mediante la Técnica de "Tierra Proyectada". Ph.D. thesis, University of Granada, Granada, Spain (2010)

19. Parreira, D.J.: Análise Sísmica de uma Construção em Taipa. Tese de Mestrado. Instituto Superior Técnico, Lisboa (2007)

20. Ashurst, J., Ashurst, N.: Practical Building Conservation. Brick, Terracotta and Earth. Gower Press, Aldershot (1988)

21. Pearson, G.T.: Conservation of Clay and Chalk Buildings. Donhead Publishing Ltd, Shaftsbury (1997)

22. Lacouture, L., Bernal, C., Ortiz, J., Valencia, D.: Estudios de vulnerabilidad sísmica, rehabilitación y refuerzo de casas en adobe y tapia pisada. Apuntes **20**(2), 286–303 (2007)

23. Pérez, C., Valencia, D., Barbosa, S., Saavedra, P., Escamilla, J., Díaz, E.: Rehabilitación sísmica de muros de adobe de edificaciones monumentales mediante tensores de acero. Apuntes **20**(2), 369–383 (2007)

24. Toumbakari, E.: Lime-Pozzolan-cement grouts and their structural effects on composite masonry walls. Ph.D. Thesis, Katholieke Universiteit Leuven, Leuven, Belgium (2002)

25. Oliver, A.: Conservations of nondecorated earthen materials, In: Avrami, E., Guillaud, H., Hardy, M. (eds.) Terra Literature Review: An Overview of Research in Earthen Architecture Conservation, pp. 108–123. The Getty Conservation Institute, Los Angeles (2008)

26. Roselund, N.: Repair of cracked walls by injection of modified mud. In: Proceedings of the 6th International Conference on the Conservation of Earthen Architecture: Adobe 90 Preprints, pp. 336–341 Las Cruces, New Mexico (1990)

27. Tolles, E.L., Webster, F.A., Crosby, A., Kimbro, E.E.: Survey of Damage to Historic Adobe Buildings after the January 1994 Northridge Earthquake. The Getty Conservation Institute Scientific Program Report, Los Angeles (1996)

28. Jäger, W., Fuchs, C.: Reconstruction of the Sistani house at Bam Citadel after the collapse due to the earthquake 2003. In: D'Ayala, D., Fodde, E. (eds.) Preserving Safety and Significance, Proceedings of the VI International Conference on Structural Analysis of Historic Constructions, vol. 2, pp. 1181–1187. Bath, UK (2008)

29. Vargas, J., Blondet, M., Cancino, C., Ginocchio, F., Iwaki, C., Morales, K.: Experimental results on the use of mud-based grouts to repair seismic cracks on adobe walls. In: D'Ayala D., Fodde E. (eds.) Preserving Safety and Significance, Proceedings of the VI International Conference on Structural Analysis of Historic Constructions, pp. 1095–1099. Bath, UK (2008)

30. On Yee, L.: Study of earth-grout mixtures for rehabilitation. MSc thesis, University of Minho, Guimarães, Portugal (2009)

31. Silva, R.A., Schueremans, L., Oliveira, D.V.:(2010) Repair of earth masonry by means of grouting: importance of clay in the rheology of a mud grout. In: Proceedings of the 8th International Masonry Conference, pp. 403–412. Dresden, Germany

32. Silva, R.A., Schueremans, L., Oliveira, D.V.: Grouting as a repair/strengthening solution for earth constructions. In: Proceedings of the 1st WTA International Ph.D. Symposium, pp. 517–535. WTA publications, Leuven (2009)
33. Silva, R.A.: Caracterização experimental de alvenaria antiga: reforço e efeitos diferido. MSc. Thesis, Guimarães: University of Minho (2008)
34. Silva, R.A., Schueremans, L., Oliveira, D.V., Dekoning, K., Gyssels, T.: On the development of unmodified mud grouts for repairing earth constructions: rheology, strength and adhesion. Mater. Struct. **45**(10), 1497–1512 (2012)
35. ASTM C 939-94 : Standard test method for flow of grout for preplaced-aggregate concrete (flow cone method, American Society for Testing and Materials. West Conshohocken, PA (1994)
36. CEN. EN 1015-11 : Methods of test for mortar for masonry—Part 11: Determination of flexural and compressive strength of hardened mortar. European Committee for Standardization, Brussels (1999)
37. Van Olphen, H.: Clay Colloid Chemistry: for Clay Technologists, Geologists, and Soil Scientists, 2nd edn. Wiley-Interscience, New York (1977)
38. Andreola, F., Castellini, E., Ferreira, J.M.F., Olhero, S., Romagnoli, M.: Effect of sodium hexametaphosphate and ageing on the rheological behaviour of kaolin dispersions. Appl. Clay Sci. **31**, 56–64 (2006)
39. Legaly, G.: Colloid clay science. In: Bergaya, F., Theng, B.K.G., Lagaly, G. (eds.) Developments in Clay Science: Handbook of Clay Science, pp. 141–245. Elsevier, The Netherlands (2006)
40. Papo, A., Piani, L., Ricceri, R.: Sodium tripolyphosphate and polyphosphate as dispersing agents for kaolin suspensions: rheological characterization. Colloids. Surf. A Physicochem. Eng. Asp. **201**, 219–230 (2002)
41. Brás, A., Henriques, F.M.A.: Natural hydraulic lime based grouts—The selection of grout injection parameters for masonry consolidation. Constr. Build. Mater. **26**, 135–144 (2012)
42. NMAC.: NMAC 14.7.4: Housing and Construction: Building Codes General: New Mexico Earthen Building Materials Code, New Mexico Regulation and Licensing Department, Santa Fe, New Mexico (2006)
43. SNZ.: New Zealand Standard 4298:1998, Materials and workmanship for earth buildings, Standards New Zealand, Wellington (1998)
44. Bui, Q.B., Morel, J.C., Hans, S., Meunier, N.: Compression behaviour of non-industrial materials in civil engineering by three scale experiments: the case of rammed earth. Mater. Struct. **42**(8), 1101–1116 (2008)
45. SNZ.: New Zealand Standard 4297:1998, Engineering design of earth buildings. Wellington: Standards New Zealand, Wellington (1998)
46. SNZ.: New Zealand Standard 4299:1998. Earth buildings not requiring specific design. Wellington: Standards New Zealand, Wellington (1998)
47. Pacheco-Torgal, F., Jalali, S.: Earth construction: lessons from the past for future eco-efficient construction. Constr. Build. Mater. **29**, 512–519 (2012)
48. Maniatadis, V., Walker, P.: A review of rammed earth construction. Natural Building Technology Group, Department of Architecture & Civil Engineering, University of Bath (2003)
49. Minke, G.: Construction Manual for Earthquake-Resistant Houses Built of Earth. Gate—BASIM (Building Advisory Service and Information Network). Eschborn, Germany (2001)
50. Gomes, M.I., Lopes, M., Brito, J.: Seismic resistance of earth construction in Portugal. Eng. Struct. **33**(3), 932–941 (2011)
51. Walker, P., Keable, R., Martin, J., Maniatidis, V.: Rammed earth: design and construction guidelines. BRE Bookshop, Watford (2005)
52. Jaquin, P.A., Augarde, C.E., Gallipoli, D., Toll, D.G.: The strength of unstabilised rammed earth materials. Géotechnique **59**(5), 487–490 (2009)
53. Jaquin, P.A., Augarde, C.E.: Earth Building: History Science and Conservation. Bre Press, Bracknell (2012)

54. Reddy, B.V.V., Kumar, P.P.: Embodied energy in cement stabilised rammed earth walls. Energy and Buildings **42**, 380–385 (2010)
55. Lax, C.: Life cycle assessment of rammed earth. MSc. thesis, University of Bath, United Kingdom (2010)
56. Escobar, M.C.: Rammed earth: feasibility of a global concept applied locally. MSc. thesis, University of Minho, Portugal (2011)
57. Cristelo, N., Glendinning, S., Miranda, T., Oliveira, D.V., Silva, R.A.: Soil stabilisation using alkaline activation of fly ash for self compacting rammed earth construction. Constr. Build. Mater. **36**, 727–735 (2012)

Characterization and Damage of Brick Masonry

Paulo B. Lourenço, Rob van Hees, Francisco Fernandes
and Barbara Lubelli

Abstract Clay brick is among the oldest used masonry materials. Given the technological evolutions since the industrial revolution, old bricks are much different from todays' bricks. This chapter provides a review on the chemical, physical and mechanical properties of mortar, brick and masonry. In addition, a discussion on the possible causes of damage and the usage of expert systems in building diagnostics is also given.

KeyWords Clay brick • Brick masonry • Mortar • Mechanical properties • Physical properties • Chemical properties

P. B. Lourenço (✉)
ISISE, Department of Civil Engineering, University of Minho,
Azurém, Guimarães 4800-058, Portugal
e-mail: pbl@civil.uminho.pt

R. van Hees
TNO, Netherlands Organisation for Applied Scientific Research,
PO Box 49, 2600AA Delft, The Netherlands
e-mail: rob.vanhees@tno.nl

R. van Hees
Faculty Architecture, TU Delft, Delft, The Netherlands

F. Fernandes
ISISE, Department of Engineering and Technologies, University Lusíada,
Largo Tinoco de Sousa, Vila Nova de Famalicão P 4760-108, Portugal
e-mail: francisco.fernandes@fam.ulusiada.pt; fmcpf@civil.uminho.pt

B. Lubelli
Faculty Architecture, TU Delft, PO Box 5043, 2600GA Delft, The Netherlands
e-mail: b.lubelli@tudelft.nl

B. Lubelli
TNO, Netherlands Organisation for Applied Scientific Research, Delft, The Netherlands

A. Costa et al. (eds.), *Structural Rehabilitation of Old Buildings*,   109
Building Pathology and Rehabilitation 2, DOI: 10.1007/978-3-642-39686-1_4,
© Springer-Verlag Berlin Heidelberg 2014

1 Introduction

Clay brick masonry, often in combination with stone masonry and timber floors, is well distributed all over the world. Clay brick, in its forms of sun dried and burnt, has been around since the beginning of civilization. Brick was easily produced, lighter than stone, easy to mould, and formed a wall that was fire resistant and durable. The characterization of old clay bricks is a hard task due to the difficulties in collecting samples, the scatter in the properties, and the lack of standard procedures for testing [1]. Still, characterization is relevant to understand damage, to assess safety, to define conservation measures, and even to make a decision on reusing or replacing existing materials, as modern materials can be unsuitable from a chemical, physical or mechanical perspective. Information about old and handmade bricks is scarce. Ancient materials generally differ from modern ones and, frequently, exhibit high porosity and absorption, and low compressive strength and elastic modulus. The mechanical properties of brick are relevant for the performance of historical constructions, as this is the main influence factor on the compressive strength and durability of masonry. Here, the properties of historical brick masonry and its components are addressed in detail.

In addition, existing buildings often exhibit damage and knowledge based diagnostic systems are much helpful for practitioners. Through a damage atlas, integrated in a diagnostic part, visual observation can be extensively used and a correct definition of the observed degradation and damage can be made, or at least one or more hypotheses can be proposed. Possible intervention techniques are not usually automatically generated by a diagnostic module, but can be available in a background information section. These aspects are also addressed next.

2 Chemical, Physical and Mechanical Characterization of Old Mortars

Mortar is a material composed of one or more inorganic binders, aggregates, water and admixtures used in masonry to provide bedding, jointing and bonding of masonry units, or used for functions like plasters and renders. Main focus here is on lime mortars, as the most diffused type of mortar found in historic buildings. Figure 1 shows a thin section of a mortar, with binder and aggregates clearly visible.

The use of lime, $Ca(OH)_2$, dates back to pre-historic times even if the Egyptians used burned gypsum ($CaSO_4.\frac{1}{2}H_2O$) as a mortar in between the limestone blocks for the construction of the pyramids [2]. The first examples of lime as a binder in mortar date back to the sixth century BC in Greece [3]. The Romans developed a new type of (pozzolanic) mortar, a sort of concrete, with hydraulic properties. Vitruvius [4] describes the Roman knowledge of lime technology, with details on different types of lime binder, process of calcination and slaking as well as recipes for mortar composition and origin of the best sand.

Fig. 1 Thin section of a lime mortar, originating from an early eighteenth century canal bridge in Amsterdam. *B* binder, *Z* aggregate

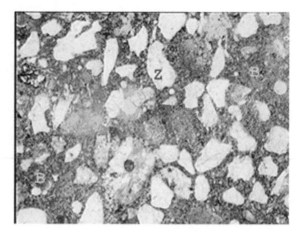

In the end of the nineteenth century, these mortars were replaced by cement based mortars. This occurred mainly because cement binders can harden and develop strength much quicker than lime binders. Incompatibility problems in the use of cement based mortars for conservation lead to the re-discovery of lime based products. Currently lime mortars are increasingly popular in conservation because of their good compatibility (physical, chemical and mechanical) with materials present in ancient buildings.

A binder can be defined as a material with adhesive and cohesive properties, which bonds mineral fragments in a coherent mass. A first distinction can be made between air-hardening (non-hydraulic) binders, which slowly harden in air, and hydraulic binders, which set and harden by chemical interaction with water. Air-hardening binders include air lime and gypsum. Hydraulic binders include hydraulic lime, lime-pozzolan, lime-cement, lime-cement-pozzolan and cement. The lime binder is obtained by a calcination process, the burning of (pure) limestone at ca. 900 °C. The obtained quicklime is slaked with water to become dry hydrated lime or, in case of an excess of added water, putty lime.

Natural hydraulic lime is obtained by calcination of limestone containing a certain amount of clay. Lime pozzolan binders are obtained by the addition of a pozzolan (natural or artificial) to the lime while mixing mortar. A natural pozzolan is a volcanic material, which originally derives from Pozzuoli, an Italian region around Vesuvius. Pozzuoli earth was used in the Roman mortars but other natural pozzolans are Santorini earth (Greece) and trass (Germany). Artificial pozzolans include metakaolin, silica fume, brick dust (preferably low fired brick) and others such as fly ash.

An aggregate can be defined as particles of rock, from natural origin or artificially crushed, with a range of particle sizes from 63 μm to 4 mm, or even 8 mm. Apart from rock aggregates, light aggregates exist such as expanded clay, vermiculite or perlite. The most common aggregate used in lime mortars is calcareous or siliceous sand, which is constituted by grains of minerals and stone. The role of sand in mortar is to make the mortar less fat, to reduce crack formation

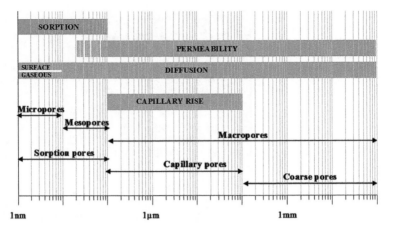

Fig. 2 Classification of pores and moisture transport mechanisms

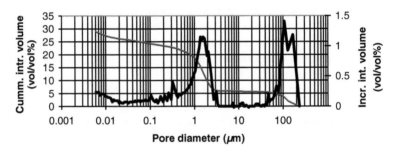

Fig. 3 Pore size distribution of a mortar, assessed by mercury intrusion porosimetry (MIP). The total porosity (ca. 29 %) can also be obtained by this technique

due to shrinkage during drying, and to give strength, hardness and porosity to the mortar. The grain size distribution of the sand has a great influence on the final porosity and pore size distribution of mortar. For example, the presence of both fine and coarse grains results in a low porosity. Porosity strongly influences hardening, mechanical strength, physical properties and durability. The ratio between the mortar components may vary depending on the quality of lime, sand and on the final use of the mortar. Historical mortars have a binder-sand ratio which may vary between 2:1 and 1:4 by volume.

Air lime hardens by reacting with carbon dioxide from the air to form a carbonate. Gypsum hardens by hydration of the hemidrate form to the di-hydrate form. Hydraulic lime contains a mix of hydrated lime, silicates and aluminates. Hardening occurs through reaction with water and by carbonation. Pozzolans, used in combination with air lime to obtain a hydraulic mortar, have in common a considerable content of silica and alumina. The knowledge of the chemical properties of a mortar is important for understanding damage processes and for designing a repair mortar, chemically compatible with the existing one. For example a gypsum based repair mortar is chemically not compatible with a dolomitic lime mortar, since, in presence of water, harmful magnesium sulphate may be formed.

Moisture transport behaviour is one of the most relevant physical properties of mortar, since it strongly influences its durability. Moisture transport behaviour mainly depends on the porosity and the pore size distribution, which can be subdivided into sorption pores (<100 nm), capillary pores (between 0.1 and 100 μm) and coarse pores (>100 μm), see Fig. 2. Small capillary pores (<1 μm) result from evaporation of water from the binder fraction. Wider capillary pores are formed by the intergranular space that is not completely filled by the binder. Figure 3 shows the pore size distribution of a mortar with a bimodal distribution.

Air lime mortars are known to have a low mechanical strength, in comparison with hydraulic lime and even more in comparison with cement based mortars. However, their capability of deformation is much higher than that of cement mortars. The strength of air lime mortars develops due to carbonation. This process may take several years, especially in very thick masonry. Therefore the strength of air lime mortar will be very low, especially in the first period after brick lying. However, if good conditions for carbonation are present, sufficient strength is developed over time guaranteeing a long service life. The strength of a mortar is greatly affected by the porosity and decreases with an increase in porosity. Compressive strength of air lime mortars ranges around 1.5–2.0 N/mm^2, whereas hydraulic lime mortars may reach ca. 10 N/mm^2. Apart from the mechanical strength of the mortar joint, the bond strength between the masonry unit (brick or stone) and the mortar joint is important. A low binder/aggregate ratio, poor grading of the aggregate or inadequate tooling of the mortar may limit the bond strength.

3 Characterization of Old Bricks

Clay bricks also exhibit different properties, which are important in the evaluation of the strength, durability and resistance to deterioration processes. These properties are closely related to the quality of the raw clay and the conditions of manufacture, namely drying and firing processes. The properties of construction materials can be grouped as chemical, physical and mechanical. Progressive ageing of bricks and permanent loads lead to material deterioration such as cracking, peeling or efflorescence, meaning that the properties exhibited currently by old clay bricks are affected in some degree and are not necessarily the original properties.

Bricks are constituted by a mixture of raw clay and water. The first step to characterize the raw clay is by means of chemical and mineralogical studies [5, 6], which are fast to perform and only require small amounts of material. This information can be also used to identify suitable raw materials for the production of missing parts or the replacement of deteriorated ones, as long as the production processes are as close as possible from the original.

The chemical composition of brick samples can be determined by x-ray fluorescence spectrometry, much used for old ceramics [7], which allows the identification of the following abundant chemical oxides and elements: silicon oxide (SiO_2), aluminum oxide (Al_2O_3), iron oxide (Fe_2O_3), potassium oxide (K_2O), titanium dioxide (TiO_2), sodium oxide (Na_2O), calcium oxide (CaO) and magnesium oxide (MgO). Silicon and aluminum oxides constitute the base elements of the clay. As an example, clay bricks from the twelfth–thirteenth century presented 38 % of silicon oxide, 21.5 % of aluminum oxide and 32.5 % of ferrous oxide, in weight [8]. In Portugal, several samples from clay bricks (Fig. 4) from monuments spread through the country and from the twelfth–nineteenth centuries were studied in [9]: Outeiro (OU, seventeenth century), Pombeiro (PO, twelfth–sixteenth century), Salzedas (SA, twelfth–eighteenth century), Tarouca (TA, twelfth–seventeenth century), Tibães (TI, seventeenth century) and Tomar (TO, eighteenth–nineteenth century). The results reported in Table 1 show that the base chemical components of the raw clay used on the bricks is relatively uniform, consisting of 54–61 % of

Fig. 4 Photographs of typical old Portuguese bricks: **a** Salzedas (*SA*), **b** Outeiro (*OU*), **c** Pombeiro (*PO*), **d** Tarouca (*TA*); **e** Tibães (*TI*) and **f** Tomar (*TO*)

Table 1 Average chemical composition of old Portuguese bricks (coefficient of variation in *square brackets*)

	SiO_2	Al_2O_3	Fe_3O_4	K_2O	Na_2O	TiO_2	CaO	MgO
OU	56.2	25.3	11.4	3.5	0.5	1.0	0.3	1.5
	[9 %]	[5 %]	[41 %]	[14 %]	[40 %]	[10 %]	[47 %]	[29 %]
PO	57.5	25.1	8.4	4.9	0.5	1.3	0.4	1.6
	[5 %]	[10 %]	[18 %]	[12 %]	[33 %]	[10 %]	[106 %]	[19 %]
SA	54.4	32.2	4.1	5.1	2.0	0.3	0.8	0.9
	[5 %]	[8 %]	[61 %]	[17 %]	[35 %]	[87 %]	[66 %]	[41 %]
TA	55.6	30.9	4.1	5.0	1.9	0.4	0.9	1.0
	[7 %]	[13 %]	[22 %]	[15 %]	[41 %]	[26 %]	[84 %]	[29 %]
TI	53.8	29.4	8.1	4.4	0.5	1.2	0.9	1.4
	[6 %]	[9 %]	[18 %]	[11 %]	[31 %]	[8 %]	[60 %]	[20 %]
TO	60.8	21.6	7.0	3.6	0.4	0.8	3.6	2.2
	[4 %]	[10 %]	[9 %]	[22 %]	[23 %]	[12 %]	[81 %]	[15 %]

SiO_2 and 22–32 % of Al_2O_3. The presence of CaO and Na_2O is often due to contamination by lime mortars or salt, respectively.

These results were processed using statistical analysis, which compares the chemical composition the bricks with the chemical composition of known ceramic samples [7]. The analysis revealed that SiO_2 contributes very little to the distinction between old samples and that no single component was found to strongly influence the provenance of the old bricks. The chemical composition of the clay found in bricks is different from ceramics and suggests that the raw clays used in the manufacture were obtained locally.

Firing of clay bricks produces a series of mineralogical, textural and physical changes that depend on many factors and influence the porosity [6, 8]. Porosity is again an important parameter concerning clay bricks due to its influence on properties such as chemical reactivity, mechanical strength, durability and quality of the brick. Generally, the quality of the brick, both in terms of strength and durability, increases with the decrease of the porosity. Commonly, historic clay bricks exhibit high porosity values, ranging between 20 and 50 % [10, 11]. The dimension and distribution of the pores are influenced by the quality of the raw clay, the amount of water and the firing temperature. If the firing temperature increases, the proportion of large pores (3–15 μm) increases and the connectivity between pores is reduced, whereas the amount of thin pores diminishes [6, 12]. This has a strong impact on the durability of the bricks as it has been shown that large pores are less influenced by soluble salts and freeze/thaw cycles. Several studies [1, 6, 13] reported that the formation of thin pores (<1.5 μm) is promoted by carbonates in the raw clay and by a firing temperature between 800 and 1000 °C. Such a pore size influences negatively the quality of bricks as their capacity to absorb and retain water increases. The density or bulk mass is related with mechanical and durability properties, and typical values range between 1200 and 1900 kg/m^3 [5, 8, 10]. Table 2 presents the results from old Portuguese clay bricks, with resulted in average values for the bulk mass of 1750 kg/m^3 and 29 % for porosity.

Table 2 Average porosity, bulk mass and compressive strength for old bricks from Portuguese monuments and the coefficients of variation between *square brackets* [7]

	Porosity (%)	Bulk mass (kg/m^3)	Compressive strength (N/mm^2)
OU	33.0 [13.9 %]	1742 [1.7 %]	8.5 [28 %]
PO	26.3 [25.5 %]	1754 [2.2 %]	9.2 [54 %]
SA	28.2 [10.6 %]	1800 [1.9 %]	14.5 [32 %]
TA	29.2 [14.5 %]	1747 [1.8 %]	8.7 [41 %]
TI	30.4 [14.7 %]	1739 [1.5 %]	6.7 [55 %]
TO	27.5 [14.2 %]	1656 [3.0 %]	21.8 [31 %]

Table 3 Typical average values for the compressive strength and modulus of elasticity of old bricks found in the literature

Date (century)	Local	Compressive strength (N/mm^2)	Elastic modulus (kN/mm^2)
1–5th	Walls, pillars, vaults and ovens from the Byzantine period	9.2–18.0 [10]	2.6–10.8 [10]
11–13th	Vaults of Our Lady Monastery, Magdeburg, Germany	13.1–14.1 [14]	–
13–17th	Siena's exterior wall, Italy	27.9 [14]	–
15th	Colle Val d'Else exterior wall, Italy	19.9 [14] 30.0 [15]	4.1 [14]
	Pienza Episcopal Palace, Italy	–	7.3 [14] 11.6–18.6 [15]
16th	Monastery of Monte Oliveto Maggiore library wall, Italy	31.1 [14]	6.3 [14]
17th	Salzedas monastery vaults, Portugal	5.2 [16]	7.3 [16]
18th	Lazzaretto de Ancona, Italy	18.5 [14]	5.8 [14]
18–19th	Centenary chimney from the ceramic industry, Spain	20.8 [15]	–

No correlation can be found between the physical or chemical properties and the mechanical properties.

The evaluation of the mechanical strength of old bricks is difficult due to their scatter. They may also be deteriorated by weather or chemical agents such as soluble salts, ice-thawing cycles or load-unload cycles. Additionally, the experimental test set-up conditions (dimensions and moisture content of the sample, boundary conditions, temperature, etc.) can also influence the results. Typical values of the compressive strength of old clay bricks are reported in Table 3, with a wide range of values (from 4 to 32 N/mm²). The average compressive strength of Portuguese old clay bricks as well as its dispersion is reported in Table 2. A large variability on the compressive strength was obtained, with coefficients of variation up to 50 %. It is possible to observe that the bricks with lower f_c exhibit also a higher dispersion. The wide range of strengths found is between 6.7 and 21.8 N/mm², with an average of 11.6 N/mm² considering the total sample and 8.3 N/mm² considering the four weakest bricks.

Another relevant mechanical parameter is the modulus of elasticity. It is not always clear how authors measured the values presented, even if most standards refer the use of the linear part of the stress–strain curve in a range of 10–50 % of the maximum stress value, which is also characterized by a large variability. The values found range from 1 to 18 GPa, which represent between 125 and 1400 f_c, where f_c is the compressive strength. Most common values are in the range of 200 f_c, with an average for the values in Table 3 of 350 f_c.

It is difficult to relate the tensile strength of the masonry unit to its compressive strength due to the different shapes, materials, manufacture processes and volume of perforations. For the longitudinal tensile strength of clay, calcium-silicate and concrete units, Schubert [17] carried out an extensive testing program and obtained a ratio between tensile and compressive strength ranging from 0.03 to 0.10.

4 Mechanical Characterization of Brick Masonry

The properties of brick masonry are strongly dependent upon the properties of its constituents. Traditional masonry is subjected to compressive stresses and the compressive strength of masonry in the direction normal to the bed joints is required for design and safety assessment purposes. Experimentally, this property can be obtained according to the European norm EN 1052-1 [18], see Fig. 5a. This

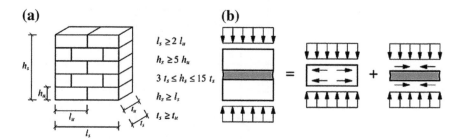

Fig. 5 Uniaxial compressive behaviour of masonry: **a** test specimen according to the European standards (for units with $l_u \leq 300$ mm and $h_u \leq 150$ mm) [17] and **b** schematic plane representation of stresses in masonry components

configuration seems to return the true uniaxial compressive strength of masonry. Mann and Betzler [19] observed that, initially, vertical cracks appear in the units along the middle line of the specimen, i.e., through the vertical joints. Upon increasing deformation, additional cracks appear, normally vertical cracks at the smaller side of the specimen that lead to failure by splitting of the prism. This persuaded researchers to investigate semi-empirical and analytical relations to predict masonry strength based on the components characteristics and on the type of masonry. Several semi-empirical relations can be gathered from the literature, e.g. [20–22], and from the codes [23, 24].

Masonry compressive failure is mainly governed by the interaction between units and mortar. A relevant factor is the difference in elastic properties between the unit and mortar. Assuming compatibility in the deformation of the components and a mortar that is more deformable than the units, the difference in stiffness leads to a state of stress characterized by compression/biaxial tension of units and triaxial compression of mortar, see Fig. 5b. In the pioneer work of Hilsdorf [25], this phenomenon was described and an equilibrium approach was developed to predict the masonry strength, assuming that failure of mortar coincides with failure of masonry. Later [26], this hypothesis is overcome by considering a limit strain criterion based on the lateral strain exhibited by brick units at failure. Other contributions were given in [27–29].

The bond between unit and mortar is often the weakest link in masonry. The nonlinear response of the joints, controlled by the unit-mortar interface, is related to two different phenomena that occur at the unit-mortar interface. One associated with tensile failure and another one associated with shear failure. Different test set-ups have been used for the characterization of the tensile behaviour of the unit-mortar interface. These include flexural testing, (three-point, four-point, bond-wrench) [30], indirect tension testing (splitting test) [31] and direct tension testing [32].

Fig. 6 Biaxial strength of solid clay units masonry [36, 37]

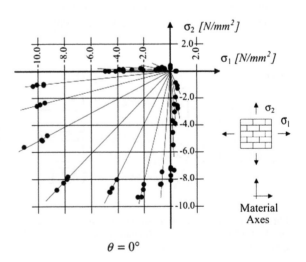

$\theta = 0°$

Experiments on the biaxial behaviour of bricks and blocks are scarce. The influence of the biaxial stress state has been investigated up to peak stress to provide a biaxial strength envelope, see Fig. 6. Basically, two different test set-ups have been utilized, uniaxial compression orientated at a given angle with respect to the bed joints [33] and true biaxial loading at a given angle with respect to the bed joints [34, 35].

5 Deterioration and Damage Mechanisms

The most important factors influencing degradation and damage to masonry are related to: environment; materials; building's design; craftsmanship in the construction of the building and its maintenance.

Environmental factors include, for example, the presence of moisture and salts, air pollution, temperature changes, dynamic loads and soil settlements. Moisture may come from sources like rain penetration, capillary rising damp or flooding. Salts may be originally present in the material (for example, a mortar which has been made using sea water or beach sand), they may come from the environment (aerosol, de-icing salts, etc.) or from the use of the building in the past (for example, chloride from salt storage, nitrates in the case of stables). Temperature variations may give rise to degradation phenomena in masonry due to differential thermal dilation, whereas dynamic loads resulting from earthquakes and vibrations provoked by wind or traffic may cause crack patterns.

Material factors are mainly related to the composition of the mortar (binder/sand ratio, grain size distribution of the aggregate) and the properties of the masonry unit/mortar combination (porosity, capillary moisture transport, adhesion, mechanical strength). Many degradation processes may only occur in the presence of water; consequently the speed at which a material absorbs and releases water has a strong influence on its risk of degradation. Therefore, moisture transport properties, which are related with porosity and pore size distribution, are of primary importance when considering the durability of a mortar and the masonry as a whole.

The design of the building, i.e. its shape, orientation and above all the details, may strongly influence the occurrence and the severity of the degradation. Also craftsmanship in the form of quality of the execution and of adequate conditions for hardening of mortars is an important factor that affects the susceptibility of the mortar to damage.

The degradation processes (chemical, physical and mechanical) exert stresses on the materials, which weaken the material until it fails and damage becomes visible. Degradation can be defined as an increase in decay, which corresponds with a decreasing performance of the material. Thus, damage can be defined as an unacceptable reduction of the performance of the material, affecting its durability. An overview of the factors influencing the durability of masonry is given in Table 4 while Table 5 gives an outline of the most important damage processes affecting masonry and damage types related to those processes.

Table 4 Overview of factors influencing the durability of mortars and masonry

Environment	Moisture supply	Rain, snow
		Ground water
		Surface water
		Floods
	Salt supply	Soil or surface water
		Use of the building (e.g. stable, salt storage)
		Air (aerosol)
		Floods
		De-icing salts
		Cleaning, surface treatments
	Air pollution	
	Exposure to fire	
	Temperature	Variations
		Extremes
	Dynamic loads	Earthquakes
		Wind
		Traffic
		Vibrations
	Differential settlements	
Materials	Mortar composition	Binder type
		Binder/aggregate ratio
		Grain size distribution of the sand
	Properties brick/stone and mortar system	Porosity
		Moisture transport properties
		Adhesion/bond
	Presence of salt in materials	
Design of the building	Original structural design of the building or modification	
	Choice of combinations of materials	
	Detailing of the building	
	Choice of repair methods and materials	
Workmanship and construction procedures	Quality of the execution	Quality of execution
		Mortar mixing on site
		Way materials are cured and curing conditions
		Protection of fresh mortar
	Lack of knowledge on (traditional) workmanship	
Maintenance	Lack of maintenance	
	Inappropriate maintenance program	

Table 5 Overview of the most important damage processes and related damage types

Physical/chemical	Most important damage types
Moisture	Biological growth
Salts	Efflorescence
Frost	Spalling
Pollution	Exfoliation
	Powdering
	(Black) crust
Structural	
Overloading, creep	Crack patterns
Settlement	Displacement/deformation
Thrust arches/vaults	
Earthquakes	

Some of the most important damage processes are discussed next. For processes in which water is involved, the crystallization of soluble salts is probably the most widespread process causing damage to historical masonry buildings. Salt damage can only occur in the presence of both salt and water. Salt moves in the capillary system of the material and accumulates where evaporation occurs. Salt accumulation and crystallization create pressures, which can exceed the mechanical strength of the material and consequently lead to damage.

As the mortar (e.g. bedding or pointing mortar, plaster, render, etc.) and brick or stone are used in masonry in combination with each other, the risk and location of salt damage will depend on the pore size distribution of the mortar/substrate combination [38]. Since moisture (and salt) transport by capillarity moves from larger to smaller pores, a fine porous mortar applied on a coarse porous material will have a larger risk of decaying than a coarse mortar applied on a fine porous substrate (this does however not necessarily imply that a fine porous mortar on a coarse porous substrate would be the wrong choice). Important damaging salts are sulphates (for example Na_2SO_4) and chlorides (for example $NaCl$). Salts precipitating in the pores of a mortar may create pressures, which may lead to damage [39, 40]. As a consequence of salt crystallization, a mortar can show damage in the form of sanding, scaling, exfoliation or crumbling, whereas the masonry units may show damages like powdering, exfoliation and spalling. Sometimes salt crystallization causes damage to a lime bedding mortar because a physically incompatible pointing mortar was chosen. This is the case of a too dense pointing applied on a more porous lime mortar (Fig. 7). Because of the hindering of the drying caused by the new cement pointing, crystallization of salts that were already present in the masonry occurs at the bedding mortar-pointing interface. This results in the detachment of the pointing (also called push-out) and also in a form of loss of cohesion (crumbling or sanding) of the underlying lime bedding mortar.

Apart from pure crystallization, the formation of expansive compounds due to the reaction of salts with mortar components may also cause considerable damage, not only to mortars, but to the masonry as a whole. Sometimes the resulting crack patterns may be mistaken for structural damage (Fig. 8a), where only after drilling

Fig. 7 Push-out of cement re-pointing due to crystallization of salts at the interface of new pointing and old bedding mortar

Fig. 8 a Crack pattern in masonry, looking like structural damage but caused by expansive reaction in the internal part of the mortar. **b** Bursting of pointing mortar due to the formation of CaO.$Al_2O_3.3CaCl_2.31H_2O$ (trichloride)

cores from the masonry it became clear that the cracks originated from swelling of the mortar inside the pier. Additional investigations with optical and electron microscopy revealed the presence of secondary ettringite concentrations, initiating the cracks. Sometimes the pointing mortar is bursting, i.e. it looks like it swells because of an increase of volume (Fig. 8b). In this last example, the damage in the form of bursting of the pointing was shown to be due to the formation of trichloride.

Other examples of such expansive compounds, which may cause damages, are thaumasite and ettringite [41]. Thaumasite and ettringite are the results of the reaction of sulphates (coming for example from the air or from bricks) with components of the hydraulic mortar. Thaumasite ($CaCO_3 \cdot CaSO_4 \cdot CaSiO_3 \cdot 15H_2O$) may form by the reaction of water with calcium carbonate, calcium sulphate and hydrated calcium silicate, which again are present in concrete or mortar mixtures as binders. The composition of hydrated calcium silicates, which may vary within a relatively wide range, is indicated by the generic formula C–S–H. Ettringite ($3CaO \cdot Al_2O_3 \cdot 3CaSO_4 \cdot 32H_2O$) may form by the reaction of water with calcium sulphate and the alumina bearing hydration products ($4CaO \cdot Al_2O_3 \cdot 13H_2O$, $3CaO \cdot Al_2O_3 \cdot 6H_2O$, $C_3A \cdot CaSO_4 \cdot 12 \cdot 18H_2O$, etc.). These products, sometimes indicated as C-A-H, are formed by hydration of Portland cement or other binders, such as hydraulic lime or mixtures of lime and pozzolan [40].

Hydrated lime (air lime) mortar cannot be affected by the reactions described above. In this case another form of expansive reaction, the one consisting in the conversion of the $CaCO_3$ into $CaSO_4 \cdot 2H_2O$ (gypsum), can take place. Sulphates present in the polluted air or coming from the brick react, in the presence of moisture, with the $CaCO_3$ in the mortar to form $CaSO_4 \cdot 2H_2O$, i.e. gypsum.

Damages due to structural causes are generally showing as cracks, often in combination with deformations. The first important step to make a diagnosis of structural damages is the survey and interpretation of the crack pattern. However, the possibility of occurring damages due to non-structural causes has also to be taken into account. The signs of damages given by the crack direction and opening have to be well evaluated.

The crack patterns may be caused by structural failures like overloading, settlements or due to extreme events like earthquakes. The main structural failures that may cause damages affecting the structural stability include: (i) dead load in heavy massive structures; (ii) soil settlements; (iii) horizontal actions due to thrust in arches and vaults; and (iv) extreme events like earthquakes or landslides. The position, the direction and the width of the cracks indicate where the local stress value reaches the strength of the material and hence, indirectly, the type of stress to which it is subjected. Knowing typical causes, which can produce damage to the structure such as vertical and horizontal actions, soil settlements, interactions between walls and floors, roofs and walls, can help understanding the visible effects (cracks, deformations, leaning, etc.) of these actions on the structure.

6 Diagnostic Systems and Expert Systems

The use of expert systems for diagnosis in building practice or for mitigation of the effects of decay and damage to buildings belonging to the cultural heritage is still not very common. The development started fifteen years ago, when the first version of MDDS (Masonry Damage Diagnostic System) was delivered as the outcome of an EU project [38, 42]. The approach concerning the assessment of damage to (historic) masonry buildings is quite comparable to the one used in medicine. In medicine it consists of three steps: anamnesis, diagnosis and therapy. In building diagnostics, generally, steps like survey, (visual) assessment, diagnosis and intervention are commonly used.

Such an approach was already adopted in the early MDDS, although this system would be very restrained in proposing interventions. The MDDS contained a damage atlas, a series of damage processes and a reasoning mechanism that made use of essential conditions to assess whether the occurrence of a certain damage process might be possible or probable. Already in the original system, the damage atlas was a very important tool, initially limited to damages concerning brick. Restricted as that was, it certainly had an important function: the use of a common language (damage terminology). Figure 9 gives an overview of the damage processes included in MDDS. A process is defined by a number of "essential conditions" that would allow the process to occur; together with a set of well-defined damage types (damage atlas), this constituted the backbone of the original system.

In the practical situation of conservation works, an assessment of the state of the building condition is the first and necessary step to properly define the problem that is to be solved. This step also includes the decision on which investigations have to be performed. An assessment will usually start with a visual inspection, or survey, of the building. A correct diagnosis is the "*conditio sine qua non*" for a proper assessment of the damage phenomenon and, subsequently, for the definition of the intervention. The part on structural damages in MDDS was initially underdeveloped, as the system main focus was on damage related to the interaction between materials and environmental factors.

Quite some additions have afterwards been made to the initial system: the reasoning has been very much refined; it is possible to introduce measurement data such as moisture content, salt content, etc. in the reasoning process in order to have a more refined hypothesis and diagnosis. Moreover, other materials have been introduced, like natural stone, mortars, renders and concrete as well as composite constructions such as masonry structures. Systems such as the current MDDS (**Monument** Damage Diagnostic System, successor of the **Masonry** Damage Diagnostic System) intend to facilitate multidisciplinary teamwork by offering a structured, transparent and consistent method for analysing and diagnosing damage [43, 44]. Additionally, several additions and improvements have been made into this extended system, such as those following from the EU COMPASS project [44]. Although MDDS does not fully cover structural damages yet, a structural damage atlas is already available on the basis of the research of de Vent [45, 46] and a more complete structural analysis module will be added.

An investigation carried out with the help of MDDS will start, like any investigation, with the survey: gathering data and performing a visual inspection of the building showing damage. The aim of this first phase of the work is to enable the inspector to acquire more insight in the situation and to structure his observations with the help of the system. The system helps in handling each situation found in

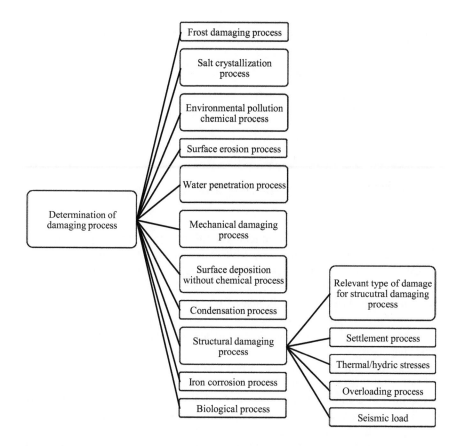

Fig. 9 Tree structure of damage processes contained in the first MDDS

a building as part of a context. Its approach is based on the way of reasoning of an expert. All information considered relevant may be inserted in the system, which is structured in such a way that at three levels, building, construction and material, descriptions can be made and data can be added. The user is free to make annotations on the building, even if they are not directly related to its decay, but will serve other purposes (e.g. statistics). Pictures and drawings can also be inserted in the consultation file, which will eventually be part of the dossier of the building.

The assessment of the type of damage found can be done at distinct levels in the system: at the level of the construction (for example a wall as a whole) and/ or at the level of each constitutive material/construction system (for example masonry unit, brick, stone, plaster, bedding mortar, paint, etc.). With the support of the damage atlas, which has been integrated in the diagnostic part, a correct definition of the observed damage type at both levels is possible (Fig. 10) and the results of the visual observation (i.e. descriptions, photos, drawings, etc.) can be included.

(a)   **(b)**

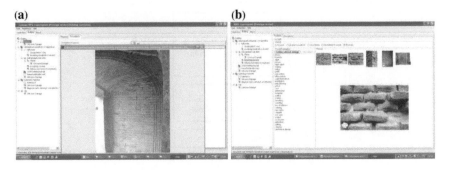

Fig. 10 MDDS damage atlas at construction level (**a**) and at material level (**b**)

The damages related to environmental conditions and material properties are mainly caused by environmental causes like water, frost, salts, pollution, flooding, etc., generally in combination with material properties. See also Sect. 5. All degradation processes that take place under those conditions are related to the presence of moisture. The system will, apart from assessing the correct damage type and coming with a hypothesis on the basis of the observations, also allow to insert measuring data, for example on moisture and salt distribution in a wall, for a more precise diagnosis.

The damages due to structural causes are generally perceived as large isolated cracks or as a diffused pattern of cracks, as addressed above. In the identification process of a structural damage pattern, the following characteristics should be taken into account: width of the crack(s) and variations over length; as far as possible: depth of the crack(s); direction of the crack(s); combination of crack(s) with deformations or displacements. Together with the visual characteristics of a pattern, the following should also be considered: the behaviour of the crack(s) in the course of time; comparisons of the actual damage found with previous damage and monitoring of its behaviour; the building materials constituting the construction; the building techniques used; the building element showing damage (e.g. wall, column, arch); and conservation measures performed over time (previous history).

For the time being, MDDS works with an integrated atlas in the diagnostic part of the system, which makes it possible to suggest one or more hypotheses, on the basis of the structural damage pattern that was assessed (Fig. 11). Investigations should also be carried out to ascertain whether the damage pattern has appeared together with other forms of deterioration: for example, a crack may occur together with a displacement or other non-structural types of damage. There are also damage types that appear to be structural, but are actually caused by salt or frost damaging mechanisms. An interesting example is that of the church tower of Noordwijk, the Netherlands (Fig. 12) [47]. The damage pattern is constituted by vertical cracks running along the corner of the tower. They are crossed by less evident horizontal cracks, running through the mortar joints. This damage pattern appears, at first sight, likely to be caused by a structural deterioration process, but it was in fact due to the formation of swelling compounds in the mortar.

Fig. 11 Structural damage atlas, integrated in the diagnostic section of MDDS

Fig. 12 Damage pattern that may at easily attributed to a structural mechanism

7 Conclusions

The present chapter addresses the properties of historical brick masonry and its components. First, historic mortars are discussed with respect to binders, aggregates, the role of porosity and mechanical strength. The strength of a mortar is greatly affected by the porosity and decreases with an increase in porosity. Compressive strength of air lime mortars ranges around 1.5–2.0 N/mm^2, whereas hydraulic lime mortars may reach ca. 10 N/mm^2. The bond strength between the masonry unit (brick or stone) and the mortar joint is also important and a low binder/aggregate ratio, poor grading of the aggregate or inadequate tooling of the mortar may limit the bond strength. Subsequently, old bricks are characterized in terms of chemical composition, average porosity, bulk mass and mechanical properties. The compressive strength of old clay bricks have a wide range of values (from 4 to 32 N/mm^2), with some concentration between 7 and 20 N/mm^2. For the modulus of elasticity, values range from 1 to 18 GPa, with an average value of 300 times the compressive strength. The ratio between tensile and compressive strength ranges from 0.03 to 0.10. Finally, the strength theories and experimental results of brick masonry under uniaxial and biaxial compression are briefly reviewed.

With respect to deterioration and damage, the most important influencing factors are discussed: environment; materials; building's design; craftsmanship in the construction of the building and its maintenance. The use of expert systems for diagnosis in building practice or for mitigation of the effects of decay and damage to buildings belonging to the cultural heritage is still not very common. In building diagnostics, generally, steps like survey, (visual) assessment, diagnosis and intervention are commonly used. With the support of a damage atlas in expert systems, a correct definition of the observed damage type is possible, providing a more objective and user-independent result.

References

1. Elbert, K., Cultrone, G., Navarro, R., Pardo, S.E.: Durability of bricks used in the conservation of historic buildings—influence of composition and microstructure. J. Cult. Herit. **2**, 91–99 (2003)
2. Boynton, R.S.: Chemistry and Technology of Lime and Limestone. Wiley, New York (1980)
3. van Balen, K.: Lime Mortar carbonation and its influence on historic structures. PhD thesis, K.U. Leuven (1991)
4. Vitruvius, P.: The ten books of architecture. In: Rowland, I.D., Noble Howe, T. (eds.) Cambridge University Press, Cambridge (1999)
5. Baronio, G., Binda, L.: Physico-mechanical characteristics and durability of bricks from some monuments in Milan. Mason. Int. **4**, 29–35 (1985)
6. Cultrone, G., Sebastián, E., Elert, K., De La Torre, M.J., Cazalla, O., Rodriguez-Navarro, C.: Influence of mineralogy and firing temperature on the porosity of bricks. J. Eur. Ceram. Soc. **24**, 547–564 (2004)
7. Castro, F., Oliveira, P., Fernandes, I.: Development of a methodology for estimation of the provenance of archaeological ceramics. Method Theor. Hist. Archaeol. **10**, 123–125 (1997)

8. López-Arce, P., Garcia-Guinea, J., Gracia, M., Obis, J.: Bricks in historical buildings of Toledo City: characterization and restoration. Mater. Charact. **50**, 59–68 (2003)
9. Lourenço, P.B., Fernandes, F.M., Castro, F.: Handmade clay bricks: chemical, physical and mechanical properties. Int. J. Architect. Herit.: Conserv. Anal. Restor. **4**(1), 38–58 (2009)
10. Maierhofer, C., Leipold, S., Schaurich, D., Binda, L., Saisi A.: Determination of the moisture distribution in the outside walls of S. Maria Rossa using radar. In: 7th International Conference on Ground Penetrating Radar, pp. 509–514 (1998)
11. Livingston, R.A.: Materials analysis of the masonry of the Hagia Sophia Basilica, Istanbul. In: 3rd International Conference on Structural Studies, Repairs and Maintenance of Historical Building, pp. 15–31 (1993)
12. Cultrone, G., De La Torre, M.J., Sebastian, E.M., Cazalla, O., Rodriguez-Navarro, C.: Behavior of brick samples in aggressive environments. Water Air Soil Pollut. **119**, 191–207 (2000)
13. Winslow, D.N., Kilgour, C.L., Crooks, R.W.: Predicting the durability of bricks. ASTM J. Test. Eval. **16**(6), 527–531 (1988)
14. Marzahn, G.A., Jahnel, R., Tue, N.V.: Finite element analysis of two ancient groined masonry vaults. In: 13th International Brick/Block Masonry Conference, pp. 147–156 (2004)
15. Bati, S., Ranocchiai, G. A.: critical review of experimental techniques for brick material. In: 10th International Brick/Block Masonry Conference, pp. 1247–1255 (1994)
16. Lourenço, PB, Ramos, LF, Vasconcelos, G, Peña, F. (2008) Monastery of Salzedas (Portugal): Intervention in the cloister and information management. In: 6th International Conference on Structural Analysis of Historic Constructions. Taylor & Francis Group, pp. 95–108
17. Schubert, P.: The influence of mortar on the strength of masonry. In: de Courcy, J.W. (eds.) 8th International Brick and Block Masonry Conference, pp. 162–174. Elsevier Applied Science, London, UK (1998)
18. CEN: EN 1052-1:1999. Methods of Test for Masonry: Determination of Compressive Strength. Brussels, Belgium (1999)
19. Mann, W., Betzler, M.: Investigations on the effect of different forms of test samples to test the compressive strength of masonry. In: 10th International Brick/Block Masonry Conference, pp. 1305–1313 (1994)
20. Kirtschigg, K.: On the failure mechanism of masonry subject of compression. In: 7th Int Brick/Block Masonry Conference, pp. 625–629 (1985)
21. Haseltine, B.: International rules for masonry and their effect on the UK. Mason. Int **1**(2), 41–43 (1987)
22. Vermeltfoort, A.: Compression properties of masonry and its components. In: 10th International Brick/Block Masonry Conference, pp. 1433–1442 (1994)
23. CEN.: EN 1996-1-1:2005. Eurocode 6. Design of masonry structures. General rules for reinforced and unreinforced masonry structures. Brussels, Belgium (2005)
24. ACI: ACI 530–11/ASCE 5–11/TMS 402–11: 2011 Building Code Requirements for Masonry Structures. Detroit, USA (2011)
25. Hilsdorf, H.: An Investigation into the Failure Mechanism of Brick Masonry Loaded in Uniaxial Compression. Designing, Engineering and Construction with Masonry Products. F.B. Jonhson, Houston (1969)
26. Khoo, C., Hendry, A.: A failure criterion for brickwork in axial compression. In: 3rd Int Brick/Block Masonry Conference, pp. 139–145 (1973)
27. Francis, A., Horman, C., Jerrems, L.: The effect of joint thickness and other factors on the compressive strength of brickwork. In: 2nd International Brick/Block Masonry Conference, pp. 31–37 (1971)
28. Atkinson, R., Noland, J., Abrams, D.: A deformation failure theory for stack-bond brick masonry prisms in compression. In: 7th International Brick/Block Masonry Conference, pp. 577–592 (1985)
29. Ohler, A.: Zur berechnung der druckfestigeit von mauerwerk unter berucksichtigung der mehrachsigen spannungszustande in stein und mortel. Bautechnik (1986)

30. Schubert, P., Hetzemacher, P.: On the flexural strength of masonry. Mason. Int. **6**(1), 21–28 (1992)
31. Drysdale, R.G., Hamid, A.A., Heidebrecht, A.C.: Tensile strength of concrete masonry. J. Struct. Div. (ASCE) **105**(7), 261–276 (1979)
32. van der Pluijm, R.: Out-of-plane bending of masonry: Behaviour and strength. PhD Dissertation. Eindhoven University of Technology (1999)
33. Hamid, A.A., Drysdale, R.G.: Concrete masonry under combined shear and compression along the mortar joints. ACI J. **77**(5), 314–320 (1981)
34. Page, A.W.: The biaxial compressive strength of brick masonry. Proc. Inst. Civ. Eng. Part 2 **71**, 893–906 (1981)
35. Ganz, H.R., Thürlimann, B.: Tests on the biaxial strength of masonry (in German). Report 7502-3. Institute of Structural Engineering, Zurich (1982)
36. Page, A.W.: The biaxial compressive strength of brick masonry. Proc. Intsn. Civ. Eng. Part 2 **71**, 893–906 (1981)
37. Page, A.W.: The strength of brick masonry under biaxial compression-tension. Int. J. Mason. Constr. **3**(1), 26–31 (1983)
38. Huinink, H.P., Petkovic, J., Pel, L., Kopinga, K.: Water and salt transport in plaster/substrate systems. Heron **51**(1), 9–31 (2006)
39. Charola, A.E.: Salts in the deterioration of porous materials: an overview. J. Am. Inst. Conserv. **39**(3), 327–343 (2000)
40. Collepardi, M.: Thaumasite formation and deterioration in historic buildings. Cement Concr. Compos. **21**, 147–154 (1999)
41. van Hees, R.P.J., Wijffels, T.J., van der Klugt, L.J.A.R.: Thaumasite swelling in historic mortars. Field observations and laboratory research. Cement Concr. Compos. **25**, 1165–1171 (2003)
42. van Hees, R.P.J., Naldini, S.: The masonry damage diagnostic system. Int. J. Restor. Build. Monum. **1**(6), 461–473 (1995)
43. van Balen, K., Mateus, J., Binda, L., Baronio, G., van Hees, R.P.J., Naldini, S., van der Klugt, L., Franke, L.: Expert system for the evaluation of the deterioration of ancient brick structures. Research report no. 8, vol. 1, European Commission (1999)
44. van Hees, R.P.J., Naldini, S., Binda, L., Van Balen, K.: The use of MDDS in the visual assessment of masonry and stone structures.In: Binda,L., di Prisco, M., Felicetti, R. (eds.) Proceedings of the 1st International RILEM Symposium On Site Assessment of Concrete, Masonry and Timber Structures, RILEM Publications S.A.R.L., Varenna, Italy, pp. 651–660 (2008)
45. de Vent, I.A.E., van Hees, R.P.J., Hobbelman, G.J.: Towards a systematic diagnosis of structural damage. Structural Analysis of Historic Construction; Preserving Safety and Significance, Bath, UK, pp. 689–696 (2008)
46. de Vent, I.E.A.: Structural damage in masonry. PhD Thesis, TU Delft (2011)
47. van Hees, R.P.J., Binda, L., Papayianni, I., Toumbakari, E.: Characterisation and damage analysis of old mortars. Mater. Struct. **37**, 644–648 (2004)

Characterization and Reinforcement of Stone Masonry Walls

**Bruno Quelhas, Lorenzo Cantini, João Miranda Guedes,
Francesca da Porto and Celeste Almeida**

Abstract Stone masonry is one of the oldest "structural materials" known to man. It is made by the superposition of stones, mortar and, very often, with infill material between leaves. The components present complex links and interactions and, in most cases, unknown geometry and high variability of their mechanical properties. These characteristics make stone masonry a highly heterogeneous material for which it is difficult to define realistic behaviour laws, a challenge that still demands further research, either through laboratory or onsite experimental campaigns. In reality, the mechanical characteristics of a stone masonry element strongly depend on the geometry and geometrical distribution of the stones along the façade and cross-section of the element and, therefore, on the layout of the interfaces, i.e. the joints. Studies developed in Italy have defined a series of parameters that try to quantify the level of fulfilment of good practice constructions rules with the expectable performance of a stone masonry wall under static and dynamic loading, in particular under seismic type loadings. This chapter discusses

B. Quelhas · F. da Porto
Department of Civil, Environmental and Architectural Engineering,
University of Padova, Via Marzolo No 9, Padova, Italy
e-mail: bruno.silva@dicea.unipd.it

F. da Porto
e-mail: francesca.daporto@dicea.unipd.it

L. Cantini
Department of Structural Engineering, Politecnico di Milano,
Piazza Leonardo Da Vinci, 32-20133 Milano, Italy
e-mail: cantini@stru.polimi.it

J. M. Guedes (✉) · C. Almeida
Department of Civil Engineering, University of Porto,
4200-465 Porto, Portugal
e-mail: jguedes@fe.up.pt

C. Almeida
e-mail: celeste.almeida@fe.up.pt

A. Costa et al. (eds.), *Structural Rehabilitation of Old Buildings*,
Building Pathology and Rehabilitation 2, DOI: 10.1007/978-3-642-39686-1_5,
© Springer-Verlag Berlin Heidelberg 2014

construction typologies and materials, assessment methodologies, earthquake induced failure mechanisms and strengthening intervention techniques on stone masonry structures, in particular walls.

Keywords Stone masonry • Assessment methodologies • Walls • Seismic behaviour • Strengthening techniques

1 Introduction

Stone masonry is one of the oldest "structural material" known to man that resisted time. It has been widely used, mainly up to the middle of the twentieth century, on a large variety of constructions, either common or monumental, some of them being today classified as local, national, or even world cultural heritage.

Stone masonry is a composite structural material made of the superposition of stones, mortar and, very often, with infill material, presenting complex links and interactions, for which the definition of realistic behaviour laws remains a big challenge; by nature, it is a heterogeneous material, whose components present, in most cases, an unknown geometry and a high variability of the mechanical properties. Thus, a great effort has been, and still is being done to gather information on this type of structures, either through laboratory [1–9] and onsite experimental campaigns [10–14] and/or including Non Destructive Techniques (NDT) and Slightly Destructive Techniques (SDT) using flat-jacks, sonic equipment, dynamic identification procedures, among other techniques [7, 15–17].

Studies carried out in Italy after major earthquakes have also allowed characterizing and classifying stone masonry walls through the assessment of the elevation and cross-section of the walls and the mechanical state of its components [18, 19]. However, and although a series of common "good construction practices" have been followed in different countries, the definition of a complete classification of the existing masonry typologies is difficult. In Portugal, some work has been done on the survey of stone masonry constructions in different regions [20], and a first attempt to create a database was done through the study of stone walls from buildings of the town of Tentúgal [21].

The next points establish and discuss construction typologies and materials, assessment methodologies, earthquake induced failure mechanisms and strengthening intervention techniques on existing masonry structures.

2 Masonry Typologies and Quality

Stone masonry is a composite structural material made of more or less regular layers of stones defining, approximately, horizontal (continuous) and vertical (discontinuous) interface lines, commonly referred to as joints, which may be filled in with mortar, small stones (wedges) and (or) other stiff material: pieces of brick, ceramics... In the case of a multi-leaf stone masonry, apart from those elements, there could be infill material within leaves, being also part of the masonry structure. This element, often considered to be secondary, is usually made of a mix of different materials, from earth to stones, broken bricks, natural fibres, etc., and placed without any particular compaction.

According to the natural resources of a specific geographical area, masonry structures were erected using the available materials and following the local constructive tradition. The construction techniques within a certain territorial ambit were established by the builders' empirical knowledge, successively transmitted through oral tradition, which afterwards became rules for that territory. Thus, since different building typologies can be found within different regions, it becomes difficult to find a complete and unique classification of the existing stone masonry. Nevertheless, the contact between cultures allows common features to be present in the different construction traditions. Since ancient civilizations, like the Hellenic one, the diffusion of this technical knowledge was supported by experts in the construction field. The treatises of architecture that appeared during the Roman Empire, or the work promoted by these and other cultural centres and present in the libraries of the mediaeval convents and monasteries are examples of that knowledge circulation. Thus, each civilization that reached a certain cultural hegemony left the traces of their technical capabilities through written or oral evidences.

A common tradition for masonry walls was developed by the Romans: from the Mediterranean coasts to the Northern regions of Europe, the Romans learnt the local building traditions and taught their own technical knowledge. During their domination, they imposed shared rules for the construction of masonry structures. In the first book of his treatise, Palladio [22] recalls Vitruvius's classification for Roman stone masonry walls, remarking that some of those building techniques were still used at his time. Five main typologies are described through short indications and graphic layouts, as follow:

a. Opus isodomum, presenting a regular texture formed by squared stones displayed along horizontal settings. Due to the variation of the proportion between the shaped stones used in the different courses, this masonry typology was also known as opus pseudo-isodomum (see Fig. 1a);
b. Roman concrete, constituted by irregular stones having limited dimensions (pebbles) bound by mixtures with pozzolanic properties (see Fig. 1b);
c. Infill walls, created through the so called Roman concrete, a composite material obtained by irregular stones of various dimensions bound by pozzolanic mixtures. The infill could be contained through regular (see Fig. 1c) or irregular external masonry texture;
d. Opus reticolatum, composed by squared stones displayed along diagonal laying planes and regularly crossed by Roman bricks forming horizontal settings (see Fig. 1d);
e. Opus incertum, dry walls formed by irregular stones without any horizontal settings; (see Fig. 1e).

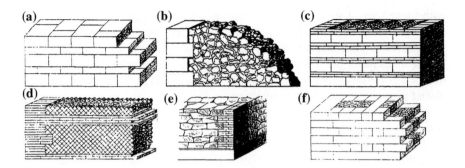

Fig. 1 Examples of the main Roman stone masonry typologies: **a** opus pseudoisodomum; **b** Roman concrete; **c** infill wall (emplecton); **d** opus reticolatum with infill; **e** opus incertum; **f** opus mixtum (layouts from Table 11 of [23])

Theses masonry typologies, still recognizable in the main Roman monuments and ruins in Europe, North Africa and Middle East, were adopted also after the Roman Empire: the mixed type with infill walls (Fig. 1f), for example, is a building technique that was largely used in Middle Age for the construction of Romanic and gothic cathedrals. Regular stones textures with an internal infill were also typical of defensive structures, like fortresses.

According to the past experience of masonry constructors, the geometry of a stone masonry, namely the distribution and superposition of the stones should respect a series of procedures that are usually referred to in the bibliography [24] as "good practice rules". The respect of such rules contributes, unquestionably, to a better distribution of forces through the masonry elements and to a better global structural behaviour. In particular, stones should be disposed along regular horizontal layers, should create discontinues vertical interfaces and, in case of a multi-leaf masonry, should guarantee a good connection and (or) interlock between different leaves.

When dealing with ordinary buildings (also known as diffused architecture), the classification appears to be more complex, since more differences are usually found due to a greater influence of more local building traditions, which, sometimes, even correspond to the introduction of new masonry typologies. The changes produced by certain historic events (like the Arabic domination in North Africa and in the Iberian Peninsula) determined a diversification of the classic masonry building techniques. Moreover, natural cataclysmic have also promoted the progressive development of the construction techniques, by showing the ineffectiveness of certain structures and forcing the investigation of more suitable building solutions. The introduction of timber frames in masonry walls (as in some seismic areas of Portugal, Greece or south Italy) was an example of a technical response meant to contrast earthquake effects.

The studies of experts, like Giuffrè [25] and De Felice [26], on diffused architectonic heritage outlined that masonry structures may present defects in the connections between their layers. With respect to the masonry walls built by the Romans, the walls constructed in further periods were characterized by infill mixtures with poorer binding properties. In fact, the monolithic behaviour of Roman walls was obtained through binding mixtures having hydraulic properties and good resistance. These characteristics were rarely obtained during the further centuries, due to the use of mixtures with low quality. As a result, multi-leaf walls are deeply influenced by the organization of their masonry sections: their mechanical behaviour is not interpretable through a simple evaluation of the external masonry texture, but it should be based on the study in depth of the characteristics of the cross-section. Considering large typological classes [27], common stone walls can be divided in:

- one leaf solid wall, composed by one stone per thickness [2] or stones well connected (usually by headers) and organized in horizontal or sub-horizontal courses (see Fig. 2a);
- two leaves wall, formed by two separate layers connected by an infill made by small rubble materials bound with mortar (Fig. 2b), or through the periodical overlapping of the stones (Fig. 2c);
- multiple leaves wall, usually constituted by external layers with a regular texture and one or more internal leaves composed by an irregular infill (Fig. 2d);
- dry stone wall, usually composed by irregular shaped blocks jointed by small stone or brick detritus and in some cases by stone wedges (Fig. 2e).

According to research carried out by Binda [28] and Borri [29] on Italian historical centres in seismic area, the quality of the stone masonry walls can be evaluated through the following combination of visual inspections and limited investigations:

- geometry: the dimension of the material components (the proportions of the stones; the height of the mortar joints, etc.), the disposal of the courses, the vertical joint staggering, the presence of headers, etc. [30];

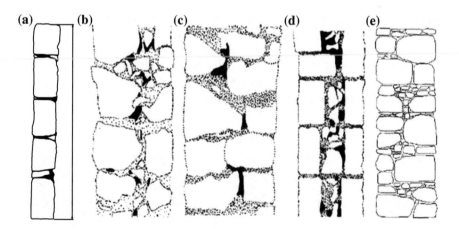

Fig. 2 Examples of different masonry typologies classification based on the characteristics of their sections (courtesy of C. Almeida and L. Binda and G. Cardani). **a** Single leaf wall. **b** Double leaves wall without connections. **c** Double leaves wall with connections. **d** Multi leaves wall (3 layers). **e** Dry wall without mortar joints

- decay status of the materials: identification of the pathologies causing the reduction of the mechanical properties of the structure [31];
- physical and chemical analysis on sampled mortar: the mortar used in the structure should be investigated for different depths and the identification of certain impurities (like hygroscopic saults or other sinterizing composts) allows to classify the mortar quality [32];
- Direct sonic tests provide qualitative indications on the connections between the components and can indicate the presence of headers [18];
- Single and double flat jack tests are able to identify quantitative parameters: respectively, the local state of stress and the deformability characteristics [33];
- Local dismantling: the direct observation of a limited area of the masonry section (usually 1/3 of its depth) allows recognizing the presence of multiple leaves and the technique used for assembling the components (regular mortar joints, rubble materials, etc.) [34].

The above mentioned study-methodology is the synthesis of the guidelines for the evaluation of the masonry quality promoted by the Italian institutions after recent seismic events. These recommendations provide indications for the interpretation of the mechanical behaviour of historical masonry structures introducing abacus of the main stone masonry textures and abacus showing the most common masonry cross sections (see Fig. 3). According to the characteristics of each stone masonry type, their quality can be evaluated according to the adequacy of the real structures to the characteristics described in the corresponding model.

2.1 Masonry Mechanical Behaviour

Studies developed in Italy have created a series of parameters that (try to) quantify the level of fulfilment of good practice constructions rules through the measurement of the deviation of the geometrical and physical characteristics of the walls from ideal conditions [36, 37]. Moreover, the Italian codes [38–40], present tables that link the fulfilment of these rules to the expectable performance of a stone masonry wall under static and dynamic loading.

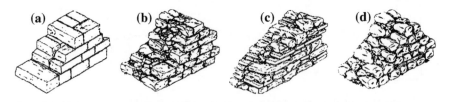

Fig. 3 Examples of different masonry typologies [35]. **a** Regular shaped stone blocks. **b** Irregular blocks and pebbles mixed type. **c.**Irregular blocks and pebbles mixed type with brick courses. **d** Mixed irregular blocks, pebbles and flakes of rocks

In reality, strength and stiffness of a stone masonry element depend on the geometry and geometrical distribution of the stones along the façade and cross-section of the element and, therefore, on the layout of the interfaces, i.e. the joints. This means that the mechanical parameters of a stone masonry depend on the mechanical characteristics of the stones, which are usually the most resisting elements, but also on the physical and mechanical characteristics of the joints and of the infill.

The contribution of the infill to the behaviour of stone masonry depends on its mechanical characteristics and on the characteristics of the cross-section, namely on the roughness of the inside face of the leaves and on the existence of transversal stones connecting the leaves. Notice that the compression of the infill induces its horizontal expansion (Poisson effect) that pushes the leaves to the outside. If the external leaves are not well connected and (or) do not present a good interlock, this phenomenon promotes an out-of-plane behaviour that may induce the vertical instability of the leaves. On the contrary, if the connection between leaves is efficiently ensured, the expansion of the infill may be avoided by the external leaves, which induces a confinement state that improves the infill mechanical response and, therefore, promotes a better global performance of the stone masonry.

In general, the mechanical characteristics of a stone masonry with mortar joints are associated to the mechanical characteristics of the stones and of the mortar, in particular to the strength and stiffness of the two materials. Many authors suggest empirical based expressions that link the compressive strength of the two materials to the compressive strength of the masonry. In particular, the following equation has been proposed [41–43]:

$$\sigma_{cn} = \gamma \sigma_{cs}^{\alpha} \cdot \sigma_{cj}^{\beta}$$

where σ_{cm}, σ_{cs} and σ_{cj} represent, respectively, the compressive strength of the masonry, of the stones and of the mortar, and γ, α and β the parameters that should be calibrated to take into account the specific characteristics of each type of stone masonry. However, the mechanical properties of a stone masonry tend to have a high scatter; the mechanical parameters depend on the particularities of the masonry that are hardly repeatable. This is particularly true for stone masonry with irregular textures, which can be considered the more general situation. Nevertheless, attempts have been made by different authors to propose ranges of values for different types of stone masonry [2, 10, 44].

As for the stiffness, in particular for the Young modulus of stone masonry (E_m), empirical based expressions have also been proposed to estimate this mechanical property, in particular through a direct proportion to the compressive strength [42, 45]:

$$E_m \big/ \sigma_{cm} = k$$

Also in this case the proposed ratio can have a very high scatter, and different values for the coefficient k have already been found, or suggested. Although most of the authors point out values for k between 500 and 1000, being the last the one usually used for stones and mortar, tests made on one-leaf irregular stone masonry walls at the Laboratory of Earthquake and Structural Engineering (LESE) of the Faculty of Engineering of Porto University (FEUP) [2] have found values between 80 and 140.

Notice that stones have, in general, high stiffness and compressive strength when compared to mortar joints. This is particularly true when dealing with old stone masonry structures with lime mortar type joints. Values of compressive strength and stiffness of, respectively, 1 MPa and 1 GPa are usually found for this type of mortar, while stones, even with some degradation, present values that are easily greater than 20 MPa and 20 GPa, respectively. This difference makes the initial behaviour of stone masonry to be mostly controlled by the mechanical characteristics of the joints, which may crush under compression forces, open under tensile forces and slide under shear forces, defining, in most cases, the main rupture lines.

However, the contribution of the stones to the mechanical properties and behaviour of the masonry is not restricted to the definition of its texture, i.e. to the delimitation of the joints through the stones interfaces. To understand the role of stones in stone masonry, firstly one should be aware that stones have a tensile fragile behaviour with a tensile strength that is commonly more than ten times lower than the compressive strength. Therefore, while crushing of stones is seldom achieved, or is confined to limited areas where localized concentration of forces occurs, cracking of stones happens more frequently and it interferes with the configuration of the rupture lines. In particular, and apart from any instability phenomenon that may occur due to slenderness and (or) the influence of infill expansion, the stones tensile rupture is the main factor that controls and restrains the compressive strength of a stone masonry. Nevertheless, the importance of the tensile strength of the stones is often neglected and the compression strength is the property that is usually underlined in the literature, even though the ultimate value of compression strength depends, itself, on the tensile strength of the stone. Notice that, although tensile strength and compression strength of the stones is related, the ratio between the first and the second depends on the quality of the stones, and a large scatter can be found even for the same type of stones. Nevertheless, typical values for this ratio are found between 0.04 and 0.10.

In reality, when a stone masonry wall is loaded by a set of vertical (perpendicular to the horizontal interface layers) and in-plane transversal forces (perpendicular to the previous), the forces flow to the basement through a set of preferential compression lines that depend mostly on the geometrical characteristics of the masonry, in particular on the way the stones are supported on each other. If a set of more or less homogeneous and continuous horizontal layers, i.e. joints, exists in between the stones, the forces flow in a more uniform way and no particular concentration of forces happens on the stones. On the contrary, if the horizontal interfaces between stones are made of small stones, or other stiff material and (or) present different mortar thicknesses, i.e. if the joints are not uniform, presenting a very heterogeneous and discontinuous stiffness distribution, the forces flow through the stiffer areas, promoting a concentration of forces in those areas and inducing a flexural behaviour on the stones between stiffer points. Since stones have a tensile fragile behaviour with a low tensile strength, especially when compared to the compressive strength, this phenomenon promotes the tensile rupture of the stones, which ends up being one of the most important factors that controls the global performance of a stone masonry wall.

Notice that stone masonry is a material that is not meant to be submitted to tensile forces. Moreover, the behaviour and performance of a stone masonry strongly depends on the level of compression force applied to the masonry. In particular, high compression axial forces induce an instable behaviour that should be avoided. However, under ordinary static conditions, stone masonry is usually submitted to levels of axial compression force that are far below its compressive strength and, therefore, these forces have a stabilizing role, in particular when transversal forces are also applied to the masonry. As an example, laboratory tests made on a three-leaves irregular stone masonry wall panel, 1.2 m high per 1.0 m wide and 0.5 m thick, have shown a quite nonlinear plastic behaviour with a vertical axial strength and an initial Young modulus of around 3.0 MPa and 2.4 GPa, respectively [4]. Notice that the high plastic behaviour is one of the main characteristics of stone masonry and that is mainly due to the crushing of the joints, a deformation that is mostly unrecoverable.

Stone masonry is, therefore, a material that is not meant to be submitted to out-of-plane forces. But loads of this type may occur due to natural, or accidental phenomena (wind, earthquakes, explosions, impacts, etc.), or static lateral impulses from the contact with other elements, and stone masonry should be prepared to face them. Exceptional dynamic load events, like earthquakes, can induce important fluctuations of the compressive axial force that may vary from zero, exposing the masonry to a very critical situation where a small lateral force could be enough to overturn the masonry, to very high values that can overcome the compressive strength of the masonry and may cause the crushing of the material. Figure 4 shows the compression behaviour curve of a three-leaf stone masonry wall.

3 The Seismic Behaviour of Existing Masonry Buildings: Morphology of Damages and Failure Mechanisms

The developments that have been implemented and tested concerning stone masonry structures, in particular walls, arches and vaults, are mainly linked to the improvement of the behaviour of that type of masonry elements under seismic type actions. In fact, throughout history, earthquakes have represented one of the main causes of damage and losses of stone masonry buildings. The post-earthquake damage surveys carried out after earthquakes affecting areas where masonry buildings had an important presence came out to be an important source of information on the recurrent damage patterns. These observations showed that one of the main sources of vulnerability for such structures is associated to local failure modes that can be essentially interpreted on the basis of two fundamental collapse mechanisms [46–53]. According to Giuffrè definition [24, 52], the most

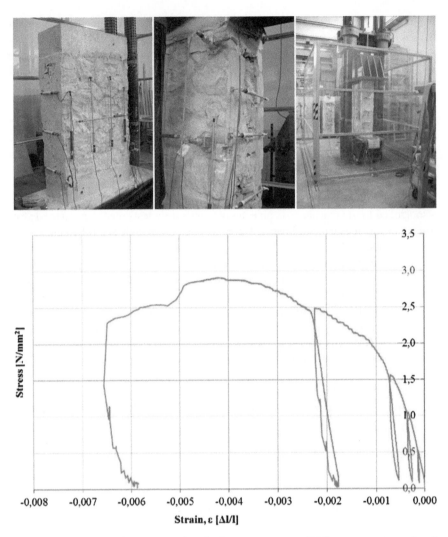

Fig. 4 Simple compression test on a three-leaf stone masonry wall [4]: test apparatus and vertical compression behaviour curve

vulnerable failure mode, referred to as the "First Damage Mode" (Fig. 5) is activated by seismic actions perpendicular to the wall, i.e. out-of-plane actions that cause the overturning of the whole panel, or of a significant portion of it. The building seismic response can be governed by such mechanisms when connections between orthogonal walls and between walls and floors are particularly poor. This is often the case of stone masonry buildings, with lack of interlocking at the connection of intersecting walls, presence of simply supported wooden floors and thrusting roofs. Only if connections are improved by proper devices, as for example tie-rods, the walls can be contained and this failure mechanism can be prevented.

Fig. 5 Deformation of a
building and typical damage
of structural walls due to a
seismic action. Example of
first and second damage mode
collapse mechanisms [54]

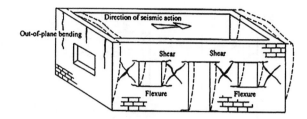

The second failure mode, referred to as the "Second Damage Mode" (Fig. 5), which can occur simultaneously, or not with the first one, is activated by seismic in-plane actions that cause the shear cracking of the panels.

Under horizontal type actions, such as those impose by earthquakes, the external masonry walls of a house can typically be subjected to out-of-plane mechanisms. This "first mode" could be considered as one the most frequent and ruinous mechanisms, as it implies the complete collapse of the wall and consequent ruin of all supported elements. The way in which it develops depends on the quality of the masonry itself and of the connections with the other structural elements [55]. In the case that the structure is not strengthened and (or) well connected to other structural elements, such as floors and roofs, the only means to restraint the overturning mechanism is the friction produced on the contact surface between the wall and the elements to which it is connected. Notice that, if the structure is strengthened, for example, by introducing ties or ring beams, then usually the simple overturning is prevented, while mechanisms relying on arch effect start to develop [56].

Out-of-plane mechanisms of masonry panels are often associated with in-plane mechanisms, either developing within the same panel, or in different panels. In fact, not only constructions have masonry walls distributed along orthogonal directions, as horizontal actions, as earthquakes, introduce horizontal loads coming from different directions. As described before, this "second mode" is caused by forces acting in the plane of the wall and it is usually characterized by diagonal cracks associated with shear forces that often result in an "X" pattern, but hardly reaching total collapse. However, when a full shear crack occurs during an earthquake, the triangular sections of the panel can become unstable, leading to collapse.

Observation of seismic damage of stone masonry walls, as well as laboratory experimental tests on stone masonry panels, showed that masonry walls subjected to in-plane loading may have two typical types of behaviour, to which different failure modes are associated:

- *Flexural behaviour*: this may involve two different modes of failure. If the applied vertical load is low with respect to the masonry compressive strength, the horizontal load produces tensile flexural cracking at the corners and the pier begins to behave as a nearly rigid body rotating around the toe (rocking). If no significant flexural cracking occurs, due to a sufficiently high vertical load, the pier is progressively characterized by a widespread damage pattern, with sub-vertical cracks oriented towards the more compressed corners (crushing). In both cases, the ultimate limit state is obtained by failure at the compressed corners, Fig. 6a.

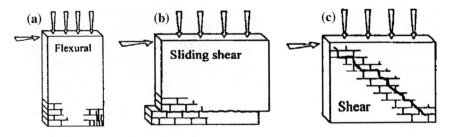

Fig. 6 Identified masonry failure mechanism [57]

- *Shear behaviour*: this may produce two different modes of failure: (i) sliding shear failure, where the development of flexural cracking at the tense corners reduces the resisting section; failure is attained with sliding on a horizontal plane bed joint, usually located at one of the ends of the pier, Fig. 6b; (ii) diagonal cracking, when failure is attained with the formation of a diagonal crack, which usually develops at the centre of the pier and then propagates towards the corners, Fig. 6c. The crack may pass prevailingly through mortar joints (assuming the shape of a 'stair-stepped' path in the case of a regular masonry pattern), or through the blocks.

The occurrence of different failure modes depends on several parameters such as: (a) the geometry of the pier; (b) the boundary conditions; (c) the acting axial load; (d) the mechanical characteristics of the masonry constituents (mortar, blocks and interfaces); (e) the masonry geometrical characteristics, namely the block aspect ratio and the in-plane and cross-section masonry pattern. In the past, many experimental tests have been carried out in order to analyse the influence of these parameters on the failure mode of masonry piers. In general, it has been assessed that rocking tends to prevail in slender piers, while bed joint sliding tends to occur in very squat piers [58, 59]. In moderately slender piers, diagonal cracking tends to prevail over rocking and bed joint sliding, for increasing levels of vertical compression [60].

For increasing levels of vertical compression [61], and increasing ratios of mortar to stone strength [62, 63], diagonal cracking propagating through stones tends to prevail over diagonal cracking propagating through mortar joints. Crushing, in general, occurs for high levels of vertical compression and is related to the compressive strength of the material [60].

It is worth pointing out that it is not always easy to distinguish the occurrence of a specific type of mechanism, since many interactions may occur between them. The damages observed in various countries due to recent earthquakes, particularly in Europe where the use of this material is very diffused in construction, (e.g. Umbria-Marche—Italy (1997-98), Azores—Portugal (1998), Andravida—Greece (2008), Abruzzo—Italy (2009), Lorca—Spain (2011), Van—Turkey (2011), Reggio-Emilia—Italy (2012), etc.—[64]) have shown that, although much knowledge already exists, and important investigation is still being carried in this field, there is still much to do, in particular in the implementation of strengthening and (or) retrofitting solutions in old stone masonry constructions to improve their performance under this type of actions. Besides, the fact that there are several buildings that were retrofitted and (or) strengthened prior to earthquakes also allowed assessing the performance, in particular the effectiveness of the applied intervention techniques; not always the adopted structural models were adequate and, therefore, the retrofitting techniques provided the expected effects. In particular, the effects of the earthquakes revealed the existence of incompatibilities between the stone masonry constructions (materials, structural systems...) and the applied intervention techniques, Fig. 7. Many failures were mainly related: to analyses performed on the basis of limited or inadequate information regarding the original structural system and the mechanical properties of the materials, to the use of unsuitable analytical tools and to the adoption of behaviour models developed for modern structures.

Although these observations call for regulatory documents that provide technicians with adequate guidance, it is very complex to define general rules and operative modalities, as it was tempted in the past. This is mainly related to the complexity of the structural typologies of existing buildings, in particular those with masonry bearing structures, but also to the fact that it is an issue that often involves social, historic, aesthetic, technical and economic aspects.

The observation of failures due to incompatibility between the original structure and the repair intervention, showed the need for developing new structural models for assessing the behaviour old stone masonry buildings, and code requirements for the intervention procedures [50, 65]. Formerly, the code requirements were oriented to the seismic adequacy of structures. In the case of cultural heritage constructions, namely those classified or protected by national or international directives, the new Italian seismic code moved from the imposition of "adequacy" to "improvement" [38–40, 66], which means more flexible, compatible and respectful interventions on existing structures. Such codes promote the improvement of the performance of old constructions through a series of intervention procedures, but without imposing the fulfilment of the (too demanding) code requirements for new constructions.

Several studies based on onsite observations after seismic events were carried out in order to define the real structural behaviour of old masonry buildings. These studies

Fig. 7 Damage caused by past interventions, Abruzzo 2009

allowed creating abacuses with the typical damages occurring to the different typologies (buildings, churches, etc.), which led to the systematization of the mechanical models able to describe their specific behaviour by kinematics models, both for in-plane and out-of-plane mechanisms.

4 Methodologies and Tools for the Assessment of Stone Masonry Structures and Control of Strengthening Interventions

Characterization of old stone masonry structures is a very complex task, which requires specific multidisciplinary methodologies of evaluation that have been improved and applied, in particular to historical constructions or constructions with high cultural value [18]. Such methodologies allow designers, such as engineers and architects, to assess the materials and structural systems of the constructions, as well as their geometry and damage state. Such information is indispensable to calibrate and assess the representativeness of the numerical models that were selected for the simulation of the construction, and that will help to understand the processes that contributed to the actual state of the construction and to measure the actual performance of the recommended intervention solutions.

These evaluation methodologies are present in the Eurocode 8 [67] and in the Italian Technical Norms for Construction [40], and specified for Cultural Heritage in the Italian Guidelines [68]. These norms define different levels of knowledge, correlated to the extent of application of the methodologies, which are then related to different confidence factors. These factors are of great importance to designers for defining retrofitting and (or) strengthening interventions.

The evaluation methodologies can be divided into two main phases: (1) knowledge phase (historical research; assessment of the structure—geometry and material; damage survey; onsite and laboratory tests) and (2) analysis phase (selection of the type of analysis, model and tools).

The knowledge phase is characterized by several steps with the ultimate goal of characterizing in a complete and detailed, as much as possible, way the structural system. This phase is essentially composed by five steps: (1a) historical investigation, paying special attention to the transformations and structural interventions performed to the structure throughout time: assessment of the efficiency of those actions and discussion on their influence to the behaviour of the structure; (1b) description of the building, through the detailed analysis of each of the architectural and structural elements and the survey of the geometrical and material characteristics of the structure; (1c) characterization of the composing material properties (physical, chemical and mechanical) through onsite (flatjacks, sonics, etc.) and laboratorial experimental tests; (1d) characterization of the more global behaviour of the structure through onsite tests, in particular by using dynamic identification tools, a technique that is becoming more and more diffuse, allowing the identification of higher and lower resistance areas, but also to assess, indirectly, the effectiveness of applied reinforcement solutions; (1e) description and definition of possible causes to the overall state of the structure. The use of different techniques and the comparison of the different results allow a more reliable and effective assessment.

The level of depth and detail involved in this 1st phase depends on the proposed objectives established for each particular case. If it intends a simplified analysis, or the analysis of only a certain part of the structure, it may not be reasonable to implement all the five steps with the same detail, or to apply them to the whole structure. This depends largely on the experience of the technicians involved in the study; a less experienced person may: (1) be extremely conservative (2) or totally neglect the knowledge phase.

The 2nd phase is characterized by a single step: (2a) define a modelling strategy for the assessment of the structures based on the previously gathered information. Within this step, and based on the objectives defined for each case, different matters have to be decided, such as the most effective: (i) modelling type analysis (limit analysis, analysis with numerical models, in particular using finite elements, etc.) and (ii) material type behaviour (linear or nonlinear).

The two phases are not always applied in a unidirectional way, i.e., after the definition of the 1st phase, and after passing to the analysis and modelling of the structure, it is often necessary to reassess the input information, entering into an iterative process.

5 Strengthening Techniques

Recent seismic events showed the ineffectiveness of some past interventions on masonry structures and, therefore, of the approaches/methodologies and tools adopted on its conception. The intervention solutions have to be designed taking into account its functional purpose, the real onsite conditions and should be validated guaranteeing the fulfilment of certain criteria [69–71], such as the requirement for: (i) structural authenticity; (ii) structural reliability; (iii) compatibility; (iv) durability; (v) non-intrusiveness (non-invasive); (vi) non-obtrusiveness; (vii) removability/reversibility, or repeatability; (viii) monitorability and controllability, all converging to (ix) a minimum intervention approach. These criteria should not be understood as absolute requirements, but as recommended conditions to achieve efficient, respectful optimal solutions, consistent with conservation principles. In fact, the fulfilment of all the criteria may be impossible in most cases, and some prioritization or choice, based on engineering judgment, is necessary.

On this validation process, the experimental studies constitute an important source of information, in particular for the development and calibration of analytical and numerical tools capable of predicting the behaviour of these structures. Having in mind that there is not a unique way of repairing, consolidating, strengthening, etc., the optimal solution should be selected among the different available techniques that respond to the safety and serviceability needs of a construction, pointing to the most friendly intervention, i.e. guaranteeing, as much as possible, the accomplishment of the previous criteria, within the available budget. At the same time, maintenance and monitoring programs should be also setup to follow the performance and physical state of the intervention solution in the long term.

As mentioned before, an important requirement to be considered in the selection of any material or technology used for repair or strengthening is the compatibility (chemical, physical, mechanical, thermal, rheological, etc.) between the new and the existing elements. A choice, regarding compatibility, is usually posed in the selection of traditional materials and techniques against modern (or innovative) ones. The first ones present, normally, longer term compatibility with the original elements due to the combination of similar properties and the absence of undesirable side-effects, as observed through past experience.

Modern and innovative materials and techniques may be considered for repair and strengthening purposes provided that sufficient scientific research and experience are available on their adequate performance and compatibility with the original elements. Some of these techniques have already shown severe incompatibility problems when used to restore or strengthened stone masonry structures. In other cases, more experience has to be gathered still before it can be said for certain that no damaging side effects may occur in the long term. The Venice Charter [69] refers directly to this subject; where traditional techniques prove inadequate, the consolidation of a monument can be achieved by the use of any modern technique for conservation and construction, provided its efficacy has been shown by scientific data and proved by experience. In turn, the ICOMOS/ISCARSAH Recommendations [70], mention that "the choice between "traditional" and "innovative" techniques should be determined on a case-by-case basis, with preference given to those that are least invasive and most compatible with heritage values, consistent with the need for safety and durability".

There is a large number of intervention techniques that have direct application to stone masonry, aiming to improve its performance under static and dynamic loadings. Among those, a set of three that have been widely used in the recent years was selected: (i) grout injection; (ii) deep re-pointing of mortar joints and (iii) application of transversal ties. These techniques can be used either independently, or combined. Nevertheless, their effectiveness continues to be studied and analysed due to the involvement of different materials and methodologies, in combination with the complexity and in-homogeneity of the masonry.

Grout Injection

The repair and strengthening by grouting of brick and stone masonry walls, Fig. 8, has been largely applied in Italy on historic buildings and dwellings in the seventies and eighties, after the main earthquakes of Friuli and Irpinia; nevertheless, no great effort was done in advance to test the effectiveness of this technique.

Even if experimental and analytical research has been carried out in the past decades on these techniques, the effectiveness was mostly assessed in terms of strength increase, rather than in terms of compatibility with the original masonry [17, 72, 73]. However, some research was carried out on the effectiveness of grout injections [6, 74–80], and more recently the research works carried by Mazzon [3] and Silva [4].

Fig. 8 Grout injection on stone masonry walls [4]: **a** drilling holes for the injection; **b** cleaning the drilled holes by blowing compressed air; **c** injection pump machine; **d** injection of a wall; **e** grout mixture leaking from a "control" hole; **f** sealing of the hole with mortar

These studies on stone masonry walls injected with grout, which involved cement or polymer-based grouts [9, 81–83] and lime-based grouts [3, 7, 83–85], allowed improving the knowledge regarding the mechanical characteristics and the structural effects of the injection technique. The conclusions recommended a careful approach and suggested a previous knowledge of the masonry wall morphology and of the masonry characteristics, since some types of walls are not injectable. Furthermore, it is a non-reversible technique and its use can raise durability and compatibility questions.

This type of intervention should be applied when there is weak cohesion between the different elements of the masonry and (or) there is an important amount of voids in the masonry cross-section, sufficiently interconnected to allow the mixture to penetrate and spread through the wall, providing a more homogeneous structural element. The technique is particularly appropriate for multi-leaf stone masonry, since a higher percentage of inner voids is expected when compared to other type of masonry. Nevertheless, the technique has been also used with success in single-leaf walls from the city of Porto, Portugal [1], although, in this case the use of this technique can be understood as deep re-pointing.

The choice of the mixture to inject is done by selecting the best characteristics for the type of wall on which to intervene, for example, the mechanical strength of the mixture and its deformation characteristics should be similar to those of the original wall. Therefore, the effectiveness of a repair by grout injection depends not only on the characteristic of the mixture used, but also on its mechanical properties, on the injection technique adopted and, once again, on the information on the wall characteristics. The technical improvements of the last years have developed new grouts with specific properties, such as low salt content and ultra-fine aggregates, and they have shown how to optimize the injection methodology, namely the injection pressure, or the distance between the injectors in function of the masonry characteristics.

The main problems connected to the grout injection technique can be summarized as follows: (i) lack of knowledge on the size distribution of voids in the wall; (ii) difficulty of the grout to penetrate into thin cracks (2–3 mm), even if micro fine binders are used; (iii) presence in the wall of fine and large size voids, together, which make difficult choosing the most suitable grain size of the grout; (iv) segregation and shrinkage of the grout due to the high rate of absorption of the material to be consolidated; (v) difficulty of grout penetration, especially in presence of silty or clayey materials; (vi) need for sufficiently low injection pressure to avoid air trapping within the cracks and fine voids, or even wall disruption.

Sometimes, in the case of disastrous events such as earthquakes, an apparent ineffectiveness of the consolidations using injection is observed. In fact, in these cases there is an inhomogeneous result of the intervention mainly due to: (1) poor design of the injection mixtures, (2) rough and uncontrolled execution of the intervention and (3) punctual distribution of the mixture due to an excessive distance between the injection holes. In fact, in most cases the ineffectiveness of the interventions injection is due to a poor execution of the technique, and not to the technique itself. Surveys after the 1997 earthquake in Umbria on damaged walls have often showed the difficulty of diffusion of the grout injection within stone masonry sections [86].

Deep Re-pointing of Mortar Joints

Deep repointing is a widely applied technique in all types of masonry, Fig. 9. This intervention consists on the partial replacement of the mortar joints with better quality mortar. It is meant to improve the masonry mechanical characteristics and it should be applied if deterioration is localized only in the mortar. This technique

Fig. 9 Deep re-pointing procedure [7]

can increase the masonry strength for both vertical and horizontal loads, but the best results are obtained especially in terms of stiffness, which is greatly increased due to the confinement effect of the joints. Actually, strength enhancement is expected only when a significant percentage of the initial weak or deeply damaged mortar is replaced by a new more compact and rather stiffer one, but still not excessively rigid and resistant to avoid creating areas in the masonry with inhomogeneous behaviour.

The main aims of deep repointing, provided that it is carried out with very good workmanship, are multiple: (i) to connect, in a rather thin section, the stones of the external leaf, substituting the original mortar in the joints when it is damaged, cracked and (or) very poor; (ii) to confine the wall at a less extent than the jacketing, but with better results, since the bond with the existing stones and mortar can be better assured; (iii) to confine better the injected material when grout injection is carried out; (iv) to provide a better penetration and distribution of the mortar compared to the random penetration and distribution of a grout injection.

Before deciding the application of the deep repointing technique an onsite investigation should be carried out in order to provide the crack pattern of the walls, the thickness of the section (it should be no more than 45.0 or 60.0 cm), the morphology of the masonry (number of leaves and stone arrangement) and the physical and chemical characterisation of the materials. Attention should be given to the choice of the mortar to avoid unwanted chemical, physical and mechanical incompatibilities. In general, cement based mortars are used, as they provide higher strength. However, this type of mortars may trigger unwanted chemical reactions in the masonry.

Sometimes, repointing is ineffective in cases where there is a poor execution of the intervention. In particular, it is frequent to found a malfunction of this technique because it wasn't well applied in depth, but limited to an aesthetic improvement of the surfaces. Furthermore, much attention must be paid not only to the depth to remove, but also to the total elimination of the original layer of mortar that is in contact with the resistant elements (stones) in order to allow the new mortar to develop bond/adherence with the elements.

In the case of consolidation using repointing, the inability to maintain the original plaster must be taken into account. As so, this type of intervention cannot be used in the presence of fine plaster or frescos, i.e., in the case of buildings with historical and artistic importance.

Often, in conjunction with the repointing operations, it is necessary to intervene on the walls also with injections and (or) transversal steel ties, to increase the effect of the improvement due to the solely introduction of new mortar in the joints. This can be particularly efficient in the case of two or three leaves stone walls reaching a thickness not higher than 60 cm.

Transversal Ties

The technique of inserting metal tie-rods perpendicular to the walls facades, a technique that is used only on multi-leaf wall panels, has the main purpose of linking the different leaves to promote a more monolithic structural element. Thus, it improves the global behaviour of the masonry, preventing the out-of-plane instability of the leaves, Fig. 10, not only under vertical compressive forces, but also

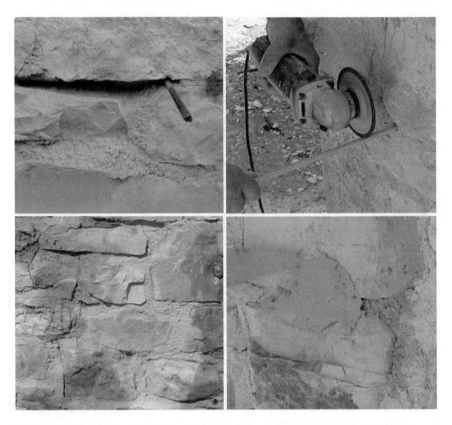

Fig. 10 Application of transversal ties in multi-leaf stone masonry walls [7]

under in-plane and out-of-plane horizontal forces, increasing the masonry global strength and stiffness. This technique has as main advantages: (i) high velocity of implementation; (ii) low cost of execution; (iii) good performance and (iv) it can be (partially) removed, in case a better consolidation solution is found.

The ties consist on simple steel bars with improved adherence, or placed inside a sleeve that is injected with grout afterwards, or treated bars with a bolted head. The holes to insert the ties are done with a rotating probe, preferably at the mortar joints. Eventually it is possible to insert the bars by hammering, taking advantage of the lesions present in the walls.

When using simple steel bars, they are fixed to the masonry by bending or injection. In this case, the action of consolidation is noticeable only when the deformation of the wall tends to increase, putting the steel bar under tensile load. When using treated bars, a slight tightening can be performed by warming the bar; the contraction that occurs due to the cooling process immediately puts the bar under tensile forces, imposing an immediately compressive transversal action on the wall leaves. To apply these contrast actions, the tie-rods can be placed not only orthogonal to the facades of the wall, but also along diagonals. For the technique to be efficient, 4 ties should be inserted per square meter, with a minimum of 2 per square meter.

In the case of pre-stressed tie-rods, special attention should be paid to the application of the actions to the bars and to guarantee that the pre-stress effect is maintained in the long term. If the holes are not injected and the anchoring systems are accessible, it is possible to perform a periodical control of the state of tension in the bars. Furthermore, in particularly aggressive environments, the protection of the bars against the corrosive action of external agents has to be ensured.

6 Final Remarks

This section discusses construction typologies and materials, assessment methodologies, earthquake induced failure mechanisms and strengthening intervention techniques on existing stone masonry structures.

The characterization of existing stone masonry structures is a very complex task. It requires specific multidisciplinary methodologies of evaluation to provide the necessary information that allows understanding the processes that contributed to the actual state of the construction, and to select the most adequate intervention solutions.

Post-earthquake damage surveys carried out after earthquakes showed that one of the main sources of vulnerability for such structures is associated to local failure modes that can be essentially interpreted on the basis of in-plane and out-of-plane collapse mechanisms. These events showed the ineffectiveness of some past interventions on masonry, in particular due to their incompatibility, in most cases physical/mechanical, with the original structure.

In this context the applicability, advantages, and limitations of some intervention techniques that have been widely used in the recent years are discussed, in particular grout injection, deep re-pointing of mortar joints and application of transversal ties, which can be used either independently, or combined.

References

1. Almeida, C., Guedes, J., Arêde, A., Costa, A.: Shear and compression experimental behaviour of one leaf stone masonry walls. In: 15th World Conference Earthquake Engineering, Lisbon, Portugal (2012)
2. Almeida, C., Guedes, J.P., Arêde, A., Costa, C.Q., Costa, A.: Physical characterization and compression tests of one leaf stone masonry walls. Constr. Build. Mater. **30**, 188–197 (2012)
3. Mazzon, N.: Influence of grout injection on the dynamic behaviour of stone masonry buildings. Ph.D. thesis, Università degli Studi di Padova: Scuola di Dottorato in Scienze dell'Ingegneria Civile e Ambientale (2010)
4. Silva, B.: Diagnosis and strengthening of historical masonry structures: numerical and experimental analyses. Ph.D. thesis, University of Padova, Italy (2012)
5. Silva, B., Guedes, J., Arêde, A., Costa, A.: Calibration and application of a continuum damage model on the simulation of stone masonry structures: Gondar church as a case study. Bull. Earthq. Eng. **10**(1), 211–234 (2012)

6. Tomaževič, M.: Laboratory and in situ tests of the efficacy of grouting and tying of stone masonry walls. In: International workshop on effectiveness of injection techniques and retrofitting of stone and brick masonry walls in seismic arean, pp. 95–116 (1992)

7. Valluzzi, M.R.:. Comportamento meccanico di murature consolidate con materiali e tecniche a base di calcea. Ph.D. thesis, University of Trieste, Trieste, Italy (2000)

8. Vasconcelos, G.: Experimental investigations on the mechanics of stone masonry: characterization of granites and behavior of ancient masonry shear walls. Universidade do Minho, Guimarães (2005)

9. Vintzileou, E., Tassios, T.P.: Three-leaf stone masonry strengthened by injecting cement grouts. J. Struct. Eng. 121(5), 848–856 (1995)

10. Corradi, M., Borri, A., Vignoli, A.: Experimental study on the determination of strength of masonry walls. Constr. Build. Mater. 17(5), 325–337 (2003)

11. Costa, A.A., Arêde, A., Costa, A., Oliveira, C.S.: Out-of-plane behaviour of existing stone masonry buildings: experimental evaluation. Bull. Earthq. Eng. (2011). doi:10.1007/s10518-011-9332-9

12. Artioli, G., Casarin, F., Dalla Benetta, M., da Porto, F., Secco, M., Valluzzi, M.R.: Restoration of historic masonry structures damaged by the 2009 Abruzzo earthquake through injection grouts. In: 9th Australasian Masonry Conference, Queenstown, New Zealand (2011)

13. Costa, A.A., Arêde, A., Costa, A., Guedes, J., Silva, B.: Experimental testing, numerical modelling and seismic strengthening of traditional stone masonry: comprehensive study of a real Azorian pier. Bull. Earthq. Eng. 10(1), 135–159 (2012)

14. Modena, C.: Interpretazione dei risultati ottenuti dalle prove in sito nell'ambito delle tre convenzioni con gli istituti di ricerca di Firenze e Milano e modellazione del comportamento strutturale dei componenti rinforzati. Convenzione di ricerca tra la Regione Toscana e il Dipartimento di Costruzioni e Trasporti dell'Università degli Studi di Padova (1999)

15. Dalla Benetta, M.: Qualificazione di murature storiche: procedure sperimentali in sito e calibrazione in laboratorio. Ph.D. thesis, University of Padova (2012)

16. Miranda, L., Rio, J., Guedes, J., Costa, A.: Propagation of elastic waves on stone masonry walls. In: Proceedings of the Eighth International Masonry Conference. Dresden, Germany (2010)

17. Binda, L., Modena, C., Baronio, G., Abbaneo, S.: Repair and investigation techniques for stone masonry walls. Constr. Build. Mater. 3, 133–142 (1997)

18. Binda, L.: Caratterizzazione delle murature in pietra e mattoni ai finidell'individuazione di opportune tecniche di riparazione. CNR-Gruppo Nazionale per la Difesa dai Terremoti – Roma (2000)

19. Binda, L., Saisi, A.: State of the art of research on historic structures in Italy. Department of Structural Engineering, Politechnical of Milan, Italy (2001)

20. Casella, G.: Gramáticas de Pedra. Levantamento de Tipologias de Construção Muraria. Centro Regional de Artes Tradicionais (2003)

21. Pagaimo, F.: Caracterização Morfológica e Mecânica de Alvenarias Antigas: Caso de estudo da Vila de Tentúgal. Faculdade de Ciências e Tecnologia da Universidade de Coimbra, Departamento de Engenharia Civil (2004)

22. Palladio, A.: I quattro libri dell'architettura. Hoepli Editore, Milano (2000)

23. Breymann, G.A.: Costruzioni in mattoni ed in pietre artificiali e naturali. Dedalo, Roma (2003). 11

24. Giuffrè, A.: Letture sulla meccanica delle murature storiche. Rome, Kappa, Italy (1990)

25. Giuffrè, A., Carocci, C.: Vulnerability and mitigation in historical centres in seismic areas. Criteria for the formulation of a Practice Code. XI World Conference on Earthquake Engineering, Acapulco, Mexico 2086 (1996)

26. de Felice, G., Giannini, R.: Out-of-plane seismic resistance of masonry walls. J. Earthq. Eng. 5, 253–271 (2001)

27. Binda, L., Penazzi, D., Saisi, A.: Historic masonry buildings: necessity of a classification of structures and masonries for the adequate choice of analytical models. In: VI International Symposium Computer Methods in Structural Masonry—STRUMAS, Rome, Italy, pp. 168–173 (2003)

28. Binda, L.: Methodologies for the vulnerability analysis of historic centres in Italy. Keynote lecture, IX International Conference on Structural Studies, Repairs and Maintenance of Heritage Architecture—STREMAH, 22–24 June, Malta, pp. 279–290 (2005)

29. Borri, A., De Maria, A.: Alcune considerazioni in materia di analisi e di interventi sugli edifici in muratura in zona sismica. XI Congresso Nazionale "L'ingegneria Sismica in Italia", ANIDIS, Genoa, Italy, pp. 25–29 (2004)
30. Ferrini, M., Melozzi, A., Pagliazzi, A., Scarparo, S.: Rilevamento della vulnerabilità sismica degli edifici in muratura. Manuale per la compilazione della Scheda GNDT/CNR di II livello, Regione Toscana, Italy (2003)
31. Binda, L., Saisi, A., Tiraboschi, C.: Investigation procedures for the diagnosis of historic masonries. Constr. Build. Mater. **14**(4), 199–233 (2000)
32. Baronio, G., Binda, L.: Experimental approach to a procedure for the investigation of historic mortars. In: IX International Brick and Block Masonry Conference, Berlin, Germany, pp. 1397–1464 (1991)
33. Binda, L., Tiraboschi, C.: Flat-Jack method applied to historic masonries. In: Proceedings of the International RILEM Workshop On Site Control and Non-Destructive Evaluation of Masonry Structures, Mantova, Italy, pp. 179–190 (2003)
34. Anzani, A., Binda, L., Cantini, L., Cardani, G., Saisi, A., Tedeschi, C.: On site and laboratory investigation to assess material and structural damage on some churches hit by an earthquake. XII Conv. Naz. L'Ingegneria Sismica—ANIDIS 2007, Pisa, Italy (2007)
35. Doglioni, F.: Codice di pratica (linee guida) per la progettazione degli interventi di riparazione, miglioramento sismico e restauro dei beni architettonici danneggiati dal terremoto umbro-marchigiano del 1997. Bollettino Ufficiale Regione Marche (1999)
36. Borri, A.: Proposta di una metodologia per la valutazione della qualità muraria. Progetto di ricerca No. 1, Valutazione e riduzione della vulnerabilità sismica di edifici in muratura, rendicontazione scientifica 1°anno, Report Reluis (2006)
37. Regione di Molise, Allegato 3D, Protocollo di Progettazione per la Realizzazione degli Interventi di Ricostruzione Post-Sisma sugli Edifici Privati, Parte I—Edifici in Muratura, 2006—'Indicazioni per la valutazione della qualità delle murature' (2006)
38. LLPP, Italian Ministry of Public Works, Norme Tecniche per le Costruzioni in Zona Sismica. Decreto Ministeriale, 16 January 1996
39. LLPP, Italian Ministry of Public Works (2009) Istruzioni per l'applicazione delle Nuove norme tecniche per le costruzioni di cui al decreto ministeriale 14 gennaio 2008. Circolare del 2/2/2009, n. 617 del Ministero delle Infrastrutture e dei Trasporti approvata dal Consiglio Superiore dei Lavori Pubblici, Suppl. ord. n. 27 alla G.U. del 26/02/2009 No. 47, Italy
40. LLPP, Italian Ministry of Public Works, Norme Tecniche per le Costruzioni. Decreto Ministeriale del 14/1/2008, Suppl. ord. n. 30 alla G.U. n. 29 del 4/2/2008, Italy (2008)
41. Bennet, R.M., Boyd, K.A., Flanagan, R.D.: Compressive properties of structural clay tile prisms. J. Struct. Eng. **123**(7), 920–926 (1997)
42. CEN, EN 1996-1-1: Eurocode 6—Design of masonry structures—Part 1-1: General rules for reinforced and unreinforced masonry structures (2005)
43. Kaushik, H.B., Rai, D.C., Jain, S.K.: Stress-strain characteristics of clay brick masonry under uniaxial compression. J. Mater. Civ. Eng. **19**(9), 728–739 (2007)
44. Corradi, M., Borri, A., Vignoli, A.: Strengthening techniques on masonry structures struck by the Umbria-Marche earthquake of 1997–1998. Constr. Build. Mater. **16**, 229–239 (2002)
45. Magenes, G., Galasco, A., Penna, A.: Caratterizzazione meccanica di una muratura in pietra XIII Convegno Nazionale—L'ingegneria Sismica in Italia. ANIDIS, Bologna, Italy (2009)
46. Binda, L., Baronio, G., Gambarotta, L., Lagomarsino, S., Modena, C.: Masonry constructions in seismic areas of central Italy: a multi-level approach to conservation. In: VIII North American Masonry Conference—8NAMC, Austin, USA, pp. 44–55 (1999)
47. Binda, L., Gambarotta, L., Lagomarsino, S., Modena, C.: A multilevel approach to the damage assessment and seismic improvement of masonry buildings in Italy. Seismic damage to masonry buildings, Balkema, Rotterdam, Netherlands (1999)
48. Borri, A., Avorio, A., Cangi, G.: Considerazioni sui cinematismi di collasso osservati per edifici in muratura. IX Convegno Nazionale "L'ingegneria Sismica in Italia", ANIDIS, Torino, Italy (1999)
49. D'Ayala, D.: Correlation of seismic vulnerability and damages between classes of buildings: Churches and houses, in Seismic Damage to Masonry Buildings. Balkema, Rotterdam, pp. 41–58 (1999)

50. D'Ayala, D., Speranza, E.: Identificazione dei meccanismi di collasso per la stima della vulnerabilità sismica di edifici nei centri storici. IX Convegno Nazionale—L'ingegneria Sismica in Italia, ANIDIS, Torino, Italy (1999)
51. Doglioni, F., Moretti, A., Petrini, V.: Le chiese ed il terremoto. LINT, Trieste, Italy (1994)
52. Giuffrè, A.: Sicurezza e conservazione dei centri storici in area sismica, il caso Ortigia. Laterza, Bari, Italy (1993)
53. Lagomarsino, S., Brencich, A., Bussolino, F., Moretti, A., Pagnini, L.C., Podestà, S.: Una nuova metodologia per il rilievo del danno alle chiese: prime considerazioni sui meccanismi attivati dal sisma. Ingegneria Sismica **3**, 70–82 (1997)
54. Tomaževič, M.: Some aspects of experimental testing of seismic behaviour of masonry walls and models of masonry buildings. ISET J. Earthq. Technol. **37**(4), 101–117 (2000)
55. Modena, C., Valluzzi, M.R., Zenere, M.: Manuale d'uso del Programma c-Sisma. University of Padova, Padova, Italy (2009)
56. D'Ayala, D., Speranza, E.: Definition of collapse mechanisms and seismic vulnerability of historic masonry buildings. Earthq. Spectra **19**, 479–509 (2003)
57. Tomaževič, M.: Earthquake-Resistant Design of Masonry Buildings. Imperial College Press, London (1999)
58. Magenes, G., Calvi, G.M.: Cyclic behaviour of brick masonry walls. In: X World Conference on Earthquake Engineering, Madrid, Spain, pp. 3517–3522 (1992)
59. Magenes, G., Calvi, G.M.: In-plane seismic response of brick masonry walls. Earthq. Eng. Struct. Dynam. **26**, 1091–1112 (1997)
60. Vasconcelos, G., Lourenço, P.B.: Assessment of the in-plane shear strength of stone masonry walls by simplified models. In: V International Seminar on Structural Analysis of Historical Constructions—SAHC06, New Delhi, India, pp. 843–850 (2006)
61. Lourenço, P.B., Oliveira, D.V., Roca, P., Orduña, A.: Dry joint stone masonry walls subjected to in-plane combined loading. J. Struct. Eng. ASCE **131**(11), 1665–1673 (2005)
62. Mayes, R.L., Clough, R.W.: A literature survey-compressive, tensile, bond and shear strength of masonry. Report No. EERC 75-15, Earthquake Engineering Research Centre, University of California, Berkeley (1975)
63. Bosiljkov, V., Page, A., Bokan-Bosiljkov, V., Zarnic, R.: Performance based studies on in-plane loaded unreinforced masonry walls. Masonry Int. **16**(2), 39–50 (2003)
64. NGDC, National Geophysical Data Center, The Significant Earthquake Database. http://www.ngdc.noaa.gov/nndc/struts/form?t=101650&s=1&d=1
65. Magenes G, Bolognini D, Braggio C. (2000) Metodi semplificati per l'analisi sismica non lineare di edifici in muratura. CNR—Gruppo Nazionale per la Difesa dai Terremoti, http://gndt.ingv.it/Pubblicazioni/Magenes_copertina_con_intestazione.htm, Roma, Italy
66. OPCM 3274, Ordinanza del Presidente del Consiglio dei Ministri n. 3274 del 20 Marzo 2003. "Primi elementi in materia di criteri generali per la classificazione sismica del territorio nazionale e di normative tecniche per le costruzioni in zona sismica". GU n. 72 del 8-5-2003, e s.m.i (2003)
67. CEN, EN 1998-3. Eurocode 8—Design of structures for earthquake resistance. Part 3: Assessment and retrofitting. European Standard. CEN, Brussels (2005)
68. D.P.C.M. 12-10-2007. G.U. n.24 of 29-1-2008, Guidelines for evaluation and mitigation of seismic risk to cultural heritage. Gangemi Editor. Rome. ISBN 978-88-492-1269-3 (2007)
69. Venice Charter International Charter for the Conservation and Restoration of Monuments and Sites. II International Congress of Architects and Technicians of Historic Monuments (1964)
70. ICOMOS/ISCARSAH Committee 2005. Recommendations for the analysis, Conservation and Structural Restoration of Architectural Heritage. ICOMOS International Committee for Analysis and Restoration of Structures of Architectural Heritage
71. ISO/TC98/SC2, ISO/FDIS 13822—Bases for design of structures—Assessment of existing structures. Final Draft. ISO, Genève (2010)
72. Modena, C., Binda, L., Anzani, A.: Investigation for the design and control of the repair intervention on historical stone masonry wall. In: VII International Conference and Exhibition, Structural Faults and Repair, Edinburgh, vol. 3, pp. 233–242 (1997)

73. Modena, C., Zavarise, G., Valluzzi, M.R.: Modelling of stone masonry walls strengthened by RC jackets. In: IV International Symposium on Computer Methods in Structural Masonry—STRUMAS, Florence, Italy, pp. 285–292 (1997)
74. Bettio, C., Modena, C., Riva, G.: The efficacy of consolidating historical masonry by means of injections. In: VII North American Masonry Conference—NAMC, Notre Dame, USA, pp. 458–471 (1996)
75. Binda, L., Modena, C., Baronio, G.: Strengthening of masonries by injection technique. VI North American Masonry Conference—NAMC, Philadelphia, USA, vol. 1, pp. 1–14 (1993)
76. Binda, L., Baronio, G., Tiraboschi, C.: Repair of brick-masonries by injection of grouts: experimental research. J. Struct. Eng. Madras, India 20(1), 29–44 (1993)
77. Binda, L., Modena, C., Baronio, G., Gelmi, A.: Experimental qualification of injection admixtures used for repair and strengthening of stone masonry walls. In: X International Brick and Block Masonry Conference—I2BMC, Calgary, Canada, vol. 2, pp. 539–548 (1994)
78. Laefer, D., Baronio, G., Anzani, A., Binda, L.: Measurement of grouts injection efficacy for stone masonry wall. VII North American Masonry Conference—NAMC, Notre Dame, USA, vol. 1, pp. 484–496 (1996)
79. Modena, C., Bettio, C.: Experimental characterization and modelling of injected and jacketed masonry walls. In: Italian-French Symposium Strengthening and Repair of Structures in Seismic Area, Nizza, pp. 273–282 (1994)
80. Tomaževič, M., Turnsek, V.: Verification of the seismic resistance of masonry buildings. In: British Ceramic Society: Load bearing Brickwork, vol. 30, pp. 360–369 (1982)
81. Miltiadou, A.: Contribution Ii l'etude des coulis hydrauliques pour la reparation et Ie renforcement des structures et des monuments historiques en maçonnerie. Ph.D. thesis, Ecole Nationale des Ponts et Chaussecs, Paris (1990)
82. Tomaževič, M., Apih, V.: The strengthening of stone-masonry walls by injecting the masonry-friendly grouts. Eur. Earthq. Eng. 6(1), 10–20 (1993)
83. Toumbakari, E.E.: Lime-pozzolan-cement grouts and their structural effects on composite masonry walls. Ph.D. thesis, Katholieke Universiteit Leuven, Belgium (2002)
84. Valluzzi, M.R., da Porto, F., Modena, C.: Behaviour and modelling of strengthened three-leaf stone masonry walls. Mater. Struct. 37(3), 184–192 (2004)
85. Vintzileou, E., Miltiadou-Fezans, A.: Mechanical properties of three-leaf stone masonry grouted with ternary or hydraulic lime-based grouts. Eng. Struct. 30(8), 2265–2276 (2008)
86. Binda, L., Cardani, G., Penazzi, D., Saisi, A.: Performance of some repair and strengthening techniques applied to historical stone masonries is seismic areas. ICPCM a New Era of Building, Cairo, Egypt, vol. 2, pp. 1195–1204 (2003)

Save the *Tabique* Construction

**Jorge Pinto, Gülten Gülay, José Vieira, Vatan Meltem,
Humberto Varum, İhsan Engin Bal and Aníbal Costa**

Abstract The main goal of this chapter consists in revising technical building aspects concerning the *tabique* construction. *Tabique* is a traditional building technique which applies raw building materials such as timber and earth, for example. External and partition walls are the main *tabique* building components, which may have a relevant structural performance in the overall stability of a building. In general, the traditional *tabique* buildings are facing high levels of degradation. This problem

J. Pinto (✉)
ECT, Engineering Department, Trás-os-Montes e Alto Douro University,
5001-801 Vila Real, Portugal
e-mail: tiago@utad.pt

J. Pinto
Portugal Laboratory Associated I3N, Aveiro University, Aveiro, Portugal

G. Gülay
Istanbul Technical University, Maslak, Instanbul 34469, Turkey
e-mail: gulayg@itu.edu.tr

J. Vieira · V. Meltem
Istanbul Aydin University, Istanbul, Turkey
e-mail: jbvieira@utad.pt

V. Meltem
e-mail: vatan@yildiz.edu.tr

H. Varum · A. Costa
Civil Engineering Department, Aveiro University, 3810-193 Aveiro, Portugal
e-mail: hvarum@ua.pt

A. Costa
e-mail: agc@ua.pt

İ. E. Bal
Institute of Earthquake Engineering and Disaster Management,
Istanbul Technical University, Maslak, 34469 Istanbul, Turkey
e-mail: iebal@itu.edu.tr

A. Costa et al. (eds.), *Structural Rehabilitation of Old Buildings*,
Building Pathology and Rehabilitation 2, DOI: 10.1007/978-3-642-39686-1_6,
© Springer-Verlag Berlin Heidelberg 2014

is mainly due to the lack of maintenance and technical knowledge. Therefore, reconstruction processes of this type of buildings also require accurate and updated technical information related to materials, to building details, to the main likely pathologies, to the knowledge of the physical and mechanical behaviour of the building components, to the reinforcement solutions, among other aspects. This research work intends to give some guidance in this context and to make a parallel analysis between the Portuguese and the Turkish *tabique* constructions.

Keywords *Tabique* • Earth construction • Timber structural elements • Rehabilitation • Traditional structures • Sustainability

1 Introduction

The fact that most of the existing old constructions are still able to perform their original functionality is an irrefutable proof of their longevity. They have stood successfully all the aggressiveness along their existence (the expected and the unexpected ones). This achievement is even more outstanding taking into account that 50 years is approximately the current expected duration (i.e. "lifetime") of a common building, nowadays. They may show a certain amount of degradation which may be seen as acceptable and natural. In fact, the natural aging effect of a construction may be valuable since it may give an identity attribute. For instance, and in a limited scale, the natural degradation of a stone masonry wall, the "patine" of the painting of a façade, the aging effect in a ceramic tile, the oxidation effect of a timber building component may be aspects that may give technical value to a building. Therefore, in the building industry, the natural aging phenomenon should be accepted and valued. At the same time, there is no other alternative because it is an irreversible phenomenon. Even in certain specific types of constructions which require robustness, this fact is unavoidable. In contrast, unexpected premature building degradation phenomena are a signal of a lack of quality, in a certain scale. In fact, buildings that face this problem are unlikely to achieve their inherent proposed functionally for a long period of time because one type of pathology may originate a trigger progressive generalized pathology phenomenon. In these cases, rapid and adequate building rehabilitation interventions are the key solution. Meanwhile, old constructions are real objects which reflect the soul of a community. In other words, they are social, economic and cultural testimonies which bring the past up to the present.

Additionally, this fabulous heritage must be seen as open technical books which are filled with a huge amount of information related to traditional building processes, and from where the technicians can learn and get inspiration. On the other hand, each ancient building must also be seen as a real scale construction sample which has been tested under real load combinations in real physical conditions and for a long period of time. The performance of the adopted structural system and building materials, and the thermal and the acoustic behaviours are some technical aspects that can be easily understood. Plus, aspects such as fatigue, second order effects, material degradation have to be included in those studies. Performing similar studies in a laboratory presents limitations, it is complex and costly.

Furthermore, each ancient building was built up according to a specific building process. That building process was included in a specific building technique type which was likely to depend on the technical knowledge, the economic potential, the intellectual status, among other aspects, of the period of time in which it was applied. Based on history, the building techniques evolve according to the introduction of new building materials and new processes, following the progress trends occurring in a specific period of time. Innovation plays a key role in these processes of progress. In this aspect, our builders who preceded us were geniuses because they were able to create magnificent pieces of works without codes or computer programs available. Bright minds, tenaciousness, spirit of adventure, empirical experience and greatness were the main tools for such impressive achievements. This research work intends to make an attempt at paying a modest tribute to all those engineers, architects, carpenters, masons and the other indispensable labourers who contributed to those building processes, directly and indirectly.

Another particularly interesting and relevant technical aspect associated to this type of construction is its sustainable value according to present day premises. In fact, they were built using simultaneously sustainable materials and sustainable building techniques. Generally, the applied building materials are natural, locally available (e.g. stone, sand or earth, among other) and, in some cases, they are also organic and, therefore, renewable (e.g. wood, straw or corn cob, among others). Meanwhile, the applied building techniques are usually simple because they do not depend on sophisticated devices or high-tech procedures. Thus, these characteristics are favourable to achieving an almost inexpressive amount of pollutant emission into the atmosphere, an enormous energy consumption reduction and also adequate water consumption. This fact is even more evident taking into account that these old buildings were built hundreds of years ago and, in the scenario of a lifetime of 50 years, they may be considered over-strengthened.

The above exposed arguments are sufficient to grasp the importance of this rich heritage which must be cherished and preserved and, above all, deserves to be passed onto the following generations. In this aspect, each generation has the duty and the responsibility of caring for it. Taking into account that most of the old buildings are private properties and the fact that conservation and rehabilitation works are costly, the governmental institutions have the obligation of being part of this maintenance process by offering efficient mechanisms and also promoting the relevance of this preservation among the community. On the other hand, the scientific and the academic communities should also join this process by delivering updated technical information and knowledge to the building industry.

Among the traditional building techniques, this research work is focused on the *tabique* building technique. There are similarities between *tabique* and wattle-and-daub building techniques. The *tabique* technique is essentially characterized by using timber structural systems filled with an earthy material (e.g. lime earth based mortar, for instance) in the building process of different types of components. The elements of the timber structural systems are connected to each other by nails. External and partition walls are the main building components built with this technique. However, chimneys and sheds are also reported as other examples of *tabique* building components. In this research work, a parallel analysis is done between Portugal and Turkey in terms of *tabique* construction. Finding similarities and differences in this traditional construction in both countries may result in a better understanding of this particular building technique and also in discoveries which may contribute to achieving better maintenance, conservation and rehabilitation building processes. It is worth to underline that both countries are seismically vulnerable and also they have similar climate conditions.

This chapter is structured as follows: Firstly, a brief retrospective of the *tabique* construction in Portugal and Turkey is done. Examples of *tabique* building typologies and some inherent specificities are provide based on Portuguese and Turkish study cases; Secondly, the applied building materials issue is emphasized; Thirdly, constructive solutions and building details concerning walls, pavements and roof are presented and described; Fourthly, structural and non-structural defects and their causes are analysed; Fifthly, some possible rehabilitation solutions are proposed and some mitigation measures are suggested; Finally, comments and recommendations for rehabilitation are presented.

2 Brief Retrospect of the *Tabique* Construction in Portugal and Turkey

2.1 Its Incidence

In Portugal, the *tabique* construction prevailed in the nineteenth century. However, there are plenty of examples dating from much earlier. It started to fall into disuse when the reinforced concrete structural elements started to be introduced in the building industry which occurred by early twentieth century. During the transition and adjusting period of time, there were constructions built according to the two building techniques in parallel and there were interesting cases in which the two techniques were combined together. Basically, the *tabique* building technique was applied all over the country. Places where the forest was scarce or the ground was essentially sandy may be exceptions. Thus, the north of Portugal is probably the richest part in terms of *tabique* heritage. For instance, the main existing period dwellings in the historic centre of the cities of the north of Portugal were built according to this technique. Furthermore, in the rural areas of this part of the country, the existing *tabique* constructions are also relevant in terms of expressiveness and heritage value and, therefore, they cannot be excluded in this work. It is necessary to underline here that concerning the *tabique* construction in the Portuguese context, this research work is mainly focused on the northeast part of Portugal, corresponding to the Trás-os-Montes e Alto Douro region. There are already some local technical building differences reported among the existing *tabique* constructions in the north part of Portugal such as in between Douro Litoral and Trás-os-Montes e Alto Douro. Related to Trás-os-Montes e Alto Douro region, an exhaustive research work has been done in this context and it was noticed that the existence of *tabique* constructions in all this territory is evident. However, its incidence is slightly higher in the Municipalities placed south.

In Turkey, the history of timber structures dates back to the Ottoman Period. Along the history, wood has been used widely as a construction material (e.g. timber framed structures with different types of fillings).Anatolia and Rumelia are rich in heritage with this characteristic and also the whole Balkan region. Timber frame structures are typical of traditional vernacular architecture that first arose in the medieval and early Ottoman period, improved in the end of the sixteenth century and became popular in the eighteenth century. First examples were not just only dwellings but also public buildings, such as mosques. Timber framed constructions were commonly used as typical house construction up until 1950–1960s when reinforced concrete became popular as a construction material in Turkey and accepted as the sole option. Although affected by the violent earthquakes throughout history, there still exist many examples of those traditional houses in Kocaeli, Yalova, Adapazarı, Düzce, Kastamonu where the North Anatolian Fault lies parallel to the region. They are mostly concentrated in Northern parts of Turkey due to the abundance of forest. Its strength, lightness in weight, workability, aesthetics as well as the fact that it was easy to find such materials are most important characteristics which make wood a functional and expansive material. Over the last half of the twentieth century, reinforced concrete frames gradually replaced those traditional timber framed structures. The main reasons for that fact are related to fire resistance and to demographic increasing. Thus, the traditional timber framed houses, which prevailed for 300 years in Anatolia, were gradually abandoned and even forgotten until the devastating earthquakes in 1999 reminded everyone the sound behaviour of traditional timber framed constructions during an earthquake.

2.2 Examples of Building Typologies and Particularities

In the Portuguese context, over three hundred *tabique* buildings have already been studied in this research work. In general, these buildings are dwellings [1], Fig. 1a. The most common featured scenarios are related to two (Fig. 1a) or three floors *tabique* buildings [2]. One floor or over three floors scenarios are rare. Perhaps, one floor *tabique* building solution is uncommon because the timber structure system of the vertical building components would be very vulnerable to the attack of insects due to its proximity to the ground and also rising damp related problems such as deterioration of the earthy render. Meanwhile, up to three floors solutions are rare because they are related to dwellings type typology and, probably, there

Fig. 1 Examples of *tabique* dwellings. **a** Portuguese (*exterior view*). **b** Turkish (*interior view*)

were structural limitations. However, the existence of five floors *tabique* dwellings in the city centre of Chaves is a fact already reported [2]. Thus, this latter typology was applied sometimes in considerably dense urban areas, for the time. Sober architecture is a characteristic associated to this type of construction. Cubic shapes and plain façades are aesthetic typical references, as Fig. 1a exemplifies. The majority of these buildings have traditional Portuguese ceramic tiles as exterior revetment of the roof (Fig. 1a). The roofs themselves are simple and normally characterized by having two or four differentiate planes. Usually, the ground floor is an open space (i.e. free of interior vertical elements such as columns or partition walls) and, usually, it is used as a storing room or a shopping store. In this case, the main structural vertical elements are external and peripheral stone masonry walls built in granite (Fig. 1a) or schist. These two types of stones are local. Meanwhile, the upper floors are for living in and, therefore, partition walls are required. The partition walls are *tabique* vertical components. In terms of external walls, there are two important possibilities which are stone masonry external walls or *tabique* external walls. *Tabique* external walls are essentially placed in the upper floors (Fig. 1a). Based on our experience, *tabique* exterior wall is a building solution scenario more applied in rural areas and in the region of Trás-os-Montes e Alto Douro. In big cities of the north of Portugal such as Oporto, this building scenario is uncommon and essentially circumscribed to the attic. Associated to the *tabique* exterior walls there are different applied options for the finishing. White wash painting (Fig. 1a), steel (Fig. 1a) or zinc corrugated plates, slate boards, and ceramic tiles are the most commonly noticed possibilities. There are interesting cases which have a combination of finishing types of the *tabique* exterior walls. For instance, the *tabique* exterior wall orientated south (which is the most susceptible to rain exposure in this region) may have steel corrugated plated as finishing and the other walls may have a white wash painting type (Fig. 1a). These presented technical building particularities also give strength to the sustainability and affordability attributes of this type of buildings. At the present time, we may conclude that the technical decisions concerning *tabique* construction were delicate and defined in order to achieve a robust, an affordable and an environmental friendly building solution. Meanwhile, in the Turkish context *tabique* construction may be related to *bağdadi* and *dizeme* constructions (Fig. 1b). This type of construction is widely used in Rumelia and Istanbul. Particularly, along the Bosporus there were/are so many mansions made by *bağdadi* construction as well as in the rest of the city that they are historic buildings which are protected by the Law of Protection of Cultural and Natural Properties.

3 Materials

As it was stated above, the main characteristic building materials of *tabique* building components are wood (in the timber structure), alloy (in the mechanical connectors which are traditionally nails) and an earthy type mortar (in the filling). Therefore, a *tabique* building component may be seen as a composite. Timber and alloy are the main structural ones. Meanwhile, the earthy type mortar material has an important constructive functionality because it works as a matrix which intends to wrap the timber and the nails elements, to fill the existing gap between them and to protect them. Some technical building aspects related to the durability of the structural elements, the thermal and the acoustic insulation performance and the fire resistance of a *tabique* building component are greatly dependent on the quality of the filling. On the other hand, this quality depends on the type of the constituents of the mixture, the proportion in which they are mixed up and the adherence ability to the support (i.e. the timber structural system of the component). Therefore, in order to better understand this traditional building technique and to be able to give appropriate technical guidance for future adequate maintenance and rehabilitation procedures, it seems fundamental to characterise and to identify these building materials previously. Therefore, several samples of the main structural materials in the buildings have been collected and experimentally analysed.

3.1 Timber

In the Portuguese context, over sixty timber element samples were tested in order to identify their wood species. The obtained results have concluded that different types of wood can be used in the timber structural system of a *tabique* component. Based on the Portuguese experience, four wood species have been identified so far. They are by order of incidence importance the *Pinus pinaster* (softwood), 65 % of incidence, the *Castanea sativa Mill* (hardwood), 25 % of incidence, the *Populus sp* (hardwood), 7 % of incidence, and the *Tília cordata* (hardwood), 3 % of incidence. *Pinus pinaster* is clearly the most applied wood in this context. These species are autochthonous of the north of Portugal. Other wood species types are likely to be found in other parts of Portugal. These results indicate that hardwood and softwood have been traditionally used in structural applications. This technical aspect is important taking into account that the wood is the main structural material in this type of building components and, therefore, its mechanical properties are relevant. Other important conclusion is related to fact that these are really successful civil engineering applications of autochthonous Portuguese wood species. Usually, the Portuguese timber is somewhat depreciated because of its considered limitations in terms of its material uniformity and related to the climate conditions of the country. In this case, by using a timber structural system, generally characterized as being a regular frame of orthogonal regular elements, it allows dissipating those possible limitations. The authors sincerely believe that this information may also give guidance to arouse the important potential that the Portuguese forests can offer, and which is often despised.

Furthermore, in the Turkish context, wood is also the main structural framework of traditional Turkish *tabique* components as horizontal, vertical and diagonal structural reinforcement. In fact, wood is hugely applied in a diverse range of building applications such as flooring, cladding for the ceiling, for cupboards, doors, windows and shutters, and as roof constructions. Depending on the climate, flora and other local conditions, various kinds of timber are utilized for different purposes. For instance, chestnut, pine, fir and oak trees are used for the load bearing structure of the roof; yellow fir is used as a substitute for fir, if fir is not available; black pine is used for producing shingles; poplar is used in producing the wood turned balustrades of staircases and windows; walnut is used for the panels on ceilings, and on the doors of cupboards and rooms; beech is used for making window casements [3]. Among these, chestnut is the most advantageous material because it is light, endurable against fungi and insects, easily found and repairable.

3.2 Nails, Connectors and Fixing Systems

In the Portuguese context, all the studied *tabique* buildings have shown nails as the mechanical connector of the fixing solution of the timber elements that form the structural system of a *tabique* component. In the region of Trás-os-Montes e Alto Douro, the most commonly used solution of this timber structural system consists in having a net formed by a set of vertical timber elements linked to each other by a set of horizontal timber elements which are applied in both faces. At the same time, in a *tabique* component, the vertical timber elements tend to have a regular cross-sections shape (in geometry and in dimension) and a regular gap in between them. Similar technical patterns occur in the horizontal timber elements. However, based on our experience, there was no evident similar pattern among the same type of *tabique* components of different *tabique* buildings. This fact was already highlighted in [4] concerning partition *tabique* walls as building components. At least one nail is applied to fix one horizontal timber element to one vertical timber element. Therefore, one horizontal timber vertical element is fixed to several vertical timber elements at the same time. On the other hand, one vertical timber element is fixed to several horizontal timber elements, in both its faces. In a previously done technical inquiry which targeted retired builders, owners of *tabique* buildings and aged people, the fact that nails had a low percentage of iron metal was one interesting resulting conclusion. The explanation for this detail consisted essentially in attempting to reduce the undesirable oxidation phenomenon of the nails. It was reported that this pathology would affect the finishing of the wall negatively by stimulating the appearance of brownish spots on the white wash finishing type. Furthermore, aiming to deliver an accurate inventory of building aspects related to this traditional building technique, several nail samples have been collected and experimentally tested. The geometry and the alloy characterizations have been the main technical aspects considered in this study. In terms of geometry, it has been concluded that it may vary significantly. In fact, we can find nails with square (constant or variable) or circular (constant) cross-section shapes and length values ranging between 3.0 and 10.5 cm. According to our research, constant square cross-section nails have been the most commonly adopted option. Nails with lengths ranging between 3.0 and 5.0 cm have been used to fix the horizontal timber elements to the vertical timber elements. On the other hand, nails with lengths ranging between 6.5 and 10.5 cm have been applied in the most important structural connections such as the connection of the vertical timber elements (e.g. board or rafter) to the horizontal timber elements placed at the ends of the vertical ones. Steel has been the alloy type experimentally found for the tested nails. Common steel or a steel poor in carbon have been the major obtained results. Furthermore, the analysed nail samples have shown an acceptable level of conservation. It is important to refer that the task of collecting building material samples has been extremely complex. In general, this task has been carried out because it has been possible to find *tabique* buildings under demolition process or to find other buildings that had already collapsed and are abandoned. These building scenarios have helped this research work but, at the same time, they also reflect the tragic reality that most of the existing *tabique* buildings have been facing during the last years in the region of Trás-os-Montes e Alto Douro. This situation has been essentially caused by an impotence attitude assumed by the owners or by the public entities in taking measures to avoid this loss of heritage Therefore, based on these conditions, in particular the ones related to the second building scenario, it is expected to have an aggravated deteriorating stage of the studied nail samples compared to the nails which are still in use. The above results may allow concluding that there is no specific characteristic associated to the nail types used in the fixing system of the timber elements. In fact, the current available nails seem adequate for this application. Anti-oxidation nails may be an interesting technical solution in order to avoid the above reported white wash pathology.

According to the Turkish experience on traditional timber framed construction, this type of construction may have an adequate seismic behaviour besides having simple joints and nailed solutions. Direct joints of traditional timber frames are generally mortise and tenons.

3.3 Filling

According to the collected data in the *tabique* buildings already studied in the Trás-os-Montes e Alto Douro region, two main technical solutions of the filling material type of *tabique* components were identified. These two solutions are a lime earthy based mortar (incidence of 70 %) and earth (incidence of 30 %) among up to sixty filling samples analysed. Basically, these two types of solutions were distinguished by performing a scanning electron microscopy/energy dispersive spectroscopy analysis (SEM/EDS) and by verifying the existence of calcium (Ca) in the identified chemical elementary composition. Since the soil of this region is essentially characterised by being granite related, the expectancy of having Ca as a chemical element of the elementary chemical composition is almost null. Therefore, a sample was considered as being a lime earthy based mortar when Ca was found in its elementary chemical composition. The identified percentage of Ca of the elementary chemical composition varied quite sharply among the studied lime earthy based samples. Values ranging between 6 and 40 % have been reported for the percentage of Ca and, at this stage, an average value of 26 % was estimated. In these cases, the Ca results from the addition of a certain portion of lime into earth, and because lime is rich in this chemical element. In contrast, elementary chemical compositions without Ca (or provided with an inexpressive amount of Ca) were considered as being related to earth samples. In this latter situation, the presence of that small amount of Ca may result from the fact that a filling sample had been intoxicated by the white wash finishing during the collecting task procedure or to the fact that Ca particles had accidentally remained in the microscopy as dust waste and resulted from previous tests. Meanwhile, the lime earthy based mortar filling type have shown higher consistence than the earth filling type. Therefore, a similar approach may be taken in the site of the *tabique* building in order to identify expeditiously the filling type and taking into consideration the above material behaviour difference. The incorporation of a certain amount of lime into the prepared earth (in terms of particle dimensions) intends to improve the plasticity of the filling material resulting consequently in a workability gain. Better scattering and easy adhesion to the support may be the two achieved attributes by applying this procedure. Furthermore, in both filling types, gravel is generally the type of earth used. Granulometric analysis was also performed in the collected filling material samples. The respective obtained results indicate that 20 % of thin particles (i.e. silt and clay) and 80 % of medium sized particles (i.e. sand) may be considered as being a traditional dimensional soil mixture for the filling material of *tabique* components. Additionally, based on the above referred inquiry, the fact that the earth was obtained nearby the building site was a consensual output. In order to confirm this information, a case study was specifically done for this purpose and the obtained conclusions give consistence to the above technical particularity [5]. In fact, samples of filling material and samples of local gravel have shown interesting similarities in terms of elementary chemical and mineralogical compositions. Thus, the exposed aspects reinforce the idea that this type of traditional building technique may be exceptionally environmentally friendly when compared to the current applied building techniques since earth is abundant, local and reusable. The amount of water content, the ratio of the constituents, the curing time, alternative organic binders and the incorporation of different types of sustainable and affordable fibre are some additional aspects which also have been under research. They aim to improve the filling material of *tabique* components in future conservation and rehabilitation processes.

In the Turkish context, a similar building pattern is verified. In a *tabique* wall (in a *bağdadi* construction), the timber structure, including the short rough pieces of timber (known as *bağdadi laths* in Turkish), is plastered in both its interior and exterior sides. The interior part of the wall is plastered with mud, sand and lime mortar which fills the existing voids between laths. Meanwhile, the exterior part of the façade walls is mainly timber plastered. Plaster mortar reinforced with organic materials such as straw is also a current traditional building technique in the Turkish *tabique* heritage.

4 Constructive Solutions and Building Details

There is still a lack of published documents related to this building technique. Additionally, the design project of a *tabique* building is generally inexistent. Thus, to report the adopted constructive solutions complemented with the respective building details of *tabique* building cases is essential. Therefore, the following sections intend to contribute in this respect.

4.1 Tabique Walls

As stated above, a timber structural system is the main structural solution of a *tabique* component such as walls. Four timber structural system typologies have been found so far. The typology described in Sect. 3.2 which is characterized by presenting an orthogonal set of timber elements (vertical and horizontal), which is designated here by typology 1, and is considered as being the most currently applied solution, Fig. 2a. This typology has been applied simultaneously in exterior and partition walls. The typology 2, see Fig. 2b, has been applied in *tabique* exterior walls placed at the roof floor level and taking into account that the timber truss is incorporated in the timber system of the wall. In this case, the timber elements are also disposed orthogonally but sloped. Typology 2 tends to present a double-faced timber structural system in which the elements of the truss are kept in the middle. Typology 3, represented in Fig. 2c, seems to be similar to typology 1 apart from including additionally 45° sloped timber elements which seem to be strategically placed. This solution was only detected in few *tabique* buildings. This fact may allow the conclusion that it is an occasionally and locally applied

Fig. 2 Timber structural system typologies of tabique walls. **a** Typology 1. **b** Typology 2. **c** Typology 3. **d** Typology 4

solution. Furthermore, based on the field work already done, this typology seems to be essentially applied on exterior walls. A modular orthogonal double faced timber structural system, typology 4 (Fig. 2d), has also been found occasionally and locally on exterior walls.

On the other hand, the building details presented in Fig. 3 are exemplificative of the fact that the typology 1 building solution may be subdivided into two sub-typologies (i.e. typologies 1.1 and 1.2). In typology 1.1, Fig. 3a, the timber elements (vertical and horizontal) are previously prepared in order to have a regular geometrical parameterization. Meanwhile, in typology 1.2, Fig. 3b, this building technical aspect seems not to have been taken into consideration. In this case, a rough irregular shape pattern is noticed in the timber elements. Location and affordability are perhaps the main reasons that justify this differentiation. In fact, typology 1.2 has been essentially detected in certain isolated rural areas of Trás-os-Montes e Alto Douro region. The fact that different types of local wood species and different types of timber finishing requirements have been noticed in the building process of *tabique* components may indicate that this building technique may have and additional attribute which is versatility.

Related to partition *tabique* walls, typology 1 has been essentially the mostly verified one among the study cases analysed in this research work. Small thicknesses have been measured for this wall type compared to exterior ones. It is

Fig. 3 Typologies 1.1 and 1.2. **a** Typology 1.1. **b** Typology 1.2

evident, that exterior *tabique* walls have a fundamental role in the overall structural system of a building since they are the main vertical structural elements, in a
certain floor of the building. However, partition walls also have an important contribution to the stability of the building because they allow a connection of the
main structural elements. They may work as a bracing system of the main structural elements. For instance, Fig. 4 presents a partition wall which is basically
supported by a timber beam. On the other hand, it is also linked to the truss of
the roof's structural system and leaned on the orthogonal exterior granite masonry
walls. The flooring and the wood ceiling boards are also transversally connected to
the partition wall at its bottom and top levels.

Furthermore, additional building details concerning *tabique* walls have been
noticed during this work. The traditionally adopted filling material described in
Sect. 3.3 is able to conserve the timber structural system of the wall [4]. Figure 5a
is related to an ancient external *tabique* wall which exemplifies this conclusion.
The respective timber elements did not suffer biological degradation or insect

Fig. 4 A partition *tabique*
wall and its connection
system

Fig. 5 Incorporation of organic products into the filling material. **a** Wood waste. **b** Corn cob

attack. The wood has even maintained its original radiance. Additionally, expedite fire resistance tests were also already performed in *tabique* wall samples and the obtained results indicate that the traditional adopted filling material may also have refractory abilities preventing the wood combustion. Meanwhile, different types of organic products have been found in the filling layer. Straw, wood waste [6] (I, Fig. 5a) and corn cob [7] (II, Fig. 5b) are some examples of these organic incorporated products. These results have inspired this research team to do a parallel research work focused on evaluating the potential of corn cob as an alternative sustainable building material. Interesting thermal insulation ability [8] and possible material properties similar to cork [9, 10] are some of the conclusions already obtained.

4.2 Pavements or Floors

The typical pavement of a *tabique* building is a timber structural system and it corresponds to the main horizontal structural element. Traditionally, this system is formed by beams (I, Fig. 6), secondary beams (II, Fig. 6) and bracing (III, Fig. 6). These structural elements are commonly disposed orthogonally to each other. They tend to show a regular geometry and a regular relative distance. Generally, the timber beam elements are supported by the vertical structural elements such as stone masonry walls or exterior *tabique* walls, in their ends. On the other hand, the timber secondary beam elements are supported by the timber beams. The secondary beams support the flooring boards and also brace the beams in order to avoid possible lateral bending instability phenomena. A timber bracing system is also applied and acts transversally on the secondary beams or on the beams. Figure 6 shows two examples of these traditional timber pavements placed on the ground floor level. In this case, there is a basement and, therefore, ceiling flooring

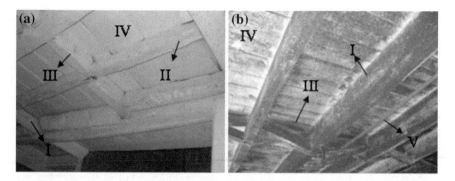

Fig. 6 Structural timber system solutions of pavements of *tabique* construction. **a** Example 1.
b Example 2. *I* beam, *II* secondary beam, *III* bracing, *IV* flooring boards, *V* steel reinforcement

is not expected. In upper floors, ceiling flooring is likely to be applied resulting
in having additional secondary timber beams placed at the bottom level of the
timber beam. The boards of the flooring (IV, Fig. 6) and of the ceiling flooring are
supported continuously on different secondary timber beams. The pavement presented in Fig. 6b does not include secondary beams and some main timber beams
are reinforced by an external steel reinforcement system (V, Fig. 6). According
to [11–13], this type of pavement may be considered as a load-sharing system.
This consideration may also be applied to the structural timber system of *tabique*
walls.

In order to complement the above information, Fig. 7 shows three frequently
applied support solution building details of the main timber beam of the pavements. At the pavement roof level, the support detail of the main timber beams
may be done similarly as the building detail shown in Fig. 7a. On the other hand,
at the pavements of intermediate floor levels, a frequently applied solution consists in considering hollows in the stone masonry wall, Fig. 7b. In both cases, the

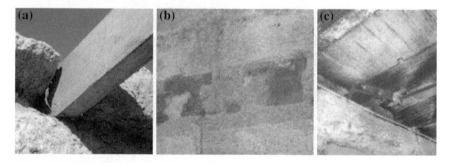

Fig. 7 Current applied support solution building details of the main timber beams. **a** Detail 1.
b Detail 2. **c** Detail 3

beam tends to be supported in a length corresponding to half of the thickness of the stone masonry wall (at that level). Plus, the hollow tends to be sized slightly higher than the respective size of the cross-section of the beam and in order to simplify the building process. When the gap between the lateral faces of the beam and the lateral faces of the hollow is filled, then lateral bending of the beam is reduced at the supports and a structural strength gain is obtained. Furthermore, an alternative solution is shown in Fig. 7c in which the main beams are directly supported on a stone continuously cantilevered. This cantilever may be obtained by contemplating a high width stone horizontal layer at the level of the pavement or considering double stickiness underneath walls.

Additionally, Fig. 8a is related to another building detail of the adopted support solution of a timber pavement on a stone masonry wall. The main beams (I, Fig. 8a) are supported similarly as detail 1 of Fig. 7a. On those beams and on the stone masonry wall, a peripheral timber beam is placed (II, Fig. 8a). On the peripheral timber beam, an exterior *tabique* wall was placed (IV, Fig. 8a). These timber elements are connected to each other by nails. In that figure, the flooring boards are also highlighted (III, Fig. 8a). Meanwhile, Fig. 8b shows the main structural system of other timber pavement of a *tabique* building. The main beam (I, Fig. 8b), the secondary beams which support the flooring boards (V, Fig. 8b), the ceiling boards (VI, Fig. 8b), the secondary beams which support the ceiling boards (VII, Fig. 8b) and the top of a partition *tabique* wall (VIII, Fig. 8b) are identified. In this particular case, the direction of the ceiling boards (A, Fig. 8b) and the direction of the flooring board (B, Fig. 8b) are perpendicular to each other. This fact explains that the respective secondary beams are also perpendicular to each other. Another interesting technical aspect is related to the fact that the secondary beams which support the ceiling boards are much sparser than the ones that support the flooring board. This aspect has to do with the loading

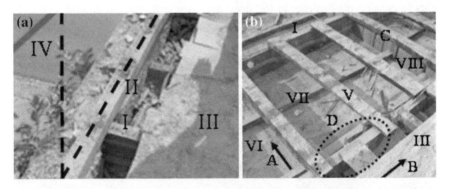

Fig. 8 Building details of the pavements. **a** Detail 1. **b** Detail 2. *I* main beam, *II* peripheral beam, *III* flooring boards, *IV* exterior *tabique* wall, *V* secondary beam (at the flooring level), *VI* ceiling boards, *VII* secondary beams (at the ceiling level), *VIII* partition *tabique* wall, *A* ceiling board direction, *B* flooring board direction, *C* top *tabique* wall finishing, *D* secondary beam arrangement

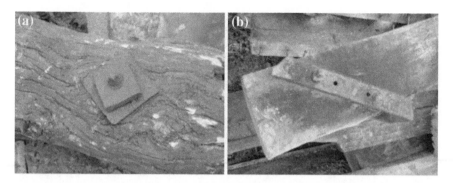

Fig. 9 Other typical metal connectors found in the survey. **a** Example 1. **b** Example 2

combinations that act on these boards. In the ceiling board case, the self-weight is basically the only load acting on it. In contrast, in the flooring board case, dead and live loads have to be considered. These boards may be intercalated placed on the main beam and according to detail D of Fig. 8b. Furthermore, the building detail C of Fig. 8b concerning the connection between a partition *tabique* wall and a pavement complements the information given in Sect. 4.1. In this case, a flooring board is strategically placed on the top of the partition wall. The partition wall connects the main beams at its ends. The ceiling and the flooring boards and the respective secondary beams are also connected to this wall at its top and assuring the referred transversal support on both sides.

As it has been noticed in the *tabique* context, the timber elements are essentially connected by metal mechanical connectors. Nails have been referred throughout this document. Figure 9 includes additional typical metal devices such as bolts and anchor plates.

4.3 Roofs

The structural roof solution of *tabique* buildings is a commonly traditional timber structural system solution, Fig. 10, made of trusses, purling, boards and rafters. Generally, the trusses (Figs. 4, 10b) are similar in terms of geometry and size, and they are placed regularly (i.e. having the same distance in between them). They also tend to be a symmetric system themselves. They are directly supported by the stone masonry or by the exterior *tabique* walls, at their ends. On the other hand, the purling may be simple or continuously supported on the trusses, and also following the above technical pattern. They support the boards which support the rafters. These last ones support the ceramic tiles. All these timber elements are nailed to each other. In certain cases, a partition *tabique* wall incorporates a truss, similarly to Fig. 2b and also to Fig. 10a (in which, remains of the horizontal timber elements attached to truss are still noticeable).

Fig. 10 Typical roof structural solution of *tabique* buildings. **a** Truss. **b** General view. *I* purling, *II* board, *III* rafter

5 Structural and Non-structural Defects and Their Causes

5.1 Decay and Non-structural Defects

In [4], degradation bipolar focus is suggested concerning vulnerability of *tabique* buildings. The roof and the ground floor are identified as the main susceptible deterioration focuses. The rain water infiltration and the direct contact between timber element and the ground are also identified in [4] as the main degradation causes. The existence of a broken ceramic tile (I, Fig. 11a) or a missing ceramic tile (II, Fig. 11a), are sufficient non-structural roof building defects able to trigger a progressive degradation collapse of the building if a maintenance process does not take place, Fig. 11b. In [14], an attempt at assessing the structural vulnerability of traditional Portuguese roofs is done, in which different vulnerable structural failure

Fig. 11 Roof depending pathological effect. **a** Portuguese *tabique* dwelling. **b** Progressive failure of the roof. *I* broken tile, *II* missing tile, *III* deterioration of the finishing, *IV* cracking of the filling, *V* lost of filling, *VI* wood degradation

scenarios were identified. The continuous presence of water in a building will damage the materials substantially. Retraction and biological attack phenomena will likely occur. Based on several study cases of really deteriorated *tabique* buildings, it was noticed that these phenomena tend to progress from the top (where there is a high pathological effect incidence) to the bottom where that effect is relatively smooth), Fig. 11a. On the other hand, the outstanding durability characteristic of timber is underlined in [15]. In fact, apart from the referred pathologic susceptibility, timber may have an impressive durability because the material degradation tends to progress gradually. In contrast, the filling of the *tabique* components tends to show a much higher susceptibility to the water presence. The finishing (such as white wash) and the filling (such as earthy based lime mortar) may be highly vulnerable to this condition and a rapid material degradation process may take place. The fact that the filling is an earthy based material will contribute to this vulnerability. For instance, in Fig. 11a an exterior *tabique* wall of a Portuguese dwelling is shown. Poor rain water roof drainage system and roof revetment deterioration have allowed the occurrence of a progressive pathological effect on the façade. Finishing degradation (III, Fig. 11a), cracking occurrence of the filling (IV, Fig. 11a), falling of the filling (V, Fig. 11a) and timber degradation (VI, Fig. 11a), is a possible deteriorating event sequence resulting from these building defects. Similar degradation sequences may occur in the inner *tabique* components.

Meanwhile, direct contact between *tabique* components and the ground is another undesirable building condition which may contribute to a premature material decay, in particular, wood biodegradation [4]. Insect attack such as termite attack will be much more likely to happen. Figure 12a presents a *tabique* Portuguese dwelling that exemplifies this fact. In this case, the combined effect of direct contact with ground and water impact is also featured. The degradation scales of Fig. 12 indicate that the intensity of the material degradation increases from the bottom to the top of the exterior *tabique* wall. In particular, it is more evident in the vertical elements of the timber structural system. This pathological

Fig. 12 Ground depending pathological effect. **a** Deteriorated façade. **b** Biological attack effect

tendency is clearly exemplified through the building detail of Fig. 12b. This evidence may justify the fact that traditionally the *tabique* components are essentially built on the upper floors and in the Trás-os-Montes e Alto Douro region context. By adopting this solution the overall durability of the building may be increased and unexpected and undesirable premature degradation may be avoided. This susceptibility may be aggravated because the vertical timber elements are not protected by the earthy filling material in their bottom face (which is the one that connects directly with the ground). It is worth adding that this biological attack may be transmitted in between timber elements and, therefore, maintenance and conservation processes are necessary in order to prevent this situation.

5.2 Structural Damages: Field Evidences After Earthquakes

Hımış, *dizeme* and *bağdadi* constructions that have experienced structural damages in recent earthquakes in Turkey can be categorized in two different periods. The first period is the non-instrumental period where the total damages are mentioned in historical documents without referring to neither the type of the structures that were damaged nor the characteristics of the structural failure. The historical documents of this sort are not helpful in getting clear insights on the vulnerability of such structures. There is a transitive period of instrumentation and documentation, which is until 1960s, where the instrumental data may be still missing but damage descriptions are better than other older historical documents. The 1943 Tosya-Ladik Earthquake (Ms = 7.2) for instance, which occurred in a period where the instrumentation was not dense, and the earthquake awareness and the organizational capabilities of the state were not as high as today. As a result, it is mentioned in the technical documents that the earthquake caused several damages in the province of Kastamonu, where *hımış* and *bağdadi* constructions constitute the largest portion of the building inventory, whilst the timber-framed residential houses did not suffer serious damages. Examples of the well-documented era can be the 1999 Kocaeli and Duzce Earthquake (Mw = 7.4 and 7.2, respectively) and 2010 Palu (Elazığ) Earthquake Mw = 6.1. According to the literature, in one district in the hills above Gölcük, where 60 of the 814 reinforced concrete structures were heavily damaged or collapsed, only 4 out of the 789 traditional buildings with two or three storey collapsed or were heavily damaged. In the 1999 Marmara earthquake, the reinforced concrete buildings accounted for 287 deaths against only 3 in the traditional so called *hımış* structures. In the heart of the damage district in Adapazari, the research showed that 257 of the 930 reinforced concrete structures were heavily damaged or collapsed and 558 moderately damaged. By comparison, none of the 400 traditional structures collapsed or were heavily damaged and only 95 of the total were moderately damaged [16]. In some buildings constructed with *dizeme* technique, laths and *dizeme* are nailed on the main posts (vertical structural elements) with small distances before plaster work. Typical *dizeme* construction is comprised of hundreds of timber elements and nailed connections. This means that in case of failure of one load path it can often be compensated with adjacent elements and joints. The purpose of using timber infill in *dizeme* construction is most probably to avoid the common early shear failure and falling out of frame as occurred in usual masonry infill. Thus, timber infill called *dizeme* provides continuous additional support to the building during the course of the earthquake shaking by "working" or energy dissipation through many cycles without loss of their integrity. Another historical record is the 1944 Gerede Earthquake, where many timber-framed buildings with *dizeme* construction behaved well during the 7.8 magnitude tremor. It is known that the damage of *hımış* and *bağdadi* structures remained limited during the past strong earthquakes in Turkey. Therefore, earthquake structural damage of traditional buildings may be essentially a result of an inappropriate construction or intervention, or the lack of maintenance. A traditional building technique may be earthquake robust.

6 Possible Rehabilitation Solutions

6.1 Non-structural Rehabilitation

Based on the technical facts presented throughout this chapter, in particular, in the previous section, additional building care concerning a reliable water-proof roof should be guaranteed in a non-structural rehabilitation process of a *tabique* building. Replacement of damaged ceramic tiles, applying new ones in non-provided areas of the roof and cleaning up of the rain water roof drainage system are tasks that should be done annually and, preferably, during the summer time, both in the Portuguese and the Turkish contexts. For the Portuguese context, July, August and September are the most reliable months to perform this task because they correspond to the dry season. However, this routine may not be straightforward. Lack of technical knowledge, cost, complexity and negligence may be some reasons that justify the fact that this maintenance routine does not take place quite as often as recommended. If the building is located in a city centre and there is no roof accessibility then the task of the revetment roof repairing may end up being even more complex and, consequently, more costly. Usually, in the Portuguese context, these buildings have more than two floors, they used to be connected to each other (i.e. they share lateral stone masonry walls) and they face the public sidewalk directly. These building constraints generally require scaffolding usage and Council House technical approval. Therefore, in these cases, embracing roof rehabilitation should contemplate a double roof revetment and adequate roof accessibility in order to avoid further future reparation. The double roof revetment solution consists in having two distinct layers of revetment materials. Ceramic tiles and corrugated plate is a building possibility. A thermal and an acoustic insulation layer may also be considered. A sandwich panel may be adequate to fulfil this technical requirement. Figure 13a exemplifies a corrugated plate application. The underneath layer (e.g.

Fig. 13 Some rehabilitation technical aspects. **a** Corrugated plate application. **b** Collapsed rain obstacle (*I*). **c** Wood decaying

corrugated plate) will work as a water-proof prevention mechanism solution. In case of ceramic tile damage occurrence during the rainy season, this solution won't allow the water infiltration. Simultaneously, it will allow reaching the dry season to perform the necessary repairing procedures without putting in risk additional pathology appearance. The corrugated plate will be supported by the existing rafters. At the same time, the plate will support the ceramic tiles. The self-weight (i.e. the dead-load) of the roof will increase slightly by adopting this building preventing solution and, therefore, it has to be considered in the structural design of the rehabilitation process. Previous research studies [14, 15] indicate that traditional timber structural elements may be oversized according to the currently applied codes, in particular [11].

The cost/benefit of adopting this prevention solution has proved to be quite convenient. In the water-proof roof context, another relevant technical aspect is concerned with the junction of two different buildings. These buildings may be different in terms of typology and/or in terms of height. A thermal dilation junction will be necessary in that junction and, therefore, an adequate water-proof solution is required there in order to avoid future rain water infiltration in the lateral walls. Generally, in old buildings, that joint is non-provided of any water-proof system. When a water-proof system exists, it tends to be technically debilitated. Thus, this building detail has to be taken into account carefully in a roof rehabilitation process of a traditional Portuguese building. Additionally, all the ornamental stones which are salient from the masonry main façade plane should also be provided with a water-proof system applied on their top. When a stone masonry wall is not protected by the roof at its top, the previous building detail should also be applied on the top stone layer of that masonry wall. In order to complement this information related to technical aspects concerning roof water-proof guarantee, the rain water roof drainage systems have to be checked and, perhaps, redesigned according to the currently applied codes. Usually, zinc or steel plates are used to materialise this rain water roof drainage system. Based on our experience, the plates or the pipes which are part of this type of drainage systems usually present an advanced stage of degradation and/or insufficient dimensions (e.g. diameter). This technical evidence indicates that these systems may require reinforcement. It is also worth referring that those pipes are likely to be incorporated in the façade or crossing it transversally. Metal oxidation will result in material degradation of the pipes which may result in water infiltration in the masonry wall. Water infiltration in the masonry wall will allow water to reach the timber pavements and the *tabique* components. If no repairing action takes place, then a progressive general building degradation process will occur. Based on our experience, these repairing actions commonly end up being unsuccessful because of the inexperience concerning this type of traditional buildings still shown by the technicians. At the same time, the inexistence of the original design project or any data related to other building intervention, add complexity to this repairing process. As it was stated above, when these buildings are located in the city centre, they tend to face the public sidewalk directly. In this case, if the pipes of the rain water roof drainage system are applied on the outer face of the façade, then they are vulnerable to mechanical damage caused by accidental loads (e.g. car impact) or vandalism acts.

A mechanical reinforcement solution of these pipes (e.g. inserting them in an additional pipe, in an approximate 3.0 m length from the ground) is recommended. On the other hand, protecting the exterior *tabique* walls from the rain impact is an additional technical building recommendation to consider in future rehabilitation processes. Keeping the original rain obstacles is highly recommended. For instance, Fig. 13b exemplifies a real case of this building scenario in which a similar obstacle has collapsed due to the lack of regular maintenance. This loss has resulted in additional roof deteriorating events (I and II, Fig. 13b) and also in water infiltration in the façade. Redesigning these types of obstacles may be necessary and convenient. The façade of the Portuguese dwelling presented in Fig. 11a apart from being non-provided with any rain water roof drainage system has also a short rain obstacle which results in an insufficient rain impact protection of the exterior *tabique* wall. In the Trás-os-Montes e Alto Douro region, exterior *tabique* walls oriented south are the most vulnerable ones. A water-proof revetment should be used in order to avoid future possible water infiltration or material and mechanical pathological effects on the walls.

The traditional exterior wall revetment solutions, in particular zinc corrugated plate and slate board, have proved to be technically efficient. Currently, applying zinc corrugated plates may be easier and more affordable. In a rehabilitation process context, if there is an already applied exterior wall revetment solution then a checking task concerning the conservation level of the material evaluation, the adequacy of the adopted fixing system (of the plates to the wall) evaluation, the adequacy of the overlapping distance of the plates evaluation and the quality of the adopted borders finishing assessment, is fundamental. Recommendation of a punctual repairing or of an extended repairing of the applied revetment may be two likely outputs. In the last case, removal of the old revetment followed by the application of a new one seems a wise decision. As it was stated earlier in this chapter, in the region of Trás-os-Montes e Alto Douro, *tabique* wall located at the ground floor level is a rare building scenario. However, in big cities such as Oporto, the existence of a basement in a *tabique* building is quite frequent. In these cases, the main timber beams of the ground floor pavement are directly supported by the stone masonry wall existing at the basement. These walls also work as a soil supporting system. Taking into account that: (i) these walls are generally non-provided with any water-proof system or any water drainage system; (ii) they are in direct contact with the soil; (iii) the basement is generally a confined dark room non-provided with any efficient natural ventilation system, then the biological wood degradation effect tends to be extremely risky. In the past, these basements were mainly used as storage room where coal, firewood, agricultural products, unused domestic products (e.g. old furniture, papers, clothing), among other types of domestic things, where kept for a long period of time. This situation certainly contributed negatively to an already poorly natural ventilation system of the basements. Adequate monitoring and maintenance tasks were also even more difficult to perform. Furthermore, the potential of storing contaminated material (e.g. firewood already under insect attack) was extremely high. Thus, the timber elements of the structural system of the pavement of the ground floor were extremely susceptible to decay phenomena. This fact explains why the pavement of the ground floor is considered as a main susceptible deterioration focus in a *tabique* building. In maintenance or rehabilitation processes, cleaning up the basement is recommended as a first step. This step will allow performing an accurate evaluation of the degradation level of the structural timber elements and, consequently, to make effective rehabilitation decisions. Improving the quality of the natural ventilation system by adding up openings (disposed preferentially perpendicularly to each other), removal of the damaged timber elements and replacing them by new ones or reinforcing them, building an adequate water drainage and water-proof of the walls supporting the soil, applying a regular preventive insect treatment, are some practical recommendations that should be taken into account in future rehabilitation processes. At this level of the building, another relevant pathological effect that is likely to occur is concerned with stone degradation. Stone material degradation may be aggravated under the above described conditions. This likelihood is even higher considering the fact that the best quality stone pieces were selected to be applied in the upper walls in which ornamental stone work would be required.

Stone replacement may be required depending on the material degradation level. Figure 13c shows an example of this pathology and also the decay suffered by a secondary timber beam of a pavement located at the ground floor. In order to finalise the contribution of this section regarding non-structural rehabilitation procedures, it is worth underlining that the occurrence of a well-defined crack in the junction of *tabique* wall/stone masonry wall or *tabique* wall/*tabique* wall is very likely. As it was explained in Sect. 4.1, regarding the building shown in Fig. 4, a partition *tabique* wall may be basically leaned on the orthogonal exterior stone masonry walls. So far, the existence of a specific connection building detail related to the junction of these two types of walls has not been verified. This building characteristic is also extended to the connection between *tabique* walls. Therefore, the well-defined crack which appears in those junctions seems to be almost unavoidable. Perhaps it should not be seen as a main structural defect. Another extremely important non-structural rehabilitation procedure is concerned with

Fig. 14 Two examples of regularly maintained *tabique* buildings. **a** Example 1 (*exterior view*).
b Example 2 (*interior view*)

the material compatibility. Adequate compatibility between new and old materials has to be guaranteed in order to avoid additional deterioration and to assure
durability of the intervention. The technical information delivered in Sect. 3 concerning the identification and characterization of the wood, the nails and the filling
may help in this respect in future rehabilitation processes. Above all, we have to
keep in mind that each *tabique* building has its specificities and, therefore, may
also require a specific non-structural rehabilitation process. It is also important to
highlight that the referred pathological technical aspects may not be specific to
the *tabique* building technique. In fact, they may also be related to the other traditional Portuguese building techniques, including even the currently applied ones.
However, a profound knowledge of a building is necessary before taking rehabilitation actions. Identifying pathology causes, understanding the pathologies,
evaluating the level of degradation, finding measures of mitigation and adequate
repairing solutions, are not straightforward tasks. Figure 14 presents two examples
of *tabique* buildings which have been maintained regularly.

6.2 Structural Rehabilitation

Structural rehabilitation of this type of structures is another important issue to
be discussed, particularly when they are located in seismically active regions.
Based on the experience concerning the past earthquakes, this type of traditional
structure has an acceptable behaviour, contrasting with some reinforced concrete
structures which had been designed poorly. For instance, Turkey has experienced
many devastating earthquakes throughout its history. Although the available
sources concerning the effects and damages in traditional timber-framed houses
are quite limited, according to the historical evidences on the existing Ottoman
and Turkish traditional houses, their seismic behaviour is quite comparable to

some engineering modern buildings. During the Marmara and Duzce Earthquakes of 1999, many of *bağdadi* and *hımış* houses from the early twentieth century remained intact or suffered minor damage, while so many reinforced concrete houses collapsed or were heavily damaged. The main reason for this good performance was due to closely spaced studs, preventing the propagation of cracks, use of weak mortar and lime or mud allowing its sliding along the bed joints during the shaking to dissipate energy. Although the mortar and masonry by themselves are brittle materials, the system rather than materials behaves ductile. Thus, while these structures do not have much lateral strength and stiffness, they do have lateral energy dissipation capacity [16, 17]. After witnessing their good seismic performance, recently some research work has focused on their structural and material behaviour for their conservation. Very scarce work can be found to illustrate examples of applied rehabilitation work on these types of structures rather than repair and restoration procedures. For mitigation purposes, care must be taken to avoid construction of timber-framed buildings on soft soils, prevention of sliding of the building frames over the foundations due to the weak connections. The use of proper sized nails, instead of metal clamps, screws or mortises, and tendon joints may avoid connections excessively stiff or smooth. The earthquakes experienced proved that lath and plaster technique (*bağdadi* frames) behaves better than brick or masonry infilled frames for being lighter and more ductile. Thus, as a rehabilitation proposal the infillings may be altered with lighter materials and connections and stud joints, tie-beams and studs should be renewed by checking the deteriorated or damaged nails, and timber posts. Recent experiments [18] and experience from earthquakes show that one of the most important points of this kind of construction is the timber-to-timber connections. Lateral cyclic displacements of the frame cause deterioration of the connection and thus results in ductile and slightly pinched cyclic loops. The timber-to-timber connections dissipate energy through the response of the nails or other connection elements. The response of the nails improves the energy dissipation not only via the plastic deformations of the nails but also the friction between the nail surface and the timber medium. Another characteristic mechanism of damage is the out-of-plane (OOP) failure of the infill material. The OOP failure transforms the infilled frame into a bare frame, something that alters the stiffness and strength properties significantly. Besides, as seen in past earthquakes, failure of the infills in random locations throughout a frame causes force localizations, thus the damage on timber around the OOP failure can be higher than that of the rest of the frame. In order to increase the seismic resistance of these structures, enlargement of openings and windows, removing of posts and braces should be strictly prevented to avoid creating soft story effect. Use of modern techniques is necessary to increase its fire resistance, to improve its energy dissipation property and to control vulnerability effect of the biological deterioration together with the traditional technique. Covering or painting timber roofs and surfaces with non-flammable materials, periodical checking of the connections and biological deteriorations, educating the local people about the details of the construction techniques of the buildings to resist earthquakes are highly recommended.

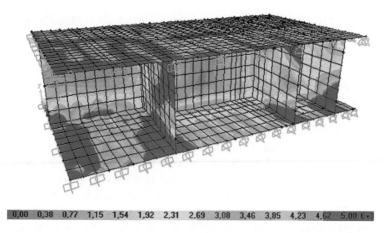

| 0,00 | 0,38 | 0,77 | 1,15 | 1,54 | 1,92 | 2,31 | 2,69 | 3,08 | 3,46 | 3,85 | 4,23 | 4,62 | 5,00 [+3 |

Fig. 15 Von-Mises stress distribution in a timber structural system of a traditional Portuguese dwelling (kPa) [19]

On the other hand, a recent research work [19] focused on evaluating the structural contribution of the partition *tabique* walls in the overall strength capacity of the timber structural system of a traditional Portuguese dwelling was done, Fig. 15. It was concluded that this contribution may increase the strength capacity significantly.

7 Final Comments

Tabique construction is a heritage that deserves to be preserved and passed on to the next generations. A *tabique* component is built efficiently because it corresponds to a simple timber structural system involved in a simple filling material. The most common existing *tabique* buildings still function which proves their inherent durability capacity. Furthermore, a *tabique* construction may be seen as a sustainable and affordable building model. Local raw materials, simple building techniques and efficient material usage are the main technical attributes that give strength to this consideration. Furthermore, this traditional building technique has also shown to be versatile because alternative timber structural systems have been found, different types of filling solutions have been identified, usage of different autochthone wood species has been verified and application of an expressive range of different revetment materials, among other verified technical building aspects. The fact that the exterior and partition *tabique* walls are possible vertical structural components also contributes to this important versatile attribute. A brief description of the overall structural system of *tabique* buildings was delivered which was focused on walls, pavements and roof. This description was complemented with inherent building details. It is worth referring that the respective stone masonry walls and timber pavement and roof structural systems are commonly applied traditional structural elements. The exterior and the partition *tabique* walls and the way that they are connected to the other major structural elements (e.g. stone masonry walls, pavements and roof) are the important specificity of the building technique under research in this work. An attempt at characterising the main pathologies (non-structural and structural) susceptible to occur in *tabique* buildings was done. Finding their causes, proposing mitigation actions to reduce their likelihood and delivering some possible reinforcement solution were also relevant technical aspects presented. This information may be valuable in future rehabilitation procedures. Roof and ground levels were identified as the main degradation focuses. Continuous presence of water, direct contact with the ground and poor natural ventilation are aspects which highly contribute to *tabique* degradation vulnerability. Therefore, regular maintenance is highly recommended. Future rehabilitation procedures should contemplate mechanisms able to facilitate regular maintenance.

References

1. Gonçalves, C., Pinto, J., Vieira, J., Silva, P., Paiva, A., Ramos, L., Tavares, P., Fernandes, L., Lousada, J., Varum, H.: *Tabique* construction in the municipalities association of the Terra Quente Transmontana. In: 3rd WSEAS International Conference on Cultural Heritage and Tourism, Corfu, Greece, ISBN 978-960-474-205-9, pp. 235–240 (2010)
2. Pinto, J., Varum, H., Cepeda, A., Tavares, P., Lousada, J., Silva, P., Vieira, J.: Study of the traditional *tabique* constructions in the Alto Tâmega region. The sustainable world, pp. 299–307. WIT Press, Southampton (2010). ISBN 978-1-84564-504-5
3. Reha, G.: Tradition of the Turkish house and Safranbolu houses. YEM Yayın, Istanbul (1998)
4. Pinto, J., Cardoso, R., Paiva, A., Cunha, S., Cruz, D., Vieira, J., Louzada, J., Varum, J.: Caracterização de paredes tradicionais de *tabique*. In: Conference "Paredes Divisórias 2011", Porto, Portugal, ISBN 978-972-8692-60-5, pp. 25–35 (2011)
5. Cardoso, R., Damião, C., Paiva, A., Pinto, J., Varum, H.: Caso de estudo de aferição do potencial da aplicação de terra local na reabilitação de construções de tabique. VII Congresso Internacional de Arquitectura de Tierra: Tradición e Innovación, Valladolid, Spain, ISBN 978-84-694-8107-3, pp. 131–140 (2010)
6. Cardoso, R., Paiva, A., Pinto, J., Murta, A., Varum, H., Nunes, L., Ramos, L.: Building details of a *tabique* dwelling in Trás-os-Montes e Alto Douro region. In: 12th International Conference on Durability of Building Materials and Components, Porto, Portugal, ISBN 978-972-752-132-6, pp. 729–736 (2011)
7. Paiva, A., Pereira, S., Sá, A., Cruz, D., Varum, H., Pinto, J.: A contribution to the thermal insulation performance characterization of corn cob particleboards. Energy Build. **2012**(45), 274–279 (2012)
8. Pinto, J., Paiva, A., Varum, H., Costa, A., Cruz, D., Pereira, S., Fernandes, L., Tavares, P., Agarwal, J.: Corn`s cob as a potential ecological thermal insulation material. Energy Build. **43**(8), 1985–1990 (2011)
9. Pinto, J., Cruz, D., Paiva, A., Pereira, S., Tavares, P., Fernandes, L., Varum, H.: Characterization of corn cob as a possible raw building material. Constr. Build. Mater. **34**, 28–33 (2012)
10. Pinto, J., Vieira, J.B., Pereira, H., Jacinto, C., Vilela, P., Paiva, A., Pereira, A., Cunha, V.M.C.F., Varum, H.: Corn cob lightweight concrete for non-structural applications. Constr. Build. Mater. **34**, 346–351 (2012)

11. CEN, EN 1995-1-1:2004: Eurocode 5: Design of timber structures. Part 1-1: General—common rules and rules for buildings. European Committee for Standardization, Brussels (2004)

12. Blass, H.J., Aune, P., Choo, B.S., Gorlacher, R., Griffiths, D.R., Hilson, B.O., Racher, P., Steck, G.: Timber engineering STEP 1: Basis of design, material properties, structural components and joints, 1st edn. Centrum Hout, ISBN 90-5645-001-8 (1995)

13. Blass, H.J., Aune, P., Choo, B.S., Gorlacher, R., Griffiths, D.R., Hilson, B.O., Racher, P., Steck, G.: Timber engineering STEP 2: Basis of design, material properties, structural components and joints, 1st edn. Centrum Hout, ISBN 90-5645-002-6 (1995)

14. Murta, A., Pinto, J., Varum, H.: Structural vulnerability of two traditional Portuguese timber structural systems. Eng. Fail. Anal. **18**(2), 776–782 (2011)

15. Murta, A., Varum, H., Pinto, J., Ramos, L., Cunha, V., Cardoso, R., Nunes, L.: Aging effect on the integrity of traditional Portuguese timber roof structures. In: 12th International Conference on Durability of Building Materials and Components, Porto, Portugal, ISBN 978-972-752-132-6, pp. 1073–1080 (2011)

16. Adem, D., İskender, T.Ö., Ramazan, L., Ramazan, A.: Traditional wooden buildings and their damages during earthquakes in Turkey. Eng. Fail. Anal. **13**, 981–996 (2006)

17. Bal, I.E., Meltem, V.: Earthquake resistance of traditional houses in Turkey: timber-frame infilled structures. In: International Symposium on Timber Structures from Antiquity to the Present, Istanbul, Turkey (2009)

18. Akyüz, U.: Seismic performance evaluation of historical traditional timber frame: *Hımış* residential houses. Scientific Project Final Dissemination Report, The Scientific and Technological Research Council of Turkey, Ankara (2010)

19. Teixeira, M.: Estudo do potencial de reutilização de elementos estruturais de madeira antiga. MSc thesis, UTAD, Vila Real, Portugal (2012)

Pombalino Constructions: Description and Seismic Assessment

Mário Lopes, Helena Meireles, Serena Cattari, Rita Bento
and Sergio Lagomarsino

Abstract This chapter describes the *Pombalino* building structures built in Lisbon downtown and other parts of Portugal during the reconstruction after the 1755 earthquake, as well as their earthquake resistant features. In particular the importance of the *Gaiola Pombalina*, a tridimensional wood truss characteristic of those constructions, in the potential seismic resistance of these buildings is discussed. The effects in their seismic resistance of the architectural and structural changes to which these buildings have been submitted since the original construction, usually with negative consequences,, is also discussed. Some strengthening and advanced analytical modelling strategies for these buildings are also mentioned. Finally, the socio-economic feasibility of strengthening this construction is briefly discussed, as well as the importance of their preservation. To be noticed that the reconstruction of Lisbon is the first time in the history of mankind that a large town was built providing widespread seismic resistance to its buildings aiming at avoiding future tragedies of the same type.

M. Lopes (✉) · H. Meireles · R. Bento
DECivil – Civil Engineering Department Instituto Superior Técnico,
Technical University of Lisbon, Av. Rovisco Pais, Lisbon 1049-001, Portugal
e-mail: mlopes@civil.ist.utl.pt

H. Meireles
e-mail: helena.meireles@ist.utl.pt

R. Bento
e-mail: rbento@civil.ist.utl.pt

S. Cattari · S. Lagomarsino
DICCA – Department of Civil, Environmental and Chemical Engineering,
University of Genoa, Via Monteallegro 1, 16145 Genoa, Italy
e-mail: serena.cattari@unige.it

S. Lagomarsino
e-mail: sergio.lagomarsino@unige.it

A. Costa et al. (eds.), *Structural Rehabilitation of Old Buildings*,
Building Pathology and Rehabilitation 2, DOI: 10.1007/978-3-642-39686-1_7,
© Springer-Verlag Berlin Heidelberg 2014

Keywords Masonry • Seismic resistance • *Pombalino* buildings

1 Introduction

Pombalino is the designation of the structural typology of the buildings built in Lisbon and other parts of Portugal during the reconstruction after the great Lisbon earthquake of 1755. The purpose of this chapter is to give an overview of the main characteristics of these buildings, their potential seismic performance and present some of the studies already performed on these buildings, including some advanced modelling techniques. In Sect. 1 a general description of this type of buildings is done, with emphasis on the earthquake resistant characteristics. In Sect. 2 the main experimental and analytical studies performed to evaluate the seismic resistance of these buildings are described. Section 3 presents some of the main changes that these buildings suffered during their already long existence and that also affect their potential seismic performance. Section 4 refers some of the main strategies and strengthening techniques to improve the seismic resistance of these buildings. Section 5 presents some advanced modelling techniques and examples of their application to *Pombalino* buildings, as well as the main results of a study at the level of an entire quarter. Section 6 presents a brief discussion on the social and economic issues that condition the feasibility of rehabilitation and strengthen of these buildings, necessary to deliver them in safe conditions to future generations but preserving their authenticity. Finally Sect. 7 presents some short notes on the historical and cultural value of these buildings.

1.1 Description of Pombalino Buildings

Lisbon and the south of Portugal were devastated by a magnitude Mw = 8.5–8.75 earthquake [1] with epicentre at southwest of the Algarve in 1755. It is estimated that more than 5 % of the population of the Lisbon region died. The reconstruction that followed was done with the concern of avoiding future tragedies by means of providing seismic resistance to the new buildings.

Of the several features of *Pombalino* buildings related with the purpose of providing seismic resistance, the most relevant and notorious is the *Gaiola Pombalina*. *Gaiola* is the Portuguese word for cage, as the *Gaiola* consists on a tridimensional wood truss that looks like a cage. The *Gaiola* is constituted by a set of plane trusses, called *frontal* walls, connected at the corners by vertical bars that belong to orthogonal *frontal* walls. Each *frontal* wall is constituted by a set of triangles, a geometry similar to the steel trusses of nowadays. Since the triangle is a geometric figure that cannot deform without variation of the length of the sides, in fact it is the only one, each *frontal* wall only needs to mobilize the axial force of its bars to resist to forces in any direction in its own plan. Therefore the

Fig. 1 *Gaiola* after removal of cover and masonry

connection between orthogonal *frontal* walls by means of common vertical wood bars yields a tridimensional truss capable of resisting forces applied in any direction. In general the space between the wood bars of the *frontal* walls is filled with weak masonry, and the surfaces are covered with a finishing material, therefore the *Gaiola* in general is not visible. Figure 1 shows photos of the *Gaiola* after removal of the masonry in a building recently demolished in downtown Lisbon, and Fig. 2 a *Gaiola* wall with the masonry filling.

Usually the *Gaiola* only develops above the top of the ground floor level in the interior walls. The façades and gable walls (between adjacent buildings) are usually built with ordinary rubble stone masonry, with some exceptions of better quality masonry, mainly at corners and some ground floor columns and walls. Their thickness may vary along the height, being between 0.60 and 0.90 m in most cases. The spandrel beams, which connect the masonry columns of the façades are of the same thickness of the columns below floor levels. When there are windows and not doors the spandrel beams extend above floor level but with a much smaller thickness, in order to allow people's access to the windows. Very often the two parts of these spandrel beams were not built simultaneously, yielding a weak horizontal surface between them. The result is that those beams have a cross-section as shown in Fig. 3.

Some interior walls, with partition purposes only, called *tabiques*, are made of one or two sets of boards or small wood bars, are thinner than *frontal* walls and have much less resistance to horizontal loads than the *frontal* walls. Figure 4 shows a photo of the interior of one of those walls, after removing the cover.

The floors are made of wood planks (typically 2 cm thick) supported on perpendicular wood joists (typically 10×20 cm^2), which are supported on the exterior walls (more often on the façades) and *frontal* walls. In buildings of better quality the wood joists are continuous from façade to façade, in others they have discontinuities on the intersection with *frontal* walls. According to the original practice the horizontal wood joists of the floors should be properly anchored inside the façades, by means of iron anchors embedded in the masonry of the façades at floor level. The *frontal* walls should also be anchored inside the façades. However, there are doubts about the quality of those connections, as well as of their widespread execution.

Fig. 2 *Gaiola* wall with
masonry filling [2]

The connections between the different wood bars of the *Gaiola* are done by
means of iron nails and cuts on the wood bars in order that they fit in each other,
as shown in Fig. 5.

Fig. 3 Location and cross-
section of spandrel beams

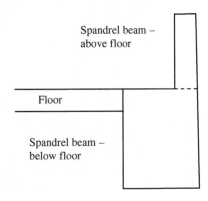

Fig. 4 Partition wall [3]

The pavements of the 1st floor (ceiling of ground floor) are usually constituted by masonry arches and vaults, as shown in Fig. 6, for two reasons: (i) create a barrier to fire, in order that possible fires on the ground floors do not spread to the upper floors, and (ii) to avoid that the soil humidity reaches the *Gaiola* wood structure above the 1st floor. At ground floor level, where the *Gaiola* structure does not exist, the arches and vaults are supported in the interior by masonry piers and walls and on the exterior by façades and gable walls.

Another characteristic of *Pombalino* construction was the standardization of the construction process, aiming at its widespread application to an entire city. For instances first the carpenters would built the wood truss, the *Gaiola*. After the bricklayers would come in and would add the masonry and the finishing to

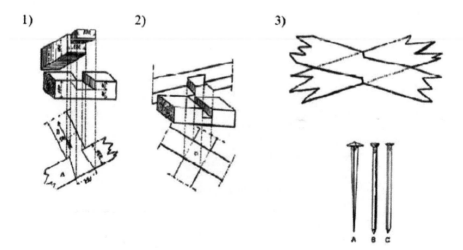

Fig. 5 Connections between wood bars of the *Gaiola* [4]

Fig. 6 Masonry arches and vaults at the ground floor ceiling [2, 5]

the *frontal* walls. The gable walls are common to both adjacent buildings (there were no expansion joints, as the flexibility of the wood pavements was enough to accommodate the effects of variations of temperature) and extend above the buildings to offer a barrier to fire propagation between buildings. Since the construction of the different buildings of each quarter was not simultaneous, there are vertical surfaces of separation of façades and gable walls, built at different periods. The standardization also extended well beyond individual buildings. The quarters of downtown Lisbon have a rectangular shape as they developed between parallel and orthogonal streets, and the buildings were all of the same height, comprising the ground floor, three upper floors and the attic. This way each quarter was constituted by a set of buildings of similar dynamic characteristics, yielding a better overall performance under seismic actions. A situation very similar to this, but with smaller buildings, can be found at the centre of Vila Real de Stº António, a southern town in Algarve also devastated by the earthquake of 1755.

In downtown Lisbon the water table is almost near the surface, as this zone is adjacent to the Tagus river. The soil is an alluvium of variable thickness, of approximately

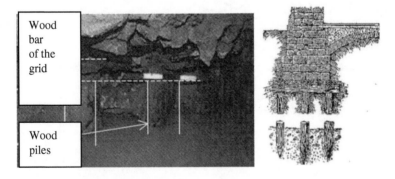

Fig. 7 Foundation scheme and piles [5]

20–30 m near the river and progressively reducing as the distance to the river increases, with a very weak load bearing capacity. Due to this the foundation system of *Pombalino* buildings is based on a tridimensional grid of wood bars on top of short length (around 1.5 m [6]) and small diameter (around 15 cm) wood piles, embedded on a large embankment made with the debris of the buildings destroyed by the 1755 earthquake and compacted by the piles. The embankment receives the loads from the structure through the wood grid and piles and distributes them by a larger area of the underlying alluvium, reducing the stresses at this level.

Figure 7 shows the constructive scheme and photos of the top of the piles at one building in downtown Lisbon, the BCP Museum. Figure 8 shows a schematic representation of a *Pombalino* building

2 Seismic Resistance

The seismic resistance of existing constructions received less attention than new constructions at the early days of modern seismic engineering, at the first half and middle of the twentieth century. In Portugal more attention was given to old buildings mainly from the decade of 1990 onwards. One of the first studies to assess the

Fig. 8 Example of a *Pombalino* building [7]

Fig. 9 In-situ tests to rupture [8]

seismic performance of an old building in Lisbon focused on a *Gaioleiro* build-ing [8], the type of buildings that were built after the *Pombalino* buildings during the late nineteenth and early twentieth centuries. This building type resulted of the progressive adulteration of their main characteristics of the *Pombalino* buildings, such as the absence of the diagonal wood bars of the *Gaiola*, weaker connections between elements and adding more floors. The studied arises from the opportunity to do tests to rupture, on site, on a building being demolished. In these tests part of the façade was used as reaction wall, in order to apply monotonic increasing hori-zontal forces in the plan of the façade strong enough to take the rest of the façade at that level to rupture. Figure 9 shows part of the façade (including instrumenta-tion details) that was divided into two unequal parts: the smallest one, that was tested, and the largest one, that is stronger and was used as reaction wall. These tests allowed to find out the stiffness for small and large displacements, as well as the respective failure loads, for the element tested to rupture.

Several tests of the same type were also performed in interior walls. This allowed calibrating a tridimensional model of the structure, that, together with the knowledge of the elements failure loads, allowed to evaluate the seismic capacity of the building. The result indicated that the building would collapse for a seismic action of approximately 43 % of the respective code (RSA, [9]) prescribed seismic action, showing the tremendous weaknesses of this type of buildings. However, this conclusion cannot be extrapolated to *Pombalino* buildings.

In the early years of the 2000 decade Rafaela Cardoso [5] analysed with detail the model of a *Pombalino* building with ground floor, 4 upper floors and attic, with the numbers 210–220 of *Rua da Prata*, whose front façade is shown in Fig. 10 and whose project was available for consultation. It is thought that this building is a late *Pombalino*, probably of the early nineteenth century period, due to the fact that it has one more floor than usual. At this time the memory of the 1755 earth-quake was starting to fade away and the strict construction rules imposed after the earthquake were being relaxed due to pressure of urban developers.

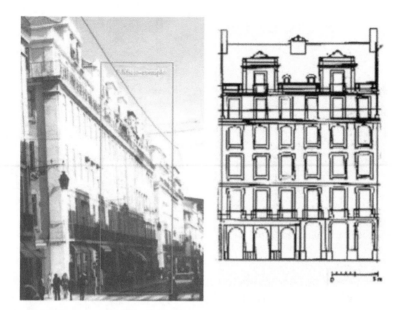

Fig. 10 Front façade of the building of *Rua da Prata* [5]

It was not possible to do an experimental characterization of the materials, therefore the structural model was based on the assumption that the materials would have average properties, estimated and calibrated from in situ and laboratory tests on specimens removed from other buildings or built in laboratory. Particular attention was devoted to the stiffness properties of the *frontal* walls, due to their relevance for the potential seismic performance of *Pombalino* buildings. Several models for the simulation of *frontal* walls were analysed [5]. In all cases each individual wood bar was represented by a linear bar hinged at both extremities, with the masonry between the wood bars represented by finite elements, as shown in Fig. 11.

The comparison between the models and the calibration of their stiffness properties was based on experimental results. Several experiments on *frontal* walls were performed at Laboratório Nacional de Engenharia Civil (LNEC) in Lisbon: (i) a *frontal* wall removed from a *Pombalino* building in downtown Lisbon, and carefully transported to LNEC, was tested under constant vertical loads and horizontal cyclic loads applied on top [10], and (ii) a set of pre-fabricated *frontal* panels tested also at LNEC [11]. Figure 12a shows schematically the dimensions of the full scale panel removed from a building and the applied loads, and Fig. 12b shows the panel after the test.

The comparison of analytical results with the results of these tests showed that the analytical stiffness always overestimated the respective experimental value. Three possible reasons for this observation were identified: (1) the connections on the extremity of the diagonal bars of the frontal wall, considered in the analytical

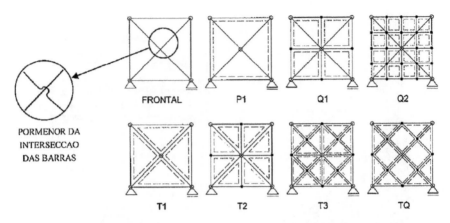

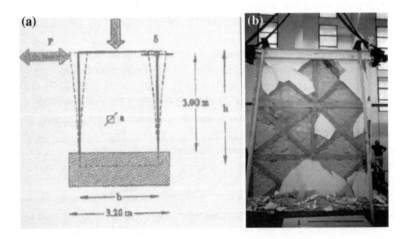

Fig. 11 Models for *frontal* walls [5]

models, could not transfer tensile forces; (2) the masonry filling contributed very little to the stiffness of the panels as it tend to detach from the rest of the panel when deformations increase; this conclusion was recently strengthened during a set of tests of panels similar to *Pombalinofrontal* walls, built and tested in the Laboratory of Structures and Strength of Materials of IST [12]. Even though, the detachment of masonry from the wood structure was clearly noticeable by eye sight at large displacements, as can be clearly seen in Fig. 13, this confirms the little importance of the characteristics of masonry for the stiffness under strong seismic actions; (3) the gaps between wood bars allow initial deformations before mobilizing the compressive strength of the diagonal of the *Gaiola*.

Figure 14a and b shows the existence of the gaps between wood bars in real buildings and Fig. 14c shows their influence on the force–displacement diagrams

Fig. 12 Scheme of test and panel after the test [10]

Fig. 13 Detachment between the masonry filling and the *Gaiola* structure [12]

of the panels [5]. The comparison of the stiffness of the analytical models with the experimental stiffness, disregarding the low stiffness branch before closing of the gaps, led to the conclusion that the masonry and the tensile diagonal of the *frontal wall* should be disregarded (except at small displacement, of little relevance under strong seismic actions). From the range of values for the wood Young's modulus given at EC7 [13] (8000–12000 MPa), the one that yielded the best match with the experimental results was the lowest value that was chosen for the analysis.

The tests on pre-fabricated panels at LNEC [11], referred above, also confirmed the conclusions about the influence of the gaps. These were diagonal compression tests on 1:3 scale models of *frontal* wall panels. Figure 15a and b shows one of the specimens and the loading shoes at top and bottom of the specimen. These tests also confirmed the detachment of the masonry filling from the surrounding wood bars of the *frontal* walls.

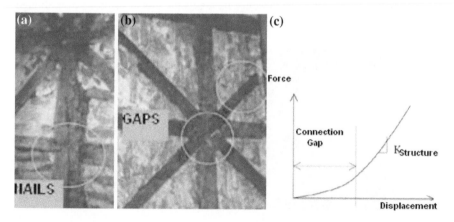

Fig. 14 a, b Connections with gaps and **c** influence on force–displacement relationship [5]

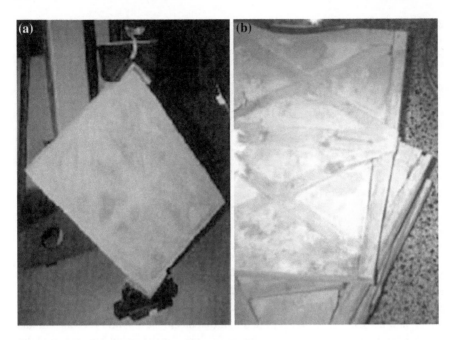

Fig. 15 **a** Panel built and **b** Panel after testing in laboratory [11]

This detachment was also observed in recent experimental studies where the influence of the masonry filling on the seismic behaviour of *frontal* walls was studied by means of the comparison of behaviour of *frontal* walls with and without masonry filling, built in laboratory [14]. The results showed that the masonry contributes to prevent buckling of the compressive diagonals, which starts at the middle section where the cross-section is reduced to half due to the crossing with the other diagonal, as shown in the details of Fig. 5. Figure 16a and b shows the specimens after testing. These tests have shown that the masonry increases the global stiffness, even though

Fig. 16 *Frontal* walls: **a** with masonry and **b** without masonry [14]

for large displacements it detaches from the wood structure. However the good quality of execution and of the materials cast some doubts on the fact that this stiffness increase would also take place on real *Pombalino* buildings or if it would be more reduced.

The global nonlinear behaviour of the building analysed by Rafaela Cardoso [5] was considered by means of an iterative procedure based on a sequence of tridimensional linear analysis by response spectra. This option allowed to use current commercial software available for structural analysis (SAP2000 [15]), as it was intended to use a methodology and software that could be used in current design practice. More details about material properties and element dimension can be found in Sect. 5.2. Figure 17 shows the plan of the structure at the ground floor.

The façades and gable walls were modelled by shell finite elements. The contribution of the *tabique* partition walls was disregarded. The foundations were modelled as built-in supports.

At each iteration the connections that failed at the previous iteration, mainly between the *frontal* walls and the façades, were removed from the model of the next iteration, as it was assumed that after a wood bar was pulled-out from the masonry the tensile strength of the connection could not be recovered. As the building was a late *Pombalino* it was conservatively assumed that the connection of the *frontal* walls with the façades were weak ($f_{tensile} \leq 5kN$). It was assumed that a façade would fail in its own plan if all the columns would fail, which corresponds to admit a limited redistribution capacity. It was concluded that failure would take place by out-of-plane collapse of the façades due to sequential rupture

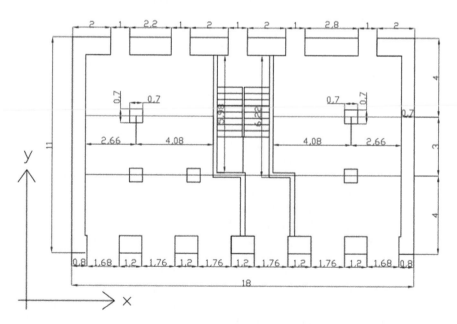

Fig. 17 Sketch of the plan view of the building: ground floor (units in metres) [16]

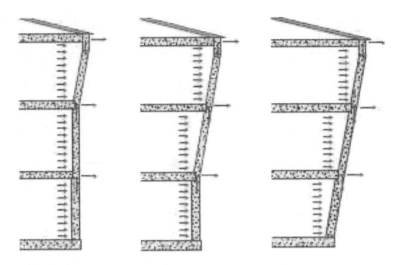

Fig. 18 Out-of-plane failure of façade walls [17]

of the connections, starting at the upper floor and propagating to the lower floors at almost the same seismic intensity, as illustrated in Fig. 18.

Rupture would take place for a seismic action of 40 % of the one prescribed for the Portuguese code RSA [9] for far field events of high magnitude, to which corresponds a value of PGA $= 0.18$ g (it was assumed a relative damping factor of $\xi = 10$ % and a behaviour factor $q = 1.5$) and a response spectra rich in high periods. However if this collapse mode was prevented because the connections were better than assumed, or had been strengthened, collapse would be triggered by failure of the ground floor columns and walls (base shear mode) but at a seismic action slightly above 100 % of the code prescribed seismic action. It is likely that many original *Pombalino* buildings would have a seismic resistance above this level, as they had one floor less and better connections. Even though there may be doubts about the strength of the connections, which should be verified for each building in strengthening projects, this result is remarkable for buildings built more than 200 years ago.

It should also be noted that the methodology is conservative, since the analysis is not done in time domain and in each iteration the spectra is the initial one, as if the earthquake would start again. However, opposed to what happens in most of Europe, where near-field events are relevant, in Portugal for many buildings as most Pombalino buildings, far-field events, such as the 1755 earthquake, condition seismic design as the zone of higher spectral accelerations extends more to the lower frequencies. These events are associated to much higher durations than near-field events, reducing the conservatism associated to the methodology used for the analysis.

The conclusion regarding the seismic resistance can be extrapolated to the *Pombalino* buildings of the present, as long as they have not been altered after the original construction, as it is thought that the *Gaiola* still continues intact in

Fig. 19 Evidence of nowadays excellent conservation state of *frontal* walls

the original buildings. It is known, from works in those buildings, that the conservation state of the *Gaiola* is good in general. For instances Fig. 19 shows parts of *frontal* walls removed from a *Pombalino* building in 2010 in excellent conditions.

The interest on *Pombalino* buildings largely exceeds their city of origin. Relevant studies were also performed the University of Minho, where the behaviour of the quarter *Martinho da Arcada*, in the corner of *Rua da Prata* and *Rua do Comércio* in downtown Lisbon, was studied [18]. Even though the buildings of this quarter had been the subject of several interventions with little concern for seismic safety, with removal of part of the interior structure and addition of steel and reinforced concrete elements, the conclusion was that the quarter would resist to a seismic action of 70 % of the one prescribed in the Portuguese code of actions RSA [9].

One of the technical issues related to *Pombalino*a building that has deserved attention from the public opinion is the possible deterioration of their foundations. During the last decade several large holes were found in the subsoil of downtown Lisbon, that fortunately caused almost little or no damage to the buildings so far. It is assumed that these holes were created by changes in the underground water flow due to the numerous underground works (basements, car parks, tube lines) that were done during the last decades. Relevant variations of the level of the water table in downtown Lisbon have also been observed in the past decade, for instances at the BCP Museum, where the photos of Fig. 7 were taken.

As it is well known the wood under strong variations of humidity tends to get rotten, which has been observed in several piles. However no relevant consequences, for instance damage in buildings due to differential settlements, has been observed so far. Together with the fact that the length of the piles is short (about 1.5 m) and do not reach the competent soil that is deeper, this shows that *Pombalino* buildings are not fully supported on the piles. It is therefore thought that the main purpose of the piles was the compaction of the superficial embankment where the buildings are supported, and that distributes the vertical loads by larger areas, transmitting much lower stresses to the weak soil below. Even though the vertical load bearing capacity seems not to have been affected, at least in the short term, to analyse better the possible consequences of deterioration of the piles, the sensitivity of *Pombalino* buildings to differential settlements that could be induced by filling the void on the rotten part of the piles, was studied. Therefore the effect of differential settlements with different profiles at the base of the building was analysed by Rafaela Cardoso [19]. The results indicate that 20 cm differential settlements from the centre to the periphery of the building would be necessary to induce collapse of connections or significant cracking, this is, only considerable settlements would have significant and noticeable effects. However, this does not mean that the deterioration of the piles is not relevant, as the voids created, even though without strong visible consequences in the short term, may induced much larger differential settlements during an earthquake due to changes in soil conditions, weakening the buildings and increasing the potential for much more damage.

3 Structural Changes

In the previous section the potential seismic resistance of original *Pombalino* buildings (without relevant structural changes after the initial construction) was discussed. However, the real seismic resistance of *Pombalino* buildings nowadays depends not only on their original characteristics but is also strongly affected by the alterations to which they were subjected during their lifetime. These were usually associated to the introduction of new facilities (for instances water, sewage or gas pipes), increasing of areas by adding more floors or changes of use and removal of columns and walls in particular at the ground floor to open large spaces for shop windows and in the interior to create larger spaces. Most of those changes were done without any concern for the seismic strength of the buildings, facilitated by a legislative gap and inexistence of technical standards applicable to works on old buildings. Several examples are shown next.

Figure 20a shows on the left hand side photo a case of water pipes introduced inside a *frontal* wall, probably during the twentieth century, cutting the wood bars of the *Gaiola* and strongly weakening the resistance of the frontal wall, specially under horizontal loads, and b) on the right hand side a photo of a case in which the pipes cross the wall in the perpendicular direction causing much less impact.

Fig. 20 a Cut in frontal wall to introduce pipes and **b** pipes crossing wall in the perpendicular direction [20]

Figure 21 shows a street in downtown Lisbon in which it is clear that there are buildings of different heights, being known that the original buildings were all of the same height, according to the reconstruction plans. The differences are due essentially to more floors added after the original construction, which strongly increase seismic effects.

Figure 22 shows one of the many buildings in which apparently façade ground floor columns were cut to create a wider shop window, weakening the building where seismic effects are stronger. Usually this is done introducing a beam on top of the columns that are cut to transfer the vertical loads to adjacent columns. However, there are cases in which entire *Gaiola* panels (above ground floor) are removed without addition of strengthening beams and without collapse, showing the excellent performance of the *Gaiola* (in the upper floors) that allows the redistribution of vertical loads from the zone that is removed to the adjacent ones.

This type of interventions took place in most of the buildings in downtown Lisbon, strongly reducing their seismic strengthening, but as the resistance to vertical loads is much less affected, the consequences will only become visible when the next strong earthquake hits Lisbon.

It can be concluded from the above that original *Pombalino* buildings possessed good characteristics of seismic resistance, considering the materials and scientific knowledge available at the time of the reconstruction. However, those

Fig. 21 Street in downtown
Lisbon with buildings of
different heights [21]

characteristics were progressively adulterated during their lifetime, leading to
buildings that nowadays have high seismic vulnerability.

4 Strengthening

In many cases it may not be economically worth or feasible, and without excessive
adulteration of the main characteristics, to strengthen old buildings to the same safety
levels prescribed for new constructions. Therefore the objective of strengthening old
buildings may be the improvement of their potential seismic performance up to minimum standards, subjected to economic restrictions and on the level of adulteration of
the original building. In this framework it may be necessary to be more selective in
the interventions on these buildings, by means of identifying the potential collapse
mechanisms and acting only upon the weakest ones. This philosophy can be illustrated
graphically as shown in Fig. 23. If a parallel between the links of the chain and the collapse mechanisms of a building is made, it is clear that strengthening the weakest link

Fig. 22 Cut of columns at ground floor level [21]

up to the strength level of another link is enough to improve the potential performance of the system (building or chain).

This issue can be exemplified with the study of the building of *Rua da Prata* [5], previously mentioned, where different strengthening strategies are discussed. In order to quantify the increase in seismic performance associated to a given strengthening solution, the parameter γ_{sis} was established; this parameter is the value that multiplied by the code prescribed seismic action (in this case the far field event prescribed by the Portuguese code of actions RSA [9]) yields the seismic action that leads to rupture. The analysis of the original structure was

Fig. 23 Strengthening strategies: analogy with chain under tensile force (adapted from [15])

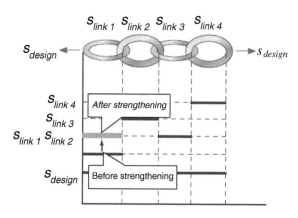

Fig. 24 Schematic
representation of
strengthening with a ring RC
beam [22]

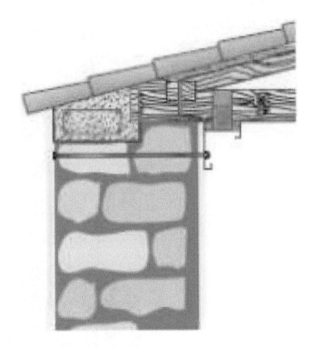

performed assuming weak connections, as already referred to in Sect. 2, yielding
a value of $\gamma_{sis} = 0.4$, associated with out-of-plane failure of the façades. If this
mechanism was prevented, for instances by strong façade-frontal connections, the
collapse would take place by the base shear mechanism, which in this situation
correspond to a value of $\gamma_{sis} = 1.05$.

One possible strengthening solution for the original building would be to build
a reinforced concrete beam at the perimeter of the top floor, connecting the inner
structure to the façades and gable walls and restricting the out-of-plane movement
of the façades and gable walls at the top. Figure 24 shows a schematic representation of a transversal cut of such a solution. However it should be noted that the
use of reinforced concrete elements in the rehabilitation of historical buildings has
been criticised by UNESCO and ICOMOS.

It should be noticed that, in this strengthening solution, the ring beams should
be properly connected to the rest of the structure so that there are no differential
movements between the masonry walls, the ring beams and the floors/roof.

The analysis of the structure after introducing the ring beams revealed that the
connections façade-frontal collapse at intermediate floors and the façade moves out-
of-plane but not as if it was a cantilever as in the original structure but as if it was a
simply supported beam, with supports on top and bottom. Due to this, collapse occurs
at a higher intensity $\gamma_{sis} = 0.7$. However, the structure would become stiffer due to
the new beam, increasing the frequencies, the spectral accelerations and therefore the
inertia forces. As the resistance to horizontal forces at the base level did not increase,

the seismic intensity associated to collapse in the base shear mechanism decreases to $\gamma_{sis} = 0.9$. The strengthening by means of ring beams can be extended to all floors. Despite the increase in stiffness and inertia forces, the resistance to the out-of-plane collapse of the façades increases to $\gamma_{sis} = 0.75$. The seismic intensity at which base shear failure takes place decreases to $\gamma_{sis} = 0.8$, since the resistance to horizontal forces at the ground floor did not increase. From this stage onwards any strengthening strategy that would increase the stiffness of the façades would be counterproductive, as the increase in the inertia forces would reduce the seismic intensity associated to base shear failure. This means that from this stage onwards the increase in the seismic resistance of the building would require strengthening both mechanisms simultaneously. This inconvenient could be eventually solved by means of an alternative strengthening strategy, for instances strengthening only the connections façades-frontals. As this consists of localized interventions that increase the strength without increasing the stiffness, it allows increasing the resistance to one collapse mechanisms without reducing the resistance to the others.

Another strategy that may be efficient is the strengthening of the pavements in their own plan by means of a set of steel angles in two orthogonal directions at 45° to the façades and side walls, in order to create an effect similar to a rigid floor that allows to transfer inertia forces to the stiffer elements of the structure, in particular the gable walls. The efficiency of this solution was studied by means of its application in a particular case study [23]. The results indicated that despite the fact that the inertia forces increased by 17 % due to the global stiffness increase, this effect was largely compensated by the redistribution of forces to the stiffer and stronger elements, reducing the action-effects in the more vulnerable elements, namely ground floor columns, façades (in the out-of-plane direction) and façades-frontal connections. Figure 25 shows (a) a scheme of the distribution of the steel angles and the deformation of one of the pavement, (b) the deformation of the same pavement without the angles, and (c) the out-of-plane deformations of the façades in both cases (the full line corresponds to the pavement strengthened with 2L200 × 200 × 20 or 2L100 × 100 × 10 angles and the dashed line to the unstrengthened pavement).

In what regards practical cases of strengthening of Portuguese old buildings, there is already considerable experience and capacity. In order to illustrate this capacity, developed essentially during the past two decades, examples of the strengthening of two *Pombalino* buildings are referred next.

The first example regards the rehabilitation and strengthening of a Pombalino building in *Rua do Comércio*, executed by the companies MONUMENTA and STAP [24], that included the following works, some of which are documented in Fig. 26:

1. Repair original wood bars and selective replacement of the bars in more advanced stage of degradation
2. Strengthening or reconstruction of interior *frontal* and *tabique* walls, filling the panels following the original techniques of the wood truss (rebuilding the *Gaiola*).

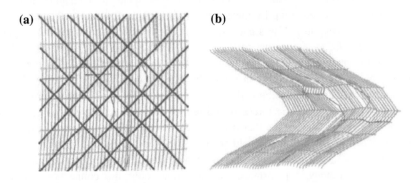

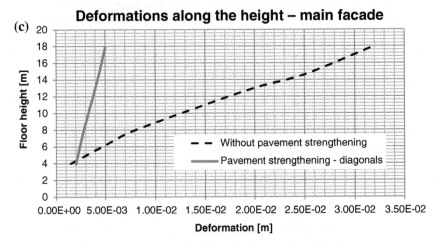

Fig. 25 Floor deformation: **a** with diagonals; **b** without diagonals; and **c** out-of-plan deformation of the façades [23]

3. Execution of a strengthening solution to increase the global resistance to horizontal forces, which consisted of:

- System of cables anchored with ductile anchor plates at the extremities, connecting the façades and gable walls to prevent independent out-of-plane movements of those walls;
- Steel plates to connect beams on the same alignment to ensure continuity;
- Devices to improve the wall–wall and wall-pavements connections.

It should be noted that the *Gaiola* continues to be part of the structural system, this is, it is not part of the problem, it is part of the solution, allowing solutions much less extensive than would be the case if the *Gaiola* was not there.

The second example is the rehabilitation of a building at *Rua Nova do Carvalho*. The structural project, by the design office A2P [25], explicitly

Fig. 26 **a** *A* selective replacement of deteriorated wood bars of the *Gaiola*; *B* selective replacement of deteriorated wood bars of the pavements; *C* steel reinforcements to ensure continuity of the wood bars of the pavements over the supports; *D* cables connecting to the façades; **b** installation of an anchorage device of a cable; **c** preparation of a façade to place anchorage plates [24]

comprised the objective of seismic strengthening. In the works on this building it was possible to preserve most of the primary elements of the building, namely:

- Foundations
- Masonry walls, strengthened with reinforced concrete layers of small thickness
- Frontal walls, repaired with new wood bars and new masonry filling at selected locations
- Ground floor columns and vaults
- Stairs and staircase
- Main bars of the wood floors
- Stones of the ground floor pavement
- Steps, horizontal platforms and ceiling of the stairs
- Part of doors and window frames

Figure 27b shows new filling of *frontal* walls made with hollow bricks and hydraulic and cement mortar. As it was already mentioned the masonry filling of frontal walls is

Fig. 27 **a** *Frontal* walls with the original filling and **b** damaged panels filled with hollow bricks with hydraulic and cement mortar

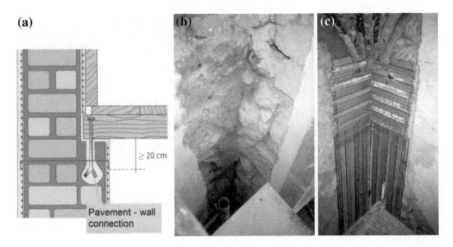

Fig. 28 a Connections wood pavements-masonry wall; and **b** and **c** between orthogonal masonry walls (adapted from [22, 26])

not as important as other parts of the structure, since what is clearly important for the seismic performance of the frontal walls is the wood structure.

In Portugal there is considerable experience in strengthening other types of masonry buildings for earthquake resistance, for example the repair and strengthening of the constructions damaged by the 1998 Faial earthquake, in Azores. Several strengthening techniques and details used in those cases can also be used in *Pombalino* buildings. Figure 28 shows two examples, the first ((a) referring to the connection between wood floors to masonry walls [22] and the second (b) and (c)) to the connection between two orthogonal masonry walls [26]. Figure 29a shows a scheme for the strengthening of a masonry wall by adding thin layers of cement mortar reinforced with steel meshes on either side [27] and, Fig. 29b, a photo of a wall strengthened in that way before adding the mortar [17].

5 Advanced Modelling

The models previously presented were intended for practical applications, therefore were based on models able to be analysed with currently available commercial software for structural analysis. In this section some advanced modelling techniques, which allow the explicit consideration of nonlinear material behaviour, are presented. These techniques make use of sophisticated computer programs, and allow performing pushover and nonlinear dynamic analysis. Actually, according to the widespread of performance-based earthquake engineering concepts, which research trends and various international and national codes [28, 29] now refer to, the possibility to perform nonlinear analyses becomes common, especially in research works: this is relevant in case of masonry buildings, that are also affected by the nonlinear behaviour. Of course,

Fig. 29 a and b
Strengthening of masonry
wall with reinforced cement
mortar on both sides (adapted
from [17, 27])

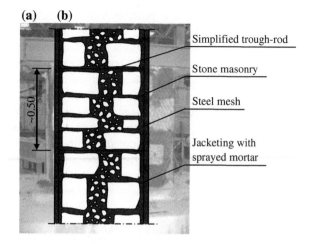

(a)   (b)

Simplified trough-rod

Stone masonry

Steel mesh

Jacketing with
sprayed mortar

~0.50

it implies the need of reliable models able to simulate the nonlinear response of various element types: for example, in case of *Pombalino* buildings, not only that of URM panels but also of *frontal* walls.

As it is known, a complete seismic assessment should include the analysis and verification of two types of response: the global one (type a), mainly related to the activation of the in-plane response of walls, and that (type b) associated to the activation of local mechanisms, which mainly involve the out-of-plane response of walls. Common assumption is to verify these two types of response separately by neglecting their mutual interaction. Type a) is usually analysed by referring to a 3D model of the structure: to this aim, among the different modelling strategies proposed in the literature, due to the regular pattern of openings in *Pombalino* building, the equivalent frame approach seems particularly suitable. One common way to analyse type b) response may be by using discrete macro-block models. In both cases, according to performance based assessment and, in particular, the use of nonlinear static analyses, the seismic verification may be: in case a), by adopting nonlinear static procedures, such as the Capacity Spectrum Method [30] or the N2 Method [31]; in case b), by referring to the nonlinear kinematic approach based on the limit analysis (e.g. as proposed in the Italian Code for Structural Design [32] and described in Lagomarsino and Resemini [33]). In the following, the attention is focused only on the global response by assuming local mechanisms are inhibited through proper constructive details: thus, for example, it is assumed that *frontal* walls are properly attached to masonry façades at reasonable distances, preventing their out-of-plane failure.

5.1 Equivalent Frame Modelling Approach

The equivalent frame approach starts from the main idea (supported by the earthquake damage survey) that, referring to the in-plane response of complex masonry walls with openings, it is possible to recognize two main structural components: piers and spandrels. Piers are the principal vertical resistant elements for both dead and seismic loads; spandrels, which are intended to be those parts of walls between two vertically aligned openings, are the secondary horizontal elements, coupling piers in the case of seismic loads. Thus, according to the equivalent frame idealisation, each wall is discretized by a set of masonry panels (piers and spandrels), in which the nonlinear response is concentrated, connected by a rigid area (nodes). Thus, by assembling 2D walls (considering only their in-plane contribution) and including the floor modelling, this approach allows one to analyse complex 3D models by performing nonlinear analyses with a reasonable computational effort; moreover, it agrees with recommendations of both national and international codes. This strategy seems particularly suitable in case of *Pombalino* buildings with good connections, as the façades are characterized by a quite regular opening pattern for which the idealisation in equivalent frame does not pose strong difficulties. Among the different models and software that work according to this approach, in the following particular attention is paid to Tremuri program which has been originally developed at the University of Genoa, starting from 2002 [34], and subsequently implemented in the software package Tremuri [35]. In fact, recently, in Tremuri program a specific element intended to simulate the response of *frontal* walls has been implemented [12, 36].

Once having idealised the masonry wall into an assemblage of structural elements, the reliable prediction of its overall behaviour mainly depends on the proper interpretation of the single element response. Different formulations, characterized by different degrees of accuracy, may be adopted. In the following, the attention is focused on a formulation based on a nonlinear beam idealization (Fig. 30): thus, the response in terms of global stiffness, strength and ultimate displacement capacity may be obtained by assuming a proper shear-drift relationship.

In case of URM panels, the formulation is based on a phenomenological representation of the main in-plane failure modes, which may occur (such as rocking, crushing, bed joint sliding and diagonal cracking); in particular, a bi-linear relation with cut-off in strength (without hardening) and stiffness decay in the nonlinear phase (for non-monotonic action) is adopted. The initial elastic branch is directly determined by the shear and flexural stiffness, computed on the basis of the geometric and mechanical properties (Young modulus E and shear modulus G) of the panel. Since the progressive degradation of the stiffness is not actually modelled, a calibration of the initial mechanical properties is necessary: in fact, they should be more properly representative of "cracked" conditions. The ultimate strength is computed according to some simplified criteria, which are consistent with the most common ones proposed in the literature and codes (e.g. in Eurocode 8—Part 3[29] and in the Italian Code for Structural Design [32]). Table 1 summarizes the strength criteria implemented in Tremuri program. Then, the failure of the panel is checked in terms of drift limit values differentiated as a function of the prevailing failure mode occurred (if shear or flexural one). This formulation is particularly suitable for nonlinear static analyses since it requires a reasonable computational effort, suitable also in engineering practice, and it is based on a few

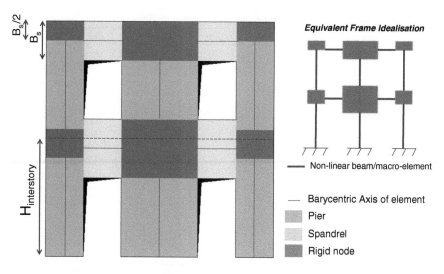

Fig. 30 Equivalent frame idealisation of a masonry wall [38]

mechanical parameters, which may be quite simply defined and related to results of standard tests. Further details on URM nonlinear beam and also on more accurate formulations implemented in Tremuri program may be found in Galasco et al. [37], Lagomarsino and Resemini [38] and Lagomarsino and Cattari [39].

To provide a reliable modelling also of *Pombalino* buildings, it is necessary to be able to describe the nonlinear response of typical *frontal* walls. In this context, the formulation proposed in Meireles et al. [36] and Meireles et al. [12] has been implemented in a nonlinear beam in Tremuri program. It aims to reproduce the hysteretic shear response of *frontal* walls and it has been formulated and calibrated on basis of the work of Meireles and Bento [42]. This work was the first experimental work to test the *frontal* walls built in laboratory under static cyclic shear testing with imposed horizontal displacements, where a specific loading protocol was used. Vertical loading was also applied to the specimen by four hydraulic jacks and rods. The objective of the experimental work was to obtain the hysteretic behaviour of *frontal* walls, by means of static cyclic shear testing with imposed displacements. Then, the hysteresis model was developed based on a minimum number of path-following rules that can reproduce the response of the wall tested under general monotonic, cyclic or earthquake loading. It was constructed using a series of exponential functions and linear functions. The hysteresis rule incorporates stiffness and strength degradations and pinching effect. It was then developed based on the experimental tests carried out [42] and the parameters are calibrated by such results. This model uses 9 parameters to capture the nonlinear hysteretic response of the wall: a first set of parameters aimed to define the envelope curve (F_0, K_0, r_1, r_2, F_u, δ_{ult}); two parameters to define the unloading curve; a last one to define the reloading curve. Figure 31 shows the assumed hysteresis model of the wall.

Table 1 Strength criteria for masonry panels implemented in Tremuri program

	Failure mechanism	Ultimate strength	Notes
Piers	Rocking/Crushing	$M_u = \frac{Nl}{0.425\,f_m}\left(1 - \frac{N}{lt}\right)$	f_m masonry compressive strength of masonry, l length of section, t thickness
	Bed joint sliding	$V_{u,bjs} = l'tc + \mu N \le V_{u,blocks}$	Mohr–Coulomb criterion with: l' length of compressed part of cross section; μ and c friction coefficient and cohesion of mortar joint, respectively. A limit value is imposed to take into account in approximate way the failure modes of blocks
	Diagonal cracking	$V_{u,dc\_1} = lt\,\frac{1.5\tau_0}{b}\sqrt{1 + \frac{N}{1.5\tau_0 lt}}$	τ_0 masonry shear strength, b reduction factor as function of slenderness [40]
		$V_{u,dc\_2} = \frac{1}{b}\left(lt\tilde{c} + \hat{\mu}N\right) \le V_{u,blocks}$	Mohr–Coulomb type criterion with: and equivalent cohesion and friction parameters, related to the interlocking due to mortar head and bed joints (such as proposed in [41])
Spandrel[a]	Rocking/Crushing	$M_u = \frac{dH_p}{2}\left[1 - \frac{H_p}{0.85\,f_{hu}\,dt}\right]$	H_p: minimum value between the tensile strength of elements coupled to the spandrel (such as RC beam or tie-rod) and $0.4\,f_{hu}\,dt$, where f_{hu} is the compression strength of masonry in the horizontal direction
	Shear	$V_u = htc$	h height of spandrel transversal section

[a]The Italian Code for structural design [32]. Differently from Eurocode 8, makes a distinction in the strength criteria to be adopted for spandrels as a function of the acting axial load: if known from the analysis, the same criteria as piers are assumed; if unknown, a response as equivalent strut is assumed. In Tremuri program, since the axial force computed for spandrels usually represents an underestimation of the actual one, the maximum value provided by these two cases is assumed as reference

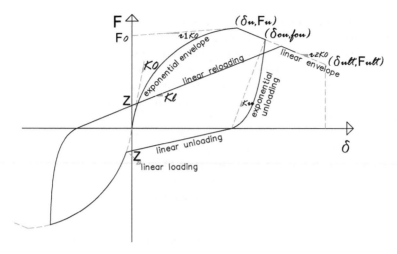

Fig. 31 Hysteresis model of frontal walls [37]

As an example, here a comparison is made between a *frontal* wall and a URM wall of equivalent dimensions (height 2.48 m; width 2.56 m; thickness 0.15 m). The masonry wall is composed of rubble masonry. The strength of the masonry panel, associated to shear failure, when subjected to a vertical stress of 20 % of the compressive capacity, is 73 kN. The ultimate drift of the masonry panel is 0.4 % (as proposed in Eurocode 8 [29] in case of a prevailing shear response). The stiffness relative to the transverse displacement between extremities of a masonry panel is calculated, according to the beam theory, considering that the panels are built-in in one (cantilever) or both extremities. By observing Fig. 32 one can see how the *frontal* walls have lower stiffness when compared to a masonry wall of approximately the same size.

Fig. 32 Comparison between a masonry wall (cantilever and fixed–fixed) and a *frontal* wall (C2x2) of the same dimensions [37]

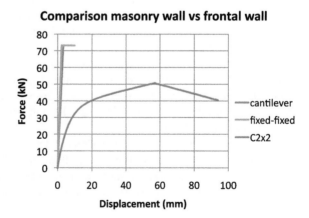

In addition to masonry and *frontal* elements, RC elements, steel and wooden nonlinear beam or tie-rods may be modelled as well.

Finally, the complete 3D model is obtained by introducing also floor elements. In particular, they are modelled as orthotropic membrane finite elements where normal stiffness provides a link between piers of a wall, influencing the axial force on spandrels; shear stiffness influences the horizontal force transferred among the walls, both in linear and non-linear phases.

In the following, a building aimed to replicate a typical *Pombalino* building was modelled and analysed by using Tremuri program. It was necessary to choose an example building. It was decided to use a modified version of the building of *Rua da Prata* previously mentioned but with some alterations in order to yield a building more similar to the original *Pombalino* buildings built after the 1755 earthquake.

5.2 Example of Equivalent Frame Modelling for a Case Study of a Pombalino Building

The building that was chosen to be analysed tries to replicate a typical *Pombalino* building. It had been the subject of research in the study by Cardoso [5] and later on in Meireles et al. [16, 43]. Its historical background and architectural drawings are also referred to and shown in the book *Baixa Pombalina: Passado e Futuro* (*Pombalino* Downtown: Past and Future) [44]. This building is recognized by the existence of a pharmacy in the ground floor, which is covered by a well-decorated panel of blue tiles, dating from 1860. Nevertheless, as it is usual in the *Pombalino* buildings of Lisbon downtown, this building has been subjected to some alterations with respect to the original layout. In this particular case one floor has been added to the original layout of 4 floors plus roof, making a total number of 5 floors plus attic. In the current study, given that it was intended to study a typical *Pombalino* building, only 4 floors plus roof were considered in the layout, so the last floor below the roof was eliminated in the drawings and modelling.

The building has six entries on the main façade and a height of approximately 15 m until the last floor (without the height of the roof). The openings have a width of 1.66 m or 1.76 m, the door on the ground floor a height of 3.5 m, the balcony on the first floor a height of 3 m and the windows on the second and third floors a height of 2 m.

At the back the openings are smaller and have a width of 1 m. At ground floor level the height of the door is 3 m and on the first, second, and third floors there are windows 1.5 m high. There are only 5 entries. The plan drawings of the building are shown in Figs. 17 and 33 for the ground floor and upper floors, respectively.

The plan of the building has dimensions of 18×11 m^2 referring to the façade and gable walls, respectively. The ground floor has 5 internal piers with dimensions of 0.7×0.7 m^2. There are stairs in the middle of the building facing towards the back façade. The staircase is made with brick masonry only on the ground

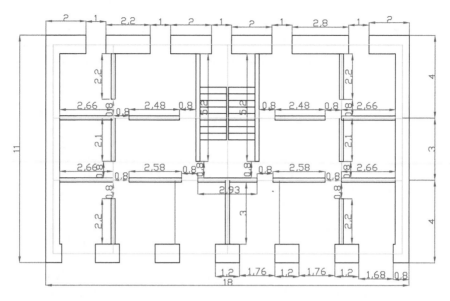

Fig. 33 Sketch of the plan view of building: upper floors (dimensions in metres) [45]

floor (on the upper floors the staircases are *frontal* walls) with a thickness of 0.24 m. On the ground floor, brick masonry walls extend up to the front of the building with a small misalignment towards the right. On the ground floor, the front and back façade piers as well as the internal piers are made of stone masonry. The gable walls as well as the front and back façades of the upper floors are made of rubble masonry.

On the upper floors (from the first to the third floor) one can find the *frontal* walls. There are two alignments of *frontal* walls parallel to the façades and five alignments (including the staircase) of *frontal* walls parallel to the gable walls. Connecting the *frontal* walls there are doors 0.8 m wide. The structural elements with their respective type of material and thickness/area can be found on Table 2. As it can be observed in Table 2, the façades (front and back) reduce in thickness the higher they are, being of 0.90 m on the ground floor and 0.75 m on the third floor.

The actions considered on the structure are the self-weight, given by the weights of the roof, the floors, the ceilings, the partition walls and the *frontal* walls, combined with the respective live loads given by Eurocode 1 [46]. The vertical loading (Table 3) to be imposed on the structure was determined based on Eurocode 1 [46] (design load = dead load + 0.3 × live load).

Table 4 summarizes the mechanical properties adopted for URM and *frontal* walls, respectively. For URM panels a drift limit value of 0.4 % and 0.8 % (as suggested also in the Italian code for structural Design [32]) has been adopted in case of prevailing shear and flexural failure modes, respectively. For *frontal* walls the value of F_{ult} (denotes failure) is taken as 80 % of the value of F_u.

Table 2 Thickness/area and material of building components [43]

Geometrical data and masonry types		
Element	Material[a]	Thickness/area
Piers (ground floor)	SM	0.7×0.7 m^2
Façades (front and back):		
Ground floor	SM	0.90 m
First floor	RM	0.85 m
Second floor	RM	0.80 m
Third floor	RM	0.75 m
Spandrels	RM	0.20 m
Gable walls	RM	0.70 m
Staircase (ground floor)	BM	0.24 m
Internal walls (ground floor)	BM	0.24 m
Frontal walls	Wood, RM	0.15 m

[a]SM, RM and BM mean stone masonry, rubble masonry and brick masonry, respectively

Regarding the floors, the joists of the floors have a section of 10×20 cm^2 and the wood pavement a thickness of 2 cm. The floors are supported by the front and back façades and by the *frontal* walls; the stairs are supported by the staircase. The floors have been modelled as orthotropic membrane elements.

The structure is modelled according to the equivalent frame model (by adopting Tremuri program) using nonlinear beams for the ordinary masonry panels and *frontal* walls according to the formulation described in 5.1. The final model of the building is presented in Fig. 34a. Here, represented in grey are the parts of the structure that are composed of rubble masonry; in purple are the parts of the structure that are composed of stone masonry; in green (dark and light depending on the size) are the *frontal* walls and in light brown are the timber beams connecting the *frontal* walls. Figure 34b identifies the alignments of the different structural elements in the plan view of the building.

Table 3 Vertical loads considered in the case study [45]

Actions considered		
Element	Location	Value[a]
–	Floors	2.0 kN/m(*ll*)
–	Stair floor	4.0 kN/m(*ll*)
Stairs	Stair floor	0.7 kN/m(*dl*)
Compartment walls	Floors	0.1 kN/m(*dl*)
Wooden floors	Floors	0.7 kN/m(*dl*)
Ceilings	Floors	0.6 kN/m(*dl*)
Frontal wall	*Frontal* walls	3.0 kN/m (*dl*)
Vaults	Masonry walls ground floor	3.5 kN/m (*dl*)
Gable walls roof	Masonry walls 4th floor	17.3 kN/m (*dl*)
Roof	Masonry walls 4th floor	4.4 kN/m (*dl*)

[a]The load type is summarized in brackets: if live load (*ll*) or dead load (*dl*), respectively

Table 4 Mechanical characteristics of masonry types and parameters of frontal walls

Masonry type	Average young modulus E (GPa)	Average shear modulus G (GPa)	Weight W (kN/ m^3)	Average compressive strength f_m (MPa)	Average shear strength τ_0 (MPa)
Stone Masonry	2.8[a]	0.86[a]	22	7	0.105
Rubble Masonry	1.23	0.41	20	2.5	0.043
Brick Masonry	1.5[a]	0.5[a]	18	3.2	0.076
Frontal wall	F_u (kN)	K_0 (kN/mm)	r_1K_0	r_2K_0	F_0/F_u
2×2[b]	50.8	6.1	0.244	−0.2745	0.728
3×2[c]	49.9	2.9	0.244	−0.2745	0.728

[a]Cracked stiffness assumed, 50 % of the value in the table was used
[b]Parameters have been calibrated on basis of experimental results obtained in Meireles and Bento [42]
[c]F_u and K_0 (as defined in Fig. 2) have been obtained for different configurations (2×3, 2×4, 3×2, 3×3 and 3×4) based on analytical models, see Meireles [45]

The mesh, that is the equivalent frame idealization, has been created by using the software package Tremuri [35] in which Tremuri has been implemented. The software creates a mesh of macro-elements for each alignment and this can be viewed for front and back façades, in Fig. 35, respectively: in red are the piers; in green are the spandrels and in blue are the parts of the façade where no damage is foreseen (rigid nodes). Furthermore, in this modelling, the foundations are modelled as built-in (no displacements or rotations allowed).

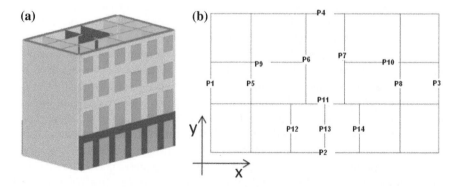

Fig. 34 **a** 3D view and **b** numbering of the alignments of the elements of the model [43]

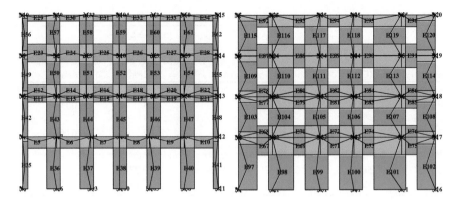

Fig. 35 Equivalent frame idealisation of front façade (*up*) and back façade (*bottom*) [45]

5.3 Example of Nonlinear Static Analysis for the Case Study Examined and Discussion of Results

The case study described in 5.2 refers to an original configuration of a *Pombalino* building. However, it should be noted, that, in reality, a considerable part of the building stock of Lisbon downtown probably is not in its original state but has been subjected to changes in its structural system, such as the ones referred to in Sect. 3. It is foreseen that these changed buildings will have a behaviour that is worse than the original building.

In the following, firstly the results of nonlinear static analyses performed on the original configuration are examined; then, the effects of some strengthening solutions on the overall response are discussed by comparing results in terms of probabilistic seismic assessment through the introduction of fragility curve concept. Pushover analyses were carried out for both *xx* and *yy* directions (see Fig. 34) and for two lateral load patterns along the height: load pattern proportional to the mass (uniform) and load pattern proportional to the mass and height (triangular). Pushover analyses enable us to have an idea of the lateral resistance of a building; the output in terms of overall base shear versus top displacement is presented in Fig. 36. At each floor the applied forces were distributed according to the distribution of mass and the top displacement refers to the average of the displacements of the nodes on the top floor. Actually, while in case of rigid floors the result of the pushover analysis is almost insensitive to the control node (usually assumed at the centre of mass), much critical is the case of the flexible ones. Actually, in this latter case, the results may be significantly affected by the control node adopted and points in the same floor may exhibit very different displacements, in particular in case of shear masonry walls characterized by very different stiffness. Thus, a reasonable compromise is to assume, for the analysis, a generic node at the level of the last floor, but to refer for the pushover curve to the average displacement of all nodes located at this level (eventually weighted with the respective pushover nodal

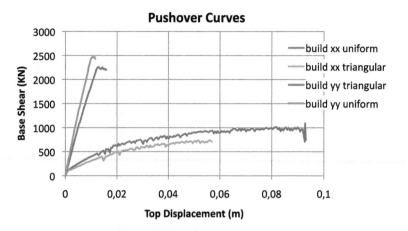

Fig. 36 Pushover curves for original building in the two directions for both uniform and triangular load patterns [45]

force) in order to consider a result representative of the whole structure and not only of local portions.

Pushover analyses performed on this basic configuration showed a significant difference between the seismic capacity of the building in *xx* and *yy* directions, in particular the stiffness and strength is much higher in the *yy* direction than in the *xx* direction; but on the other hand, the ductility of the system is much higher on the *xx* direction and is practically non-existing in the *yy* direction. In fact, in *xx* direction the piers are very slender (due to the opening's configuration) and with a very moderate coupling provided by spandrels (which show a "weak" behaviour): thus, a prevailing flexural response occurs associated to higher drift than in case of the shear failure. In general the structure exhibits a soft storey failure mode; moreover, since floors are quite flexible, a very moderate redistribution of seismic loads may occur among masonry walls. Indeed, in neither of the two directions the building seems to provide a reliable system against the earthquake.

Starting from the study of the response of the basic configuration of the *Pombalino* structure, the following strengthening solutions have been analysed mainly based on engineering judgement.

Due to this, the following retrofitting schemes have been proposed and analysed:

a. Increase the in-plane stiffness of floors (transforming flexible floors into rigid floors);
b. Increase the in-plane stiffness of floors plus inclusion of four shear walls on the ground floor;
c. Increase the in-plane stiffness of floors plus inclusion of eight steel frames on the ground floor;
d. Increase the in-plane stiffness of floors plus inclusion of tie-rods at front and back façades.

The first one is the one that will be seen to be the most effective and crucial improvement to the structure. The last three are seen to be added improvements to the structure if one wishes to increase the earthquake resistance of the building even more. The first intervention may be reversible or not, depending however on the type of intervention. All the other interventions are reversible.

As regards to the *increase of the in-plane stiffness of floors* (case a), traditional timber floors are typically flexible. The increase of the in-plane stiffness of floors is an evident and most effective method of improving the seismic behaviour of old masonry structures. This is mainly because the increase of in-plane stiffness of floors enables the horizontal forces to be redistributed between the failing walls to the adjacent remaining walls and the structure behaves like a box. A significant role in the stability of the entire building is assigned to the floors. These structures are required, in addition to an adequate performance level, an adequate rigidity and an efficient connection to the supporting walls, especially in what concerns seismic actions. For this reason, the restoration of a floor is an opportunity to improve the behaviour and efficiency of the entire structure.

Starting from the original configuration, mechanical parameters of orthotropic membranes aimed to simulate floors have been increased to simulate such type of intervention (e.g. obtained by the insertion of plywood or horizontal bracing composed of steel ties and arranged in crosses). Figures 37 and 38 show the resultant pushover curves, in *xx* and *yy* directions, respectively. The contribution that each alignment (walls) has to the base shear of the building was also evaluated in both directions. For this purpose, and taking the *xx* direction as an example (Fig. 37), a graph was plotted with, firstly, the total base shear as a function of the top displacement ("Building" legend), secondly, the base shear corresponding to the façade masonry walls (P2 and P4 alignments) as a function of the respective top displacement of that alignment ("P2" and "P4" legend) and, thirdly, the base shear

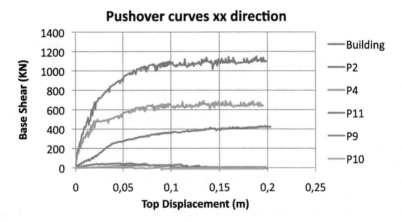

Fig. 37 Pushover curves, contribution of each wall to the base shear, *xx* direction [45]

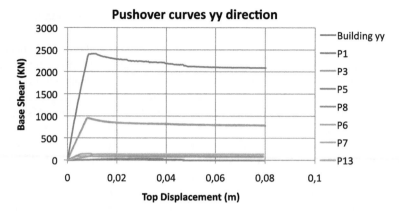

Fig. 38 Pushover curves, contribution of each wall to the base shear, *yy* direction [45]

corresponding to the alignments of the"frontal" walls as a function of the respective top displacement of that alignment ("P11", "P9" and "P10" legend).

Based on the previous graphs, the highest contribution to the base shear comes from the outside masonry walls. The contribution to the base shear given by the internal walls/columns is not negligible but is small. In other words, the *frontal* walls/internal walls alignments contribute very little to the total base shear of the respective alignments, the majority of base shear being a contribution of the surrounding masonry walls. This is because the *frontal* walls do not have continuity in height, they are interrupted at ground floor (above the first floor the contribution of the *frontal* walls to the horizontal shear may be not so low), and because they have a lower stiffness when compared to the masonry walls. Indeed, from the comparison between a single URM panel and a *frontal* wall illustrated in 5.2, one can conclude that the stiffness of the *frontal* wall is lower than the stiffness of the thick (see Table 2) surrounding masonry walls of the *Pombalino* buildings.

As regards to the *increase the in-plane stiffness of floors plus inclusion of four shear walls on the ground floor* (case b),the inclusion of shear walls is a typical procedure to improve the seismic resistance of a building. The modelled shear walls are 48 cm thick and are composed of brick masonry. It was decided that the shear walls should only be placed in the *xx* direction since this direction is the most vulnerable one (after the strengthening of the diaphragms and given the presence of the gable walls with no openings on the *yy* direction).

As regards the *increase the in-plane stiffness of floors plus inclusion of eight steel frames on the ground* (case c), the inclusion of eight steel frames on the ground floor arises from the idea that including shear walls with no openings on the ground floor is not a very much welcoming idea from the architectural and functional perspective. The ground floors of these buildings are often used as restaurants, cafés or stores facilities and the inclusion of shear walls here is not very convenient from the point of view of the owners. The eight steel frames (pillars and beams) are each one composed of four HEA140 cross sections. Again, it was decided that the steel frames should be placed only in the *xx* direction for the reasons previously described.

Finally, as regards to the *increase the in-plane stiffness of floors plus inclusion of tie-rods at front and back façades* (case d), the model was prepared for the case of tie-rods at the front and back façades. In the model bar elements with prestressing were introduced. The tie-rods are placed at the top of the piers (placed along the spandrels), connecting the piers between each other. They are prestressed, prestressing the spandrels. The idea is to couple the piers with the prestressed spandrels. The modelled tie-rods are 2.4 cm in diameter and made of steel. An initial strain of 20 % the yielding strain of the steel was used. The tie-rods were only placed in the *xx* direction, where we have spandrels.

Figure 39 shows the comparison among the pushover curves obtained for all the different configurations examined. Different configurations vary in terms of strength, stiffness and ductility. Since all these three aspects play a fundamental role in the seismic assessment, a more effective comparison is discussed in the following in terms of probabilistic assessment through the introduction of fragility curves.

To this aim, firstly pushover curves have been converted in the equivalent SDOF oscillator (according to criteria proposed in Eurocode 8—Part 3 [29]); then proper damage states (from—slight damage—to 4—collapse) have been defined on the resultant capacity curves by adopting the criteria proposed in Lagomarsino and Giovinazzi [47]. Figure 40 shows the fragility curves obtained by assuming a β value (that is the standard deviation of the natural logarithm of spectral displacement associated to different damage states) equal to: 0.53, 0.54, 0.51 and 0.49 from damage state 1–4, respectively. These values summarize the uncertainties associated to errors in the model, input parameters, definition of limit states and variability of the seismic input; they have been computed according to the proposal of Pagnini et al. [48]. Moreover further details may be found in Meireles et al. [43]. The seismic input has been assumed as the earthquake type 1 (far-field

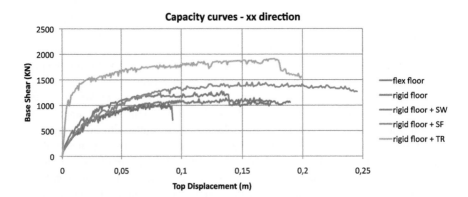

Fig. 39 Pushover curves for all the examined configurations

Fig. 40 Fragility curves
for earthquake type 1:
xx direction (*top*) and *yy*
direction (*bottom*) [43]

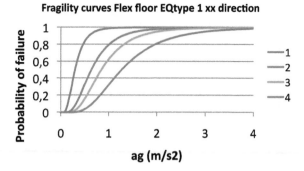

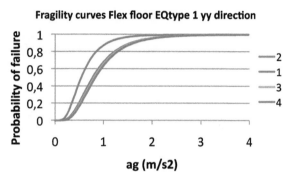

event of high magnitude, richer in larger periods) recommended in the Portuguese
national annex of Eurocode 8 [29], for Lisbon.

Finally, Fig. 41 illustrates the damage probability for earthquake type 1(high
magnitude, far field event, with low frequency contents) in the *xx* direction for all
the studied cases. Discrete damage-state probabilities can be calculated as the difference of the cumulative probabilities of reaching, or exceeding, successive damage states (as computed from the fragility curves):

$$P_0 = 1 - P[L_1|S_d]; \quad P_k = P[L_k|S_d] - P[L_{k+1}|S_d] \ for$$
$$k = 1,2,3; \quad P_4 = P[L_4|S_d]$$

In Fig. 41 Pr0 represents the probability of having "no damage", Pr1 the probability of having "slight damage", Pr2 the probability of having "moderate damage",
Pr3 the probability of having "heavy damage" while Pr4 the probability of reaching "collapse".

Based on the results obtained, it is clear that building without retrofitting presents the highest value of probability of damage Pr4 ("collapse"). Retrofitting
the building by stiffening the floors enables reducing this value significantly.
Retrofitting the building by stiffening the floors and including shear walls or steel
frames does improve slightly the situation, reducing the value of Pr4 and spreading it more through Pr3 to Pr1. The retrofitting scheme that mostly improves the
seismic performance of the building, with respect to the previous cases, is the case

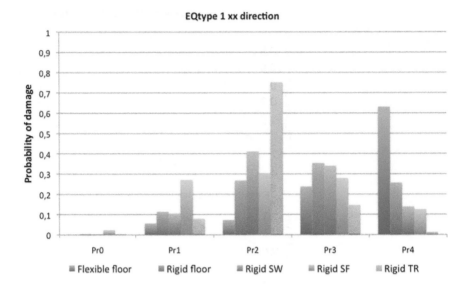

Fig. 41 Probability of damage for earthquake type 1 in the *xx* direction

of the inclusion of tie-rods in the front and back façades. This reduces significantly the damage probability Pr4. Nevertheless, this retrofitting possibility seems to increase very much the damage probability Pr2 when compared to the other retrofitting strategies.

5.4 Analysis of a Pombalino Quarter

Besides modelling buildings and structural elements, it is also worth mentioning a study of an idealized *Pombalino* quarter [49]. The purpose was to get some insight on how the buildings interact with each other, since the gable walls are common to adjacent buildings. For this purpose the model of an idealized quarter was developed based on the design of three real *Pombalino* buildings that constitute one quarter of the entire quarter of buildings, that was then replicated twice, giving rise to a model of a quarter with double symmetry. The wood floors were simulated by a set of bars with axial stiffness in both directions, since it is not expected that the floors exhibit any relevant distortion stiffness, specially under strong seismic actions. The main conclusion is that, even though the floors have no distortional stiffness, they have enough axial stiffness to force the buildings in one band (alignment of buildings) of the quarter to move together, with similar horizontal displacements at each floor level, for the modes with lower frequencies. Figure 42 shows the deformed shapes of the 1st and 2nd modes, that illustrated the band effect.

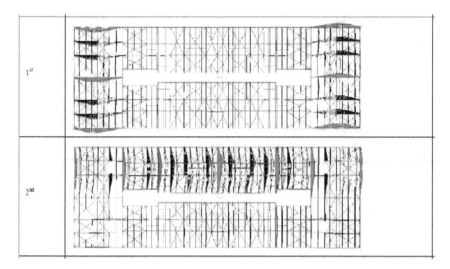

Fig. 42 First and second mode shapes of a *Pombalino* quarter [47]

However, for higher frequency modes, buildings in the same band exhibit deformed configurations with different horizontal displacements at each floor in the direction of the alignment, thus with relevant axial deformation of the floors. These effects show that strengthening one building in the direction of a band may lead to shared improvements on seismic behaviour with other buildings in the same band, at the expenses of less improvements of that building.

As it should be expected, the analysis of the quarter confirms that the buildings cannot rotate freely, as this is restricted due to the fact that they share the gable walls. In this context it does not make sense to consider accidental eccentricities in strengthening design of these buildings, such as the ones prescribed by several codes for the design of new buildings (assumed isolated).

6 Economic Feasibility of Strengthening

The economic feasibility of maintain and strengthen *Pombalino* buildings depends on the capacity to adapt the buildings to new functions or to the same functions but with different demands. For instances *Pombalino* buildings had very small rooms with 5 or 6 m^2, no lifts, etc. Some of these characteristics are not compatible with nowadays activities and architectural requirements. For instances office buildings require larger areas, and are not a solution for the whole downtown, as it would become desert at night and weekends, which is not desirable. Local authorities also want downtown Lisbon to have a life of its own, therefore part of the *Pombalino* buildings should be used for housing purposes. It is therefore important to adapt these buildings to modern uses and functions according to modern standards. Besides, this adaption is critical for the buildings to provide some income and stimulate the private sector to contribute to rehabilitation and strengthening works.

The adaptation to new uses and functions requires larger rooms, pointing to solutions that may imply to remove some interior walls. Even though this is a debatable issue, it is the opinion of the authors that economic of preservation fundamentalisms are not the best option, but one must be very demanding and assertive and try to compatibilize both criteria. In this framework it may be arguable that the secondary may have to be sacrificed to preserve the essential. The issue is what is essential and what is secondary. The *Gaiola* can be considered essential, in general. This is due not only to its symbolism and associated cultural value, that will be discussed further in the next section, but also due to the fact that due to its good state of conservation and its structural capacity it can still contribute significantly to the buildings seismic capacity. This may be relevant to reduce or avoid the need for widespread works throughout the buildings, as it could be the case without the *Gaiola*, for instances in some or many *Gaioleiro* buildings. On the other hand the partition walls called *tabiques*, as the one shown in Fig. 4, can be considered as secondary, both from the structural as well as historic and cultural heritage points of view: they have no special characteristics that distinguish them from partition walls in other types of masonry buildings, and have much less strength and stiffness than the frontal walls. Therefore the removal of interior walls to create larger spaces should follow strict criteria for the preservation of the most relevant characteristics of the buildings (from the point of view of cultural heritage) and should not be done only according to architectural criteria related to future uses of the buildings. The removal of some partition walls may weaken slightly a building, what can be compensated, and allow an architecture more compatible with the uses and functions of nowadays without reducing the historic and cultural value of the buildings [50]. However this issue deserves a deeper analysis and debate, not only by architects, engineers and promoters, but by the whole society, as urban rehabilitation and the preservation of the cultural heritage is an issue that interests the whole society, and not only the main economic and technical agents involved in design and construction.

A common problem that arises from the adaptation of old buildings to new functions is the cut of façade columns at ground floor levels aiming at creating larger spaces for shop windows. This is not acceptable, given the potential consequences. Therefore the actual and future owners of shops in zones and/or buildings of relevant cultural value should assume that if they want to keep or set-up a business in such zones they must accept some restrictions to the changes that can be done in those buildings, namely that cut ground floor columns is not acceptable. This does not seem to be a problem difficult to solve or that creates incompatibilities with most modern uses. There are in downtown Lisbon several good examples of integration of shop windows with ground floor columns, for instances by using the columns as supports for shelves or just leaving the columns between exterior accesses. Figure 43 shows two examples of compatibility between original structure and modern uses and functions.

Fig. 43 External accesses and window shops with integration of original columns [21]

7 Cultural Heritage

The reconstruction of downtown Lisbon after the destruction caused by the great Lisbon earthquake of 1755 was the first time in the History of mankind that a large town, a European capital, was built with techniques aiming at the explicit purpose of providing seismic resistance. These include for instance the *Gaiola*; the fact that the buildings were built in blocks and had similar characteristics such as the same number of floors and were generally symmetrical; the fact that they had regular openings in the front and back façades, were robust and had good quality of construction and besides they had thick masonry exterior walls surrounding the *Gaiola*. The *Pombalino* architecture was also austere with no useless decorative features, especially on the façades.

Therefore Lisbon downtown is a part of mankind's cultural heritage, a landmark that must be preserved and transferred to future generations in safe conditions and preserving the authenticity of its buildings. This is of the interest of the Portuguese people and authorities that should also promote the international recognition of downtown Lisbon historical and cultural value.

The need to preserve/improve safety standards in what regards the earthquake resistance of nowadays *Pombalino* buildings is not incompatible with the interests on the preservation of the original structure, as it continues to offer a relevant contribution to the buildings earthquake resistant capacity, as it was previously referred to: the seismic resistance of original *Pombalino* can be above the actual Portuguese code prescribed value of the seismic action, a noticeable fact for 250 years old buildings.

However, the compatibility of the preservation of the most important characteristics of the original structure with the requirements of modern uses and functions requires the art and skill of architects and engineers, and the acceptance some restrictions to architectural changes by financial institutions, public authorities, owners, tenants and all other agents involved in urban rehabilitation.

8 Synopsis

The reconstruction of Lisbon and other Portuguese southern towns after the Great Lisbon Earthquake of 1755 was performed with the major concern of avoiding future catastrophes of the same kind. The buildings where designed with a tridimensional wood truss embodied in the interior masonry walls, that provides resistance to horizontal forces in any direction. The wood truss, called *Gaiola* (cage) *Pombalina*, is a characteristic of these buildings, and its widespread application during the reconstruction of Lisbon was the first case in history of an entire town built with the purpose of providing seismic resistance to its buildings.

Other characteristics of these buildings are also described: their regular distribution in quarters within a rectangular mesh of streets, the same number of floors for all buildings, the foundation system that includes short wood piles, the lack of the *Gaiola* in the ground floor to isolate it from the water in the soil as the water table is very near the surface, the wood pavements in all floors except on the first one, that is made of masonry for fire protection, the gable walls belonging simultaneously to two adjacent buildings and the industrialization and systematization of the construction process for mass production, etc. However after the generation that lived the earthquake was gone, progressive adulteration of these characteristics took place, such as the addition of more floors, addition of heavy decorative elements of the façades, removal of the diagonals of the *Gaiola* and poorer workmanship and poorer connections between elements. This process continued and was aggravated more recently, during the twentieth century, with the removal of ground floor columns for window shops, insertion of water and gas pipes inside the *Gaiola* walls, removal of entire *Gaiola* panels to create larger spaces, etc.

Recently the scientific interest by this type of buildings has increased and several studies to evaluate their seismic resistance were performed. It was concluded that the original buildings would probably possess the capacity to resist the seismic action prescribed for new buildings by the current Portuguese code for actions in structures, without considering the 1.5 seismic factor prescribed by the code. Even though this is less than the resistance of many modern buildings, it is a remarkable result considering the materials and knowledge available 250 years ago. However the reality is that most *Pombalino* buildings don't meet these standards due to the negative alterations during the nineteenth and twentieth centuries. Some strengthening techniques used to increase the seismic resistance of these buildings, taking advantage of the *Gaiola Pombalina* are described. Some advanced modelling techniques, able to simulate the nonlinear behaviour of this type of buildings, are described and some results shown. Since this buildings share the gable walls, there are no expansion joints between buildings and they interact within each quarter. Results of the analysis of these effects on the dynamic behaviour of an entire quarter are presented. The economic feasibility of strengthening these buildings, as well as the interest in their preservation due to their historical value are briefly discussed.

References

1. Chester, D.K.: 1755 The Great Lisbon Earthquake (in Portuguese), vol. 1, Published by Público, Houston (2005)
2. Appleton ,J.: Characterization of the building stock (in Portuguese), Cap. 9 Earthquakes and Buildings. Mário Lopes, Orion Editions, Coordinator (2008)
3. Appleton, J.: Rehabilitation of old Buildings. Pathologies and Intervention Techniques (in Portuguese). Orion Editions. (2003)
4. Segurado, J.: Civil carpentry works (in Portuguese), Professional Instruction Library, (date unknown)
5. Cardoso, R.: Seismic vulnerability of old masonry structures—application to an old *Pombalino* building (in Portuguese). MOPTC Prize 2003, MSc thesis in structural Engineering, IST, Lisbon (2002)
6. Mascarenhas, J.: Construction systems. V—The *Pombalino* rentable building of Lisbon downtown, (in Portuguese), Livros Horizonte, 2nd edn (2005)
7. Mascarenhas, J.: A study of the Design and Construction of buildings in the *Pombalino* Quarter of Lisbon. PhD Thesis, University of Glamorgan, U.K. (1996)
8. Lopes, M.: Evaluation of the seismic performance of an old masonry building in Lisbon. In: Proceedings of the 11th World Conference in Earthquake Engineering, paper no 1484, Acapulco, México (1996)
9. RSA.: Code for safety and actions in buildings and bridges, (in Portuguese). Decree-Law n°235-83 INCM, Lisbon (1983)
10. Ramos, J.S.: Experimental and numerical analysis of historical masonry structures (in Portuguese). MSc thesis, School of Engineering of University of Minho, Guimarães (2002)
11. Cruz, H., Moura, J.P., Machado, J.S.: The use of FRP in the strengthening of timber reinforced masonry load-bearing walls. In: Proceedings of Historical Constructions, possibilities of Experimental and Numerical Techniques, Guimarães (2001)
12. Meireles, H., Bento. R., Cattari, S., Lagomarsino, S.: A hysteretic model for "frontal" walls in *Pombalino* buildings. Bulletin of Earthquake Engineering, DOI 10.1007/s10518-012-9360-0, Online publishing (2012)
13. Eurocode 5: Design of timber structures Part 1–1: General—Common rules and rules for buildings. ENV 1995-1-1, CEN: Brussels (2005)
14. Teixeira, M.J.: Rehabilitation of *Pombalino* buildings. Experimental analysis of frontal walls (in Portuguese). MSc dissertation, IST (2010)
15. SAP2000: Structural Analysis Software. Computers and Structures Inc, Berkeley (2006)
16. Meireles, H., Bento, R., Cattari, S., Lagomarsino, S.: Formulation and validation of a macro-element for the equivalent frame modelling of internal walls in *Pombalino* buildings. Paper 2853, Proceedings of 15thWCEE, Lisbon, Portugal (2012)
17. Costa, A.: Repair and strengthening of constructions (in Portuguese). Cap. 11, Earthquakes and Buildings, Coordinator: Mário Lopes, Orion Editions (2008)
18. Ramos, L.F., Lourenço, P.B.: Moddeling and vulnerability of historical city centers in seismic areas: a case study in Lisbon. Eng. Struct. **24**, 1295–1310 (2004)
19. Cardoso, R., Bento, R., Lopes, M.: Foundation differential settlement effects on the seismic resistance of *Pombalino* buildings. Conference of the 250 anniversary of the 1755 Great Lisbon Earthquake, Lisbon (2005)

20. Lopes, M., Bento, R., Cardoso, R.: Structural safety of Lisbon downtown (in Portuguese). Magazine Monumentos, no 21, DGEMN176-181 (2004)
21. Monteiro, M., Lopes, M.: Negative interventions and execution errors (in Portuguese). Cap. 10 Earthquakes and Buildings, Coordinator: Mário Lopes, Orion Editions (2008)
22. Carvalho, E.C., Oliveira, C.S., Fragoso, M., Miranda, V.: General rules for rehabilitation and reconstruction of current buildings affected by the seismic crisis of Faial, Pico and S. Jorge triggered by the 9th July 1998 earthquake (in Portuguese). Regional Government of Azores (1998)
23. Neves, S.: Seismic analysis in downtown Lisbon (in Portuguese). MSc Dissertation, IST, Lisbon (2008)
24. Cóias, V.: The rehabilitation of Lisbon buildings and the seismic risk (in Portuguese). Conference Reabilitar, Lisbon (2010)
25. Appleton, J.: Rehabilitation of old buildings: a sustainable choice (in Portuguese). II Quercus Meeting—Sustainable Architecture (2010)
26. Guedes, J., Costa, A.: Stabilization of the façade of the church of Ponte da Barca. Workshop The intervention on the built heritage. Design practice on conservation and rehabilitation", Porto (2002)
27. Cóias, V.: Structural rehabilitation of old buildings. Low intrusiveness techniques (in Portuguese). Argumentum/GECoRPA (2007)
28. ASCE/SEI 41/06.: Seismic Rehabilitation of Existing Buildings. American Society of Civil Engineers, USA (2006)
29. Eurocode 8- Part 3.: Design of structures for earthquake resistance. Part 3: Assessment and retrofitting of buildings. ENV 1998-3, CEN: Brussels (2005)
30. Freeman, S.A., Nicoletti, J.P., Tyrell, J.V.: Evaluation of existing building for seismic risk- A case study of Puget Sound Naval Shypyard. Brementon, Washington. In: Proceeding of the U.S. National Conference on Earthquake Engineers, pp. 113–122, EERI, Berkeley (1975)
31. Fajfar, P.: A nonlinear analysis method for performance-based seismic design. Earthquake Spectra 16(3), 573–591 (2000)
32. Italian Code for Structural Design (D.M. 14/1/2008, Official Bulletin no. 29 of 4/02/2008, 2008 (in Italian)
33. Lagomarsino, S., Resemini, S.: The assessment of damage limitation state in the seismic analysis of monumental buildings. Earthquake Spectra 25(2), 323–346 (2009)
34. Galasco, A., Lagomarsino, S., Penna, A., Cattari, S.: TREMURI program: seismic analyses of 3D masonry buildings. University of Genoa, Genoa (2009)
35. 3Muri Program, release 4.0.5 http://www.stadata.com
36. Meireles, H., Bento, R., Cattari, S., Lagomarsino, S.: The proposal of a hysteretic model for internal wooden walls in *Pombalino* buildings. In: Proceedings of the 2011 World Congress on Advances in Structural Engineering and Mechanics (ASEM11plus), Seoul, South Korea (2011)
37. Galasco, A., Lagomarsino, S., Penna, A., Resemini, S.: Nonlinear seismic analysis of masonry structures. In: Proceedings of 13th World Conference on Earthquake Engineering, Vancouver 1-6 August, paper no 843 (2004)
38. Lagomarsino, S., Resemini, S.: The assessment of damage limitation state in the seismic analysis of monumental buildings. Earthquake Spectra 25(2), 323–346 (2009)
39. Lagomarsino, S., Cattari, S.: Nonlinear seismic analysis of masonry buildings by the equivalent frame model. In: Proceedings of 11th D-A-CH Conference, 10–11 September 2009, Zurich, (invited paper), Documentation SIA D 0231, ISBN 978-3-03732-021-1 (2009)
40. Turnsek, V., Cacovič, F.: Some experimental results on the strength of brick masonry walls. In: Proceedings of the 2nd International Brick Masonry Conference, Stoke-on-Trent, 149–156 (1970)
41. Mann, W., Müller, H.: Failure of shear-stressed masonry—an enlarged theory, tests and application to shear-walls. In: Proceedings of the International Symposium on Load bearing Brickwork, London, 1–13 (1980)

42. Meireles, H., Bento, R.: Cyclic behaviour of *Pombalino* "frontal" walls. In: Proceeding of the 14th European Conference on Earthquake Engineering, 325, Ohrid, Macedonia (2010)

43. Meireles, H., Bento, R., Cattari, S., Lagomarsino, S.: Seismic assessment and retrofitting of *Pombalino* buildings by fragility curves. In: Proceedings of the 15th WCEE, Lisbon, Portugal, paper 2854, September 24–28 (2012)

44. Santos, M.H.R.: *Pombalino* Downtown: Past and Future (in Portuguese). Livros Horizonte, Lisbon (2000)

45. Meireles, H.A.: Seismic Vulnerability of *Pombalino* buildings. PhD Thesis, IST, Technical University of Lisbon, (2012)

46. Eurocode, 1.: Actions on Structures—Part 1–1: General actions—densities, self-weight, imposed loads for buildings, ENV 1991. CEN, Brussels (2001)

47. Lagomarsino, S., Giovinazzi, S.: Macroseismic and mechanical models for the vulnerability and damage assessment of current buildings. Bull. Earthquake Eng. **4**, 415–443 (2006)

48. Pagnini, L.C., Vicente, R., Lagomarsino, S., Varum, H.: A mechanical model for the seismic vulnerability of old masonry buildings. Earthquakes Struct. **2**(1), 25–42 (2011)

49. Monteiro, M., Lopes, M., Bento, R.: Dynamic behaviour of a *Pombalino* quarter. In: Conference of the 250 anniversary of the 1755 Great Lisbon Earthquake, Lisbon (2005)

50. Mira, D.: Analysis of the *Pombalino* constructive system. Rehabilitation of a building (in Portuguese). MSc dissertation in Architecture, IST (2006)

Analysis and Strengthening of Timber Floors and Roofs

Jorge M. Branco and Roberto Tomasi

Abstract In many countries, traditional construction comprises floor and roof systems in wood. Current knowledge assumes the need to preserve and to protect existing wood systems as a cultural value with important advantages to the overall behaviour of the building. In this chapter, after a description of the most common systems used in traditional wooden floors and roofs, strengthening techniques are presented. For that, the analysis and design of the existing structures is discussed and the effectiveness of the strengthening techniques is evaluated through experimental results obtained in laboratory.

Keywords Floors • Refurbishment • Trusses • Traditional connections Strengthening

1 Introduction

The importance of the preservation of timber structures of historical and cultural interest has increased its importance in the recent years, as can be attested by different guidance documents for intervention issued by national or international technical committees (e.g. international ICOMOS IWC, RILEM RC 215 AST, RILEM TC RTE and Italian UNI-NORMAL WG20 [1]). These documents

J. M. Branco (✉)
Civil Engineering Department Campus de Azurém, ISISE, University of Minho,
4800-058 Guimarães, Portugal
e-mail: jbranco@civil.uminho.pt

R. Tomasi
Department of Structural and Mechanical Engineering, University of Trento, Via Mesiano
70, 38100 Trento, Italy
e-mail: roberto.tomasi@ing.unitn.it

A. Costa et al. (eds.), *Structural Rehabilitation of Old Buildings*,
Building Pathology and Rehabilitation 2, DOI: 10.1007/978-3-642-39686-1_8,
© Springer-Verlag Berlin Heidelberg 2014

recognise the importance of timber structures from all periods as part of the cultural heritage, whereas some standards recommend the assessment and the conservation of existing timber structures also for technical and structural reasons: for instance, since 2003 the Italian technical codes indicate the saving of timber diaphragms as a strategic issue in the reduction of seismic vulnerability of existing buildings, do not supporting the practice of substituting these structural elements with heavier and more invasive diaphragms, such as concrete slabs.

The assessment of existing timber structures requires, and relies, on the determination of the mechanical properties of the individual timber members as well as the behaviour of joints and, of structural system effects. In existing timber structures, the first step towards the safety assessment is the evaluation of the actual mechanical properties of the material. Despite significant effort in the development of non-destructive testing, true strength of a timber member can only be directly determined in a destructive test, which is often unacceptable in the case of historic buildings and other existing timber structures. In practice, it is the lack of knowledge about the material and its structural behaviour that normally leads to the replacement of existing wood structures, instead of their retrofitting.

In other hand, when the decision is the strengthening of the existing structure instead of its substitution by a new one, the misunderstanding of the overall behaviour of the timber structure can result in unacceptable stress distributions in the members because of inappropriate joints strengthening (in terms of stiffness and/or strength).

This chapter aims to fulfil this gap of knowledge by presenting the state-of-the-art on the strengthening of wooden floors and roofs.

2 Timber Floors

2.1 Traditional Floor Systems

Strengthening and stiffening of old timber floors are often needed as they were designed to bear moderate loads and may suffer from excessive in-plane and out-of plane deflections with respect to current requirements.

The structural refurbishment of traditional timber floors can be achieved, in order to increase the bending stiffness of the main elements, by including other elements, such as a concrete slab, or timber planks. The structural behaviour of the resulting timber composite structure is governed by the strength and stiffness of the mechanical joints adopted to connect the existing timber beams to the new element.

Another important aspect to be keenly considered is the wooden floor diaphragm in-plane stiffness, which may affect the structural performance of a traditional masonry building subjected to lateral loads: the common configuration of existing timber floors with a crossly arranged single layer of wooden planks could possibly need an in-plane shear strengthening, in order to assure an efficient redistribution of lateral seismic load through the bearing walls.

2.2 Out-of-Plane Structural Behaviour

The reinforcement techniques commonly adopted in the practice consist in coupling the existing beams with concrete or wooden slabs: different configurations are possible depending on the slab material and connection system (Fig. 1).

Design of composite sections requires the consideration of partial composite action, due to the impossibility of achieving an extremely rigid shear connection between web and flange.

The slip modulus of the connection system is a key parameter for the mechanical characterization of the global behaviour of refurbished traditional timber floors, because it affects not only the effective bending stiffness of the composite structure, but also the internal stresses distribution. As described in Fig. 2, the real behaviour referred to case (b) is in between cases (a) and (c).

The connectors in a timber composite floor are disposed along the beam according to the shear force distribution, typically with a spacing in the central zone of the beam (for a simple supported configuration) higher than the one in the support zones.

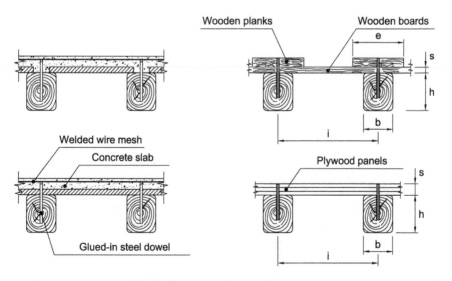

Fig. 1 Examples of reinforcement techniques of timber floors

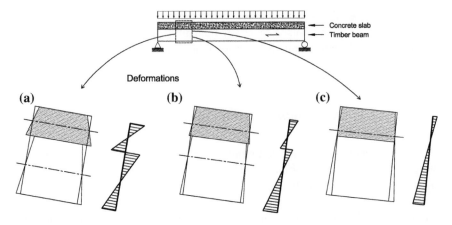

Fig. 2 Influence of the connection stiffness in a timber composite structure subjected to flexure: **a** no composite action; **b** partial composite action; **c** complete composite action (adapted from [2])

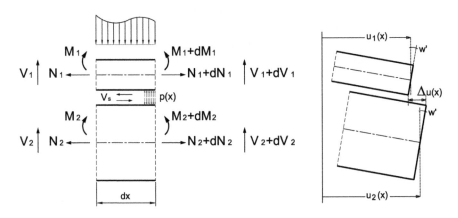

Fig. 3 Equilibrium and deformation in an incremental element

Eurocode 5 [3] adopts the approximate 'gamma method' where an effective flexural stiffness $(EI)_{ef}$ for the composite section is calculated as a function of the stiffness (slip modulus) of the shear connection, under the following hypotheses:

- the girder and the slab are considered as two Bernoulli beams (shear displacement are neglected), which are connected by means of a continuous linear behaving interface (Fig. 3);
- the static configuration is that of a simple supported single span beam with a sinusoid-like distributed load along its length.

For this simple case it is possible to find an analytical solution by means of the differential equation of equilibrium [4] (Fig. 3).The effective moment of inertia can be expressed by a Steiner-like expression (parallel axis theorem) given by:

$$(EJ)_{ef} = \sum_i E_i J_i + \gamma_2 \cdot E_2 A_2 \cdot a_2^2 + \gamma_1 \cdot E_1 A_1 \cdot a_1^2 \tag{1}$$

referring to the upper component as "1" and to the lower component as "2"; being the coefficient γ_i:

$$\gamma_1 = \left[1 + \frac{\pi^2 \cdot E_1 A_1 \cdot s_{eq}}{K \cdot l^2} \right]^{-1} ; \gamma_2 = 1 \tag{2}$$

and a_i the distance between the barycentre of the i-component and that of the composite section. K is the slip modulus of the connection.

If fastener spacing is not uniform along the length of the beam, but varies according to the shear force, an equivalent spacing is assumed:

$$s_{eq} = 0.75 s_e + 0.25 s_m \quad \text{with } s_e < s_m < 4 s_e \tag{3}$$

where s_e is the spacing at the ends of the beam and s_m is the spacing at the middle of the beam.

The parameter K is conventionally estimated for the serviceability limit state (K_{ser}), being used for the ultimate limit state a reduced stiffness $K_u = 2/3\ K_{ser}$ (maintaining the hypothesis of linear behaviour of the connection). According to Eurocode 5 [3], for steel-to-timber or concrete-to-timber connections, K_{ser} should be based on mean value of the density ρ_m for the timber member and has to be multiplied by 2.0.

It can be observed that the effective bending stiffness value $(EJ)_{eff}$ depends not only on the properties of the cross section, but also on the length of the beams [5].

According the expression based on the "gamma" method reported in Eurocode 5 [3], the maximum normal stress in each i-componente should be taken as:

$$\sigma_i \pm \sigma_{m,i} = \frac{N_{i,d}}{A_i} \pm \frac{0.5 \cdot h_i \cdot M_{i,d}}{J_i} = \left(\frac{\gamma_i \cdot E_i \cdot a_i}{(EJ)_{ef}} \pm \frac{0.5 \cdot E_i \cdot h_i}{(EJ)_{ef}} \right) \cdot M_d \tag{4}$$

The maximum shear stresses occur where the normal stresses are zero. The maximum shear stress in the web member is (Fig. 4):

$$\tau_{2,max} = \frac{0.5 \cdot E_2 \cdot h^2}{(EJ)_{ef}} \cdot V \tag{5}$$

The load on a fastener should be taken as:

$$F = \frac{\gamma_1 \cdot E_1 \cdot A_1 \cdot a_1 \cdot s_{eq}}{(EJ)_{ef}} \cdot V \tag{6}$$

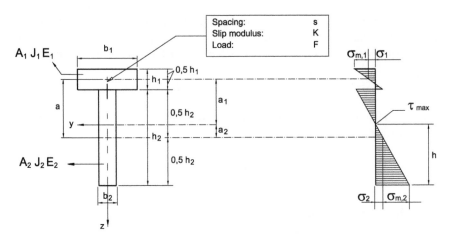

Fig. 4 Internal stress distribution on a timber composite structure subjected to flexural load

The following efficiency parameter η is introduced in order to synthesize the capacity of the connection system to limit the internal slip of the composite section, theoretically ranging from $\eta = 0$ $(EJ_{ef} = EJ_0)$ $to \eta = 1$ $(EJ_{ef} = EJ_\infty)$:

$$\eta = \frac{(EJ)_{ef} - (EJ)_0}{(EJ)_\infty - (EJ)_0} \tag{7}$$

In the refurbishment techniques the parameter η expresses the efficiency of the connection system adopted, and for the common system it practically ranges from 0.4 to 0.7.

2.3 In-Plane Structural Behaviour

Historic city centres of most countries all around the world are comprised of UnReinforced Masonry (URM) buildings with timber diaphragms. While in Mediterranean countries such as Italy or Portugal, traditional floors are made of relatively squat joists with a layer of flooring boards nailed orthogonally to the main elements, in other countries or regions (New Zealand, Australia or Northern Europe) joists tend to be slender. Although each country has its own peculiar typology of timber floors and different refurbishment techniques, it is possible to identify common behaviours and similar problems related to the in-plane stiffness of diaphragms.

Whereas the floor is not satisfactorily connected to the adjacent walls, or the in-plane stiffness is inadequate (Fig. 5), different collapse modes involving overturning of the walls may be observed, as masonry walls by themselves have, generally, insufficient resistance to lateral loads acting out-of-plane [6].

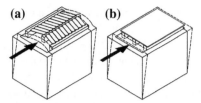

Fig. 5 Diaphragm role in preventing the overturning modes of masonry walls: **a** an inadequate in-plane floor stiffness causes the overturning of the walls perpendicular to the seismic action; **b** a stiff diaphragm allows forces to be transmitted to the walls parallel to the seismic action [6]

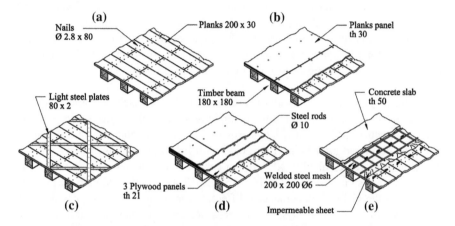

Fig. 6 Different timber floor in-plane shear strengthening techniques: **a** existing simple layer of wood planks on the timber beams; **b** second layer of wood planks crossly arranged to the existing one and fixed by means of steel studs; **c** diagonal bracing of the existing wood planks by means of light steel plates or FRP laminae; **d** three layers of plywood panels glued on the existing wood planks; **e** a stud-connected reinforced concrete slab (all measures in mm) [6]

In order to adequately model the global behaviour of a masonry skeleton, it is fundamental to characterize the in-plane stiffness of horizontal diaphragms, which plays an undeniable key-role in distributing seismic lateral loads to the resisting walls. As a matter of fact it is expected that the more the in-plane stiffness grows, the more the collaboration between systems of piers increases. In addition, earthquake damage has demonstrated that the in-plane stiffness of horizontal diaphragms often influences the out-of-plane wall response, by determining the type of local mechanism occurring (II mode mechanisms).

The poor earthquake-performance of URM buildings has been widely recognized for decades: the recent Christchurch earthquake in New Zealand (22nd February 2011) that severely damaged the URM buildings and killed more than 150 people, demonstrated once more their vulnerability during seismic events [7]. In those cases the floor diaphragms made of light timber elements and the lack of adequate connection to the masonry perimeter influenced substantially the seismic performance of the complete masonry structure. Diaphragm in-plane stiffness of timber floors is therefore a key parameter, whose accurate prediction is essential in masonry buildings' seismic assessment.

The need to increase the in-plane stiffness has induced, in the recent past, some strengthening solutions which recent earthquakes have demonstrated to be inadequate or, in some cases, even unfavourable. As a matter of fact, measures like the substitution of timber floors with concrete ones, the insertion of a concrete curb "inside" the thickness of the masonry walls, could imply a significant self-weight increase and a physical (stiffness) incompatibility with the existing masonry walls, inferring negative repercussions on the global behaviour of the constructions.

Therefore, after the 1997 Umbria-Marche earthquake (Italy), some floor refurbishment techniques have been reconsidered: from 2003 the current Italian standard code on existing buildings bans the possibility to insert a concrete curb in the depth of the existing masonry walls, and suggests new alternative strengthening techniques for the horizontal diaphragms. Some of them, presented in the next figure, have been investigated in a previous campaign at the University of Trento (Fig. 6):

As far as guidelines or standards are concerned, not much is available regarding the in-plane behaviour of timber diaphragms. In Italy, some suggestions for determining the stiffness of a single square sheathed diaphragm (in the original built condition and after being reinforced with FRP laminae) are provided by the National Research Council (CNR [8]) but cannot be considered thorough. In North America, provisions (FEMA 356 [9], ASCE SEI 41/06 [10]) contain formula for calculating the yield displacement of different types of diaphragms, providing tabular values of the equivalent shear stiffness. Recently, in New Zealand, a considerable effort (involving different research groups and laboratories) has been done to improve the understanding of the behaviour of URM buildings, with particular attention to the role played by timber diaphragms on the global building response under seismic action and on the determination of the local failure mechanisms. In 2011 a supplement document of the national standard NZSEE 2006 was proposed [11], containing detailed methods for stiffening and strengthening flexible timber floor and roof diaphragms. The procedure for the diaphragm assessment, in terms of strength and deformation, is based on a series of correction factors that take into account the many different parameters influencing the diaphragm response.

2.4 Design of Possible Intervention: Case Studies

Hereinafter is summarized a case study about the out-of plane reinforcement of the timber floor of Belasi Catle (Italy), as described in [12]. The floor was made of twenty Scot Pine timber joists with variable reciprocal distance ranging from 50–99 cm, span length from 6.0–7.2 m, and average cross section of 150 × 180 mm. The structural analysis of the floor, performed on the assumptions made regarding geometry, conditions at the supports, materials and loads, confirmed that the mechanical properties (modulus of elasticity, MoE, and modulus of rupture, MoR) of the timber members, whose effective resistant cross section was reduced by decay, were incompatible with the level required in the limit state verifications, ultimate and serviceability, according to Eurocode 5 [3].

The choice of coupling a new timber plank to the existing beams with inclined X–crossed screws to strengthen the floor of Belasi Castle was motivated by the outcomes of the preventative evaluation.

The existing beams were coupled with glulam planks strength class GL 24 h according to EN 1194:1999 [13], 80 mm thick. A T-beam compound section, whose web (original beam) and flange (new plank) were made of wood, with deformable connections between flange and web, was thus obtained. Web and flange were separated by a new boarding, with a thickness of 30 mm, which replaced the existing decayed one.

The connection system consisted of self-tapping double thread screws, steel grade 10.9 ($f_{uk} = 1000$ MPa), X-crossed with an angle $\alpha = 45°$.

Assessment of the stiffness of the connection system is crucial for determining the mechanical parameters of the composite structures. The slip modulus k_s (which is equivalent to the K_{ser} instantaneous slip modulus provided by the current limit state standards) of the adopted connection configuration was determined according to the calculation method proposed by Tomasi et al. [14] on the basis of the results from a series of push-out tests for different screw inclinations and joint configurations.

At Belasi Castle, the technique was applied from the floor's upper side (to which access was required), after adequate propping of the joists. In general, existing floorboards can be maintained. However, in the case reported, they were removed because of their poor conditions and replaced with new ones. Glulam planks were laid above and parallel to the existing beams. The two screw threads, with a different pitch but of equal length, were designed so that one penetrated and the other tightened the two members. The screws were inserted without predrilling, thus yielding further advantages in terms of simplicity, ease and quickness of assembly, and improvement of the withdrawal resistance of the connector.

Screws were unequally spaced along the beam, according to the shear force, and they were staggered in order to avoid the occurrence of splits along the fibres.

Tests were performed in the laboratory on one dismantled beam from the Belasi Castle floor (Fig. 7). A series of four point bending tests according to EN 408:2010 [15] were carried out on three consecutive configurations of the dismantled beam (unreinforced, with unconnected boards and reinforced).

The specimen was loaded using an actuator with a 25kN load cell, while the beam's deformation was monitored with four LVDTs connected to an acquisition system.

The effectiveness of the reinforcement was assessed by comparing the stiffness of the timber beam, before and after intervention, considering the testing configurations in Fig. 8.

Figure 9 reports the experimental results, in terms of the load to slip relationship, for the reinforced beam tested until failure in the laboratory. The experimental response is compared with the theoretical curves for both the case of rigid connection system $(EJ)_\infty$ and no connection system $(EJ)_0$.

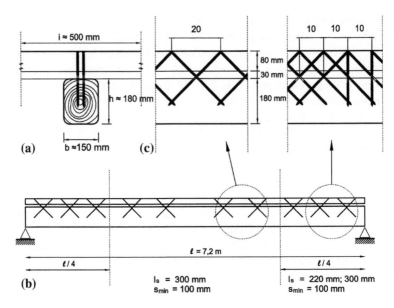

Fig. 7 **a** transversal section and **b** longitudinal section of the composite structure; **c** detail of the geometry of the connection system: *l* beam span, l_s screw length and *s* spacing of the fasteners

Fig. 8 Four-points bending test setup (EN 408:2010 [15])

The calculated value of the experimental efficiency parameter was $\eta = 0.74$. This was determined using the values of the bending stiffness EJ_{ef} (reinforced beam), EJ_0 (beam with unconnected board) derived from the test data carried out on the single elements, and EJ_∞ (rigid connection system) deduced from the experimental data adopting the "parallel axis theorem".

Fig. 9 Load to slip relationship from the 4-point bending test: theoretical curves with *1* infinitive stiffness of connection ($\mu = 100$ %); *2* experimental curve up to failure ($\mu = 74$ %); *3* zero stiffness of connection ($\mu = 0$ %)

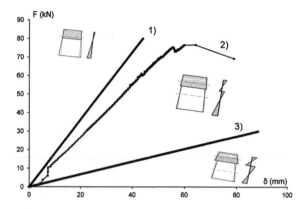

3 Traditional Timber Roofs

3.1 Historic Evolution

Roofs represent the most traditional timber structures with an extraordinary importance in the construction progress. The use of these structures is old as mankind. In fact, the first timber roofs were used by primitive men as protection against the atmospheric agents, and since then, its presence in Architecture is permanent. Although the low robustness of those structures hindered that few examples survived, there are proofs that in Ancient Greece (fourth century BC) a well-defined technique of construction of inclined timber roofs existed. However, these roofs structure did not use trusses but were formed by simply supported rafters.

The timber truss structural configuration is based on the Roman techniques of the fourth century, developed when Christian churches were covered by king-post trusses. It is believed that the most ancient example of timber trusses, built by architect Stephen of Aila in the sixth century, is located at the St. Catherine church in Mount Sinai. These timber trusses are formed by two rafters connected to a tie beam and one king post, all with the same cross-section and with two struts of minor section, Fig. 10.

This structural example, innovative for its time, represents unfortunately one isolate example. In fact, for centuries different configurations were used, as can

Fig. 10 Timber truss of the St. Catherine church in Mount Sinai [16]

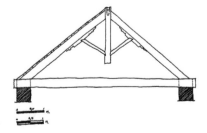

be concluded from the observations of different manuals of architecture. For example, in fourteenth century timber trusses were constructed with precise techniques, but without a clear idea of the structural scheme to use. In the beginning of the Renaissance, 1427–1433, Mariano di Jacopo designs diverse timber trusses, in particular: two with the king post stiffly connected to the tie beam reinforced at mid-span; one without tie beam and another without king post and with the tie beam made of steel. Moreover, the solution of connecting the tie beam to the king post has not precise date. For instance, in 1539 Giovan Battista Da Sangallo draws one truss with three posts and two rafters where the tie beam is not connected to the posts, while Andrea Palladio in 1570 presents diverse trusses with the tie beam connected to the king post. However, for the first time, in "The Four Books on Architecture" by Andrea Palladio published in 1570 [17], the configuration of the truss became defined by the static system that better fits (safety) the roof architecture.

In 1584, Serlio, in his "Seven Books of Architecture" [18], presents a large range of possible structural schemes for timber trusses. However, the selection of the best solution and the design of the different elements is left to the carpenter experience. In Fig. 11 some trusses proposed by [18] are shown. In these examples, the suggestion to increase the number of posts for bigger spans, and the necessity to reinforce the tie beam to bending stress is clear.

The evolution of the truss configuration differs significantly from region to region. While in Mediterranean regions a type of truss with slight pitch is consolidate, in Central and Northern Europe a more complex and Gothic geometry has been developed.

For example, in Westminster Hall (London) in 1394, in order to give greater strength to the framing, a large arched piece of timber is carried across the hall, rising from the bottom of the wall to the centre of the collar beam. The latter is also supported by curved braces rising from the end of the hammerbeam. Those elements are the portions remaining of the tie beam when this is cut through in order to give greater height in the centre (Fig. 12).

In Northern Europe, timber roofs structures with a three-dimensional structural behaviour are common, in particular in roofs with significant pitches. In fact, in this case, lateral wind loads are higher and bracing elements in all three directions are of greater importance. This type of solution is also present in Mediterranean regions, in particular in structurally demanding roofs. In Centre and Northern Europe, the timber truss is part of the structure, having habitable lofts, and tie beam has the additional function of supporting the pavement. In some cases, as

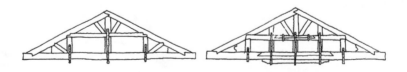

Fig. 11 Examples of timber truss configurations following Serlio [18]

Fig. 12 Representation of the timber roof of Westminster Hall [16]

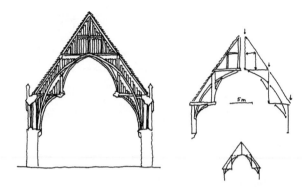

the medieval stave churches (Northwester Europe), timber trusses are integrate in a post and beam construction related to timber framing. The wall frames are filled with vertical planks and, for lateral bracing. Additional wooden brackets are inserted between the rafters, resulting in a very rigid construction.

Until the nineteenth century, timber trusses were designed as simple supported beams or as reticular structures. In these structures, elements, in particular the tie beam, are subjected to considerable bending. Only after this period, structures were designed so that axial forces (compression or tension) were the main stresses. This was achieved using the three-hinged static scheme, detaching the tie beam from the post. After this, construction manuals are based on more consistent knowledge on the structural behaviour of timber trusses and of wood as material.

The range of possible configurations in which it is practical to assemble different timber elements to form a truss is vast. Obviously, the selection depends of many factors: span, loads, roof inclination, etc. Nevertheless, the disposition of the different timber elements, normally, follows the rule that the shorter elements should work in compression and the longer in tension, thus preventing problems of buckling.

3.2 Trusses Systems

King-post truss is certainly the most wide spread timber truss configuration. They represent the simple timber truss geometry that has been used in monumental constructions, like churches and monasteries, and in rural architecture. This truss is characterized for having a horizontal member (tie beam), two principal inclined rafters connected to the tie beam to form the roof, a vertical member at the middle (king post) and two inclined struts, connecting the king post to each principal rafter, Fig. 13. However, some variations of this geometry are known, for small spans or when the purpose of the truss is to support one point load, the struts can be removed. Sometimes the tie beam is removed and replaced by a collar located in an upper level.

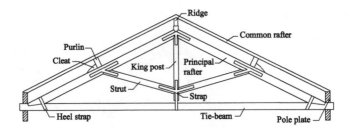

Fig. 13 King-post truss

King-post trusses are the most basic structural configuration of inclined timber roofs and can have three different static behaviours: three-hinged arch, simply supported beam and reticular beam. It is the connection between the king post and the tie beam that determines the overall static behaviour of the truss:

- Three-hinged arch (Fig. 14a)—in theory the best behaviour, only possible when the tie beam is detached or connected without the introduction of bending stiffness. Therefore the tie beam is essentially tensioned, and bending is associated, majorly, with its self-weight;
- Simply supported beam (Fig. 14b)—cannot be considered a real truss. The loads applied will be directly transmitted through the king post to the tie beam, which behaves as a simply support beam with a three point bending diagram;
- Reticular beam (Fig. 14c)—adds to the isostatic triangle, composed by the struts and the tie beam, the king post connected to the tie beam with different possible joints. Therefore, the tie beam becomes an element subjected to tension and bending, in which the bending stress can be important.

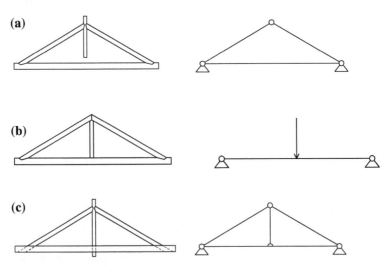

Fig. 14 Possible basic static behaviours of the king-post truss: **a** three-hinged arch; **b** simply supported beam; **c** reticular beam

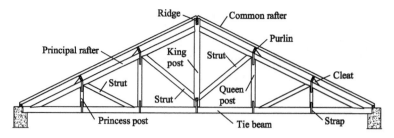

Fig. 15 Howe truss with Princess posts (*Princess-post truss*)

Below the purlins, two oblique elements, the struts, leaving from the base of the king post, support the rafter. Consequently the rafters bending provided by the purlins point loads, is minimized. Moreover, now, the king post is submitted to tension provided by the vertical component of the compression force acting on the struts. In Portugal, in opposition to other Mediterranean countries, trusses are made by rafters incapable to resist to bending moments. In fact, traditional Portuguese timber trusses are characterizes by slender elements, which should only be subjected to axial stresses. In Italy, for example, timber trusses have more robust cross-sections, since significant bending moments in the rafters are common. In some cases, double rafters are employed to increase the resistance of the structure. The difference comes from the fact that in Portugal, when more purlins are needed to support the covering structure, posts are added to the king-post truss originating a more complex truss geometry (Howe truss). For spans over 7-8 m, king-post configuration no longer is acceptable. In fact, more purlins are required and, as already pointed out, the rafters are not prepared to suffer significant bending stresses. As consequence more struts supporting the rafters are included, resulting in trusses as that shown in Fig. 15. To balance the vertical component of the compression force acting in the struts, a post, connecting the tie beam to the rafters, is needed. Suspending the tie beam from the post reduces its bending moments and self-weight bending deformation. The first post is the king, the second and third are called queens and the fourth and fifth are the princess posts. It is common to call the Howe truss by the name of the posts, see Fig. 15.

Howe truss allows longer spans by simply adding few elements to the king-post configuration, keeping slender elements. Usually the spans do not exceed 13 m and the Howe truss more common is the so-called Princess-post truss.

3.3 Traditional Connections

During the timber roofs structure gold period, between the thirteenth and nineteenth centuries, joints between members were made by notches or woodworking joints aiming to prepare the connections between two or more elements. In twentieth century, the necessity to fulfil industrial criteria (large production

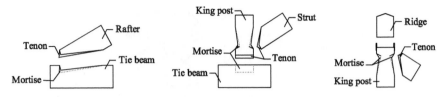

Fig. 16 Examples of step joints of timber trusses presenting tenon and mortise [19]

scale, prefabrication, standard quality, etc.) and the wide spread of metallic elements, such as nails, screws, bolts and straps, the need for carpenter art declined. As a consequence, less carpentry work was needed and metallic elements became essential to correct and improve the contact between the connected members. In any case, the conception and execution of joints has always been the most complex task for the designer. In fact, most frequent pathologies, or failures of timber roofs structures are caused by inefficient joints, wrongly designed and/or wrongly executed. Usually, joints between roof structural members are made by notches, called single step in the case of one notch, double step joint when it has two notches, etc. Stresses are directly transferred by compression and/or friction between the surfaces in contact. In some cases, in particular in the oldest examples, notched joints present tenon and mortise (Fig. 16). Those elements were used to ensure the perfect contact between the connected elements and to prevent the out-plane instability.

The most relevant connections to the timber roof structural behaviour are those of the tie beam to rafters and posts. However, the most stressed connection is the tie beam-rafter joint. Stresses can be so high, that in some cases, in these connections the notched should have double steps. In addition, in terms of durability, they represent one of the most vulnerable points of a timber roof structure.

In the more recent examples, as a cause-effect of the reduction of available carpenters, joints are usually strengthened with metal elements. The use of those elements, not only counteracts out-of-plane actions, but also ensures the adequate contact between connected elements. In addition, the strengthening through metal elements provides an important improvement in the timber truss structural behaviour under seismic actions by ensuring safety under reversal loads.

3.4 Structural Behaviour and Design

For a correct design of any element or structure, it is essential to understand is structural behaviour. Designers must be aware of the material properties and must feel comfortable in the interpretation of the structural system performance. The misunderstanding of the global behaviour of traditional timber trusses systems, in particular, the joints behaviour, can result in unacceptable stresses distribution in the members caused by inappropriate joints strengthening (in terms of stiffness and/or strength). An incorrect intervention on the joints can change the overall behaviour of the system with important consequences in the stresses distributions as it will be pointed out.

In this work, attention will be focused on the interpretation and analysis of the structural behaviour of traditional timber trusses. Works of Bodig [20], Dinwoodie [21] and Giordano [22] about the wood material properties are suggested.

The design of traditional timber trusses is a common topic in timber structures manuals (e.g. Alvarez [23], Pfeil [24] and Piazza [25]). Usually, attention is focused on the overall static behaviour of trusses and on the joints design. The truss timber elements are designed following the Eurocode 5 [3], and some verification rules for the connections, derived from the possible failure modes, can be found in literature [23], [25–27], because traditional carpentry joints are not directly analysed in this code. In the definition of the static model, loads are assumed as point loads applied in the nodes and the joints are modelled as perfect hinges, despite their semi-rigid behaviour even when not strengthened.

3.4.1 Truss Static Behaviour

In a plane structure, like traditional timber trusses, submitted to concentrated loads on the joints, without bending of the members, stress distribution in the structure results directly from its geometry. However, this behaviour can be easily modified if the static model is changed. Assessment of constructed timber trusses shows various differences on their structural model. In fact, despite construction recommendations, intuitively developed over centuries by carpenters, it is common to find examples where they were not taken into account. After an evaluation of the variations in the truss behaviour that can be achieved as result of the model assumed in the design [28], a few considerations are made:

- The application of concentrated loads out of the joints, for example caused by a wrong positioning of the purlins, can compromise the structural global safety;
- The eccentricity of the supports, relatively to the tie beam-rafter connection must be minimized. It is recommended that the reaction force passes by the intersection point of the tie beam and rafter axis;
- The tie beam must be suspended to the posts. Iron strap shall be used, nailing it only in the post, suspending the tie beam with a connection without bending stiffness and preventing out-of-plane deformations;
- When the tie beam-post connection is rigid, the natural frequencies and modal shapes of the truss are clearly modified;
- For non-symmetric loads, e.g., snow, earthquakes and wind, the influence of the joint stiffness become relevant;
- The performance of the tie beam-rafter connection is crucial, not only in consequence of the high level of stresses concentrated there but also because they represent zones where biological deterioration is more frequent;
- The supports must be able to resist horizontal movements. Friction forces are insufficient to resist do horizontal movements caused by earthquakes.

Although the importance of these observations and recommendations, the key for an adequate structural behaviour is the selection of the adequate truss configuration. Indeed, as example, traditional Portuguese timber trusses were thought to behave as plane structures submitted to point loads on the joints causing essentially axial stresses in the members. When the number of posts is less than the total number of purlins, the structural safety is severely reduced. In this case rafters are subjected to bending moments, despite being slender. As a practical rule, the truss configuration must have a number of posts equal to the total number of purlins used to support the covering structure.

3.4.2 Traditional Connections

For a better understanding of the mechanical behaviour of traditional timber connections, it is essential to point out the resistant mechanisms of wood under compression. The wood compression strength differs with the stress direction relatively to the fibres orientation. Parallel to fibres, the maximum resistance is achieved and the failure mode consists on the buckling instability of cellulose. In the normal direction of fibres orientation, the resistance is minimum and the failure mode is associated with the instability of cellulose under radial compression. For compression, intermediate resistance values are obtained, which can be quantified following Hankinson formula and adopted by various standards. Some standards propose different expressions, taking into account the reduction of resistance caused by the possibility of crushing of the fibres in consequence of the difference between the strength of earlywood and latewood. For instance, [26] suggests reducing in 20 % the resistance of compression parallel to the fibres. However, such difference between those two methods became off small account for compression at an angle range of 30–60° to the fibres direction, which represents the most realistic values. In addition, some standards take into account the increase in resistance to compression perpendicular to the fibres resulting from the ratio between the loaded surface and the total surface.

The design of traditional timber connections comprehends essentially the verification of the compression transmitted between the contact surfaces of the connected elements. Behind this simple definition, it is important to point out that the contact surface, through which the forces are transmitted, is normally smaller than

Fig. 17 Main resistant mechanism in a traditional timber connection [25]

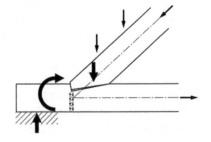

the cross-section of the element and is not orthogonal to the fibres direction of any connected element. In the schematic drawing of Fig. 17, reporting a connection between the tie beam and the rafter, which can be assumed as a general example, the main resistant mechanism is presented.

In the design of traditional timber connections, it is essential to understand the force equilibrium in the joint and, therefore, to identify all critical zones verifying for each one its resistance capacity. Analysing the forces mechanism that occurs inside the joint it is important to draw attention to:

- Existence of not negligible tension perpendicular to the fibres often followed by significant shear stresses;
- Concentration of stresses caused by the particular shape of the notches of each connected element;
- Possibility that cracks begin in the step edges;
- Eccentricity in force transmission.

3.5 Degradation

Nowadays, a considerable number of timber roofs require structural intervention in consequence of natural degradation of the material (aging), improper maintenance of the structure, faulty design or construction, lack of reasonable care in handling of the wood or accidental actions. A correct assessment of the degradation extension and their effects in the overall structural behaviour is urgent. Only after that, accurate restoration actions, which can vary from simple works in the covering structures to large substitution of structural members with joints and elements strengthening, are possible.

The most common degradation processes in timber roofs structures can be divided in three groups. The first, of biological origin, results in a reduction of the resistant cross-section of the elements caused by wood-degrading organisms. This type of degradation is more frequent near the supports because of the presence of high level of moisture, normally related with bad construction details incapable to keep exterior wood dry and accelerate runoff.

The second degradation process is related with structural requirements, in particular, with insufficient load-bearing capacity of structural members, normally translated in significant deformations and, with deficient bracing systems. Changes in the nature of the use of the construction, resulting in different service loads, scarce cross-sections of the structural elements to resist the stress level applied, insufficient elements in the bracing system, inadequate design, or construction of the connections and the presence of natural wood defects are the main causes of this type of anomalies.

Finally, bad structural systems can be responsible for stresses, in an element or in various, that can compromise the local and overall safety of the structure. The most frequent examples of this faulty design of timber roofs are: wrong positioning of the purlins; tie beam-rafter connections eccentric relatively to supports; and incorrect connection between tie beam and posts, in particular with the king post that must suspend the tie beam without introducing bending stiffness. The application of concentrated loads out of the joints, originated by a wrong positioning of the purlins, can compromise the structural global safety of the roof system. Indeed, it is demonstrated by [16] that the consideration of an eccentricity of 20 cm, a value smaller than the ones reported in the preliminary survey undertaken, is enough to compromise the rafters safety of typical Portuguese king-post truss.

3.6 Joints Strengthening

Frequently, restoration works requires the strengthening of timber structures. In the case of timber roofs structures, this strengthening normally involved the connections between the truss structural elements. Joints strengthening can be done in different forms: from simple replacement or addition of fasteners, to the use of metal plates, glued composites, or even full injection with fluid adhesives. Each solution has unique consequences in terms of the joint final strength, stiffness and ductility.

A series of monotonic and cyclic tests on unstrengthened specimens were performed in order study the primary behaviour characteristics of the connection, as well as its sensitivity to a few parameters. Subsequently, connections strengthened with basic metal devices were tested under monotonic and cyclic loading. The purpose of these tests was to uncover any advantages and deficiencies in the behaviour of the joint and of the device itself, as well as to determine a need for different types of strengthening [27].

The four basic types of intervention considered in this study are modern implementations of traditional strengthening techniques: the binding strip, the internal bolt, the stirrup and the tension ties (Fig. 18).

Four strengthening schemes, representing modern implementations of traditional techniques using metal elements, were studied. Three of them, stirrup, bolt and tension ties were analysed for two values of the skew angle, 30 and 60°, while the fourth, the rigid binding strip, was only evaluate in the case of a 30° skew angle. This last technique was abandoned because of the difficulties that it practical implementation and use revealed.

Comparing the experimental force–displacement curves, obtained for the unstrengthened and strengthened connections (Fig. 19 and 20) it is recognize that all the strengthening schemes improve the behaviour of the originals connections. In the case of connections with 30° skew angle (see Fig. 19), the strengthening techniques analysed increases the stiffness, in particular, in the positive loading direction and the maximum resistance for both directions.

The elasto-plastic behaviour with limited ductility evidenced by the unstrengthened connections under negative loading direction is substituted by full nonlinear curves exhibiting high ductility in the strengthened connections. Comparing the strengthening techniques evaluated, the less efficient (maximum force and stiffness), is the tension

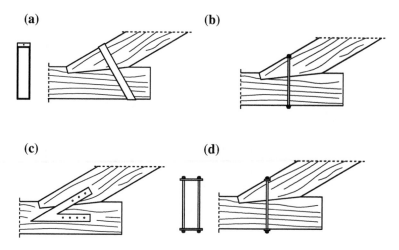

Fig. 18 Traditional strengthening techniques evaluated: **a** binding strip; **b** internal bolt; **c** stirrup; **d** tension ties

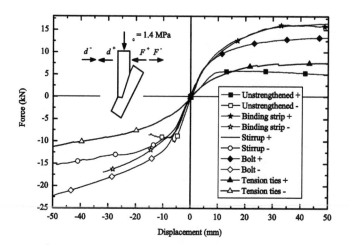

Fig. 19 Force-displacement curves for unstrengthened and strengthened connections with a 30° skew angle under monotonic loading

ties. Connections strengthened with stirrups and binding strip attained the same range of maximum force; however, this later scheme has a lower ductility capacity under negative forces. In particular, the maximum resistance for the strengthened connections with stirrups and internal bolts is achieved near the end of the test. However, in the strengthened connections with binding strip, when the tests were interrupted, the force value had already decreased. Therefore, between the internal bolt and the binding strip, the first one is more efficient in terms of ductility capacity with the goal to assure a better seismic behaviour of the connection. The effect of the strengthening schemes in the negative loading directions of the monotonic tests is obvious: the increase of the

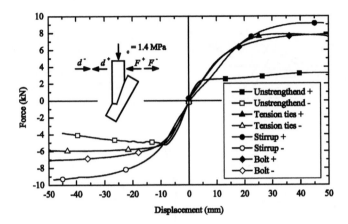

Fig. 20 Force-displacement curves for unstrengthened and strengthened connections with a 60°
skew angle under monotonic loading

maximum force and of the ductility capacity. The benefits to the stiffness are not significant (the stiffness exhibit by the tension ties technique is even lower). However, the brittle behaviour exhibited by unstrengthened connections under negative forces disappears
in all strengthened specimens. Therefore, the main profit of adding a metal device to the
joints is the improvement of ductility, with clear advantages for seismic behaviour. Only
the binding strip showed limitations in terms of maximum displacement under negative
forces.

Extending the comparison in the case of connections with a 60° skew angle, the
main conclusions are kept (see Fig. 20). The more evident benefit, which is normally
the main goal of a strengthening intervention, is the improvement of the connection
ductility. In particular, the original behaviour of the unstrengthened connections under
monotonic loading in the negative direction, characterized by limited ductility, is substituted, when strengthened, by a full non-linear behaviour with considerable ductility.

The strengthening techniques evaluated do not cause changes in the stiffness,
but result in an increase of the maximum force achieved in the tests in both directions, in particular, under monotonic loading in positive direction. Between the
strengthening techniques studied, the differences in the force–displacement curves
are much more visible when under monotonic loading in the negative direction.
In this case, the use of stirrups is the most efficient technique (maximum load
achieved), while the response of the tension ties and the bolts are quite similar.

The improvements in the connections behaviour under monotonic loading, provided by the evaluated strengthening techniques are highlighted by the response
under cyclic loading (for more details, see Ref. [29]). Without any strengthening
device, the connection is not able to prevent the failure caused by load reversals
(detachment of the connected elements) even when the rafter compression stress is
augmented (from 1.4–2.5 MPa).

The unstrengthened connections showed limited capability to dissipate energy,
in particular for the 30° skew angle (hysteretic equivalent viscous damping ratio

equal to 2.5 and 14.2 % for 30 and 60° skew angles, respectively). All strengthening techniques adopted were efficient in the improvement of the hysteretic behaviour of the connections. Hysteretic equivalent viscous damping ratios evaluated from tests results are considerable (higher than 10 % for a 30° skew angle and higher than 25 % for 60° skew angle—for more details, see Ref. [29]). The best results under cyclic loading are achieved when the stirrup, bolt or tension ties techniques are used. The binding strip provides the strongest connections, but the equivalent damping ratio is nearly half of the values presented by the others strengthened connections.

4 Final Discussion

Traditional old buildings, in most regions, comprise timber floors and roofs. Therefore, interventions in such buildings, with or without particular historic and/ or cultural values, require knowledge on timber structures.

Any intervention in an existing timber structure starts with the safety assessment involving the evaluation of the mechanical properties of the material and the geometrical characteristics of the elements and joints, as well as of the interaction with the supporting elements and of the participation to the global behaviour of the construction. Non-destructive methods have been used to assess the condition and state of degradation of timber structures, despite being recognized that only with destructive tests the correct strength of a timber element can be determined. Wood is a natural cellular material sensitive to aging and with important creep behaviour, which properties depends on defects and moisture content.

In other hand, connections are normally the weak element in a timber structure. High stresses concentration can exist while reliable and practical models of the joint of traditional timber structures are missing.

In consequence of those difficulties, timber structures have been substituted by new ones made of concrete or steel. Technicians have difficulties to characterize the material, to understand the structure and to define an accurate strengthening intervention.

This chapter aims to fulfil this gap of knowledge by presenting the state-of-the-art on the strengthening of timber floors and roofs. First, the most common timber systems are described and analysed. Then, possible strengthening techniques are presented and discussed based on past experience, scientific knowledge and tests results, underlining the importance and viability of such procedures.

References

1. ICOMOS International Wood Committee. Principles for the preservation of historic timber buildings. In: 14th Symposium in Pátzcuaro, Michoacán, Mexico, 10–14 Nov 2003
2. Ballerini, M., Crocetti, R., Piazza, M.: An experimental investigation on notched connections for timber—concrete composite structures. In: Proceedings of the 'WCTE 2002 World Conference on Timber Engineering', Shah Alam, Selangor Darul Ehsan, Malaysia (CD-Rom)

3. EN 1995-1-1: 2004/A1:2008. Design of timber structures. Part 1- 1: general—common rules and rules for buildings. European Committee for Standardization, Brussels
4. Kreuzinger H.: Mechanically jointed beams and columns, Lecture B11 Timber Engineering, Step 1, STEP/Eurofortech, Centrum Hout, Almere (1995)
5. Thelandersson, S., Larsen, H.J.: Timber Engineering. Wiley, London (2003)
6. Piazza, M., Tomasi, R., Baldessari, C., Acler, E.: Behaviour of refurbished timber floors characterized by different inplane stiffness. Structural Analysis of Historic Construction: Preserving Safety and Significance. In: VI International Conference on Structural Analysis of Historic Construction, Bath, 2–4 July 2008, vol. 2, pp. 843–850. Taylor & Francis, London, 2008, ISBN: 978-0-415-48107-6
7. Wilson, A.W., Quenneville, P.J.H., Ingham, J.M.: Assessment of timber floor diaphragms in historic unreinforced masonry buildings. In: International Conference on Structural Health Assessment of Timber Structures, Lisbon, Portugal, 16–17 June 2011
8. CNT-DT-201. Istruzioni per Interventi di Consolidamento Statico di Strutture Lignee mediante l'utilizzo di Compositi Fibrorinforzati. National Council of Research (2005)
9. ATC. Prestandard and Commentary for the Seismic Rehabilitation of Buildings. FEMA-356, Federal Emergency Management Agency, Washington, D.C. (2000)
10. Standards ASCE/SEI 41/06. Seismic Rehabilitation of Existing Buildings. American Society of Civil Engineering (2007)
11. NZSEE: Assessment and Improvement of Unreinforced Masonry Buildings for Earthquake Resistance. New Zealand Society for Earthquake Engineering, Wellington (2011)
12. Riggio, M., Tomasi, R., Piazza, M.: Refurbishment of a traditional timber floor with a reversible technique: the importance of the investigation campaign for the design and the control of the intervention. Int. J. Archit. Heritage, **2012**, 1–24 (2012). doi: 10.1080/15583058.2012.670364
13. EN 1194 (1999) Timber Structures. Glued Laminated Timber. Strength Classes and Determination of Characteristic Values. European Committee for Standardization, Brussels
14. Tomasi, R., Crosatti, A., Piazza, M.: Theoretical and experimental analysis of timber-to-timber joints connected with inclined screws. Constr. Build. Mater. **24**, 1560–1571 (2010)
15. EN 408:2010. Timber Structures. Structural Timber and Glued Laminated Timber. Determination of Some Physical and Mechanical Properties. European Committee for Standardization, Brussels
16. Tampone, G.: Il restauro delle strutture di legno. Ulrico Hoepli, Milan (1996)
17. Palladio, A.: I quatro libri dell'architettura. In Venetia: Apresso Dominico de Franceschi. Ulrico Hoepli, Milan (1945)
18. Serlio, S.: Il sette libri dell'architettura. Venezia. Re-printed Forni Editions, Bologna (1584)
19. Appleton, J.: Reabilitação de edifícios antigos. Patologias e tecnologias da intervenção, Orion, Lisbon (2003)
20. Bodig, J., Jayne, B.: Mechanics of wood and wood composites. Van Nostrand, New York (1982)
21. Dinwoodie, J.: Wood: Nature's Cellular Polimeric Fibre-Composite. The Institute of Materials, London (1989)
22. Giordano, G.: Tecnica delle construzioni in legno. Ulrico Hoepli, Milan (1999)
23. Alvarez, R., Martitegui, F., Calleja, J.: Estructuras de madera—Diseño y cálculo (1996)
24. Pfeil, W., Pfeil, M.: Estruturas de madeira. LTC (2003)
25. Piazza, M., Tomasi, R., Modena, R.: Strutture in legno. Materiale, calcolo e progetto secondo le nuove normative europee. Ulrico Hoepli, Milan (2005)
26. SIA 265. Constructions en Bois. Swiss society of engineers and architects (sia) norm (2003)
27. Branco, J.M.: Influence of the joints stiffness in the monotonic and cyclic behaviour of traditional timber trusses. Assessment of the Efficacy of Different Strengthening Techniques. PhD thesis, University of Minho and University of Trento, (2005)
28. Branco, J., Cruz, P., Piazza, M., Varum, H.: Behaviour of traditional Portuguese timber roof structures. In: Proceedings of World Conference on Timber Engineering, Portland (2006)
29. Branco, J.M., Piazza, M., Cruz, P.J.S.: Experimental evaluation of different strengthening techniques of traditional timber connections. Eng. Struct. **33** (8), 2259–2270. http://hdl.handle.net/1822/13592

Advancements in Retrofitting Reinforced Concrete Structures by the Use of CFRP Materials

José Sena-Cruz, Joaquim Barros, Mário Coelho and Carlo Pellegrino

Abstract Reinforced concrete is nowadays one of the major structural materials used in infrastructures' construction. As the years passed since the beginning of its use, technology has changed and today it is known that a large number of constructions do not verify some of the most recent safety requirements in addition to the durability problems that several structures face. This calls for the need for corrections and/or strengthening activities. Fibre Reinforced Polymer materials (FRP) have been used in Civil Engineering applications due to: lightweight, good mechanical properties (stiffness and strength), corrosion-resistant, good fatigue behaviour, easy application and virtually endless variety of shapes available. Recent advancements in retrofitting reinforced concrete structures by the use of CFRP materials promoted by the authors of the present chapter are herein summarized. Firstly a brief overview on the most common properties of old reinforced concrete constructions those who might need for interventions, is presented. Then, the most common repairing and/or strengthening techniques are addressed. Finally, a deeper look is taken on the new repairing and/or strengthening techniques using FRP reinforcements.

Keywords Reinforced concrete · Old constructions · Strengthening · FRP

J. Sena-Cruz (✉) · J. Barros · M. Coelho
ISISE, Civil Engineering Department, School of Engineering, University of Minho,
4800-058 Guimarães, Portugal
e-mail: jsena@civil.uminho.pt

J. Barros
e-mail: barros@civil.uminho.pt

M. Coelho
e-mail: mcoelho@civil.uminho.pt

C. Pellegrino
Depertment of Civil, Environmental and Architectural Engineering, University of Padova,
Via Marzolo 9, 35131 Padova, Italy
e-mail: carlo.pellegrino@unipd.it

A. Costa et al. (eds.), *Structural Rehabilitation of Old Buildings*,
Building Pathology and Rehabilitation 2, DOI: 10.1007/978-3-642-39686-1_9,
© Springer-Verlag Berlin Heidelberg 2014

1 Introduction

Reinforced concrete (RC) can be simply described as the combination of concrete and a reinforcement element, typically steel. The use of RC in building construction remounts to the beginning of the twentieth century. Nowadays it is probably the most used construction material.

As its use has more than 100 years, the technology around the application of RC is very mature and robust nowadays. Nevertheless, the evolution of RC technology was made at the same time as RC construction. This means that large percentage of existing RC structures are under-designed or designed under outdated regulations or construction practices and need for intervention. But the motivation for the intervention can be related to other causes. Wrong or inaccurate execution, deterioration and damage due to time and environmental factors, or simply the need for changing the use type/load.

In attempts to solve these problems several repair and strengthening techniques have been proposed. Over the last two decades, extensive research has been developed on the strengthening of RC structures with fibre reinforced polymer (FRP) materials. The high stiffness and tensile strength, low weight, easy installation procedures, high durability (no corrosion), electromagnetic permeability and practically unlimited availability in terms of geometry and size are the main advantages of these composites.

In this chapter an overview about recent advances on the use of FRP materials for the strengthening of RC beams, columns and beam-column joints is presented.

2 Beams

Regardless of the causes that lead to the need for repair or strengthen reinforced concrete beams, up to the 90s this has been done using traditional construction materials and techniques. Those include mainly the jacketing of the element to be strengthened with steel plates or with new RC overlays. In the case of steel plates, epoxy adhesives, mechanical anchors or a combination of both, is normally used to assure stress transfer from the RC beam to the steel plates. If the strengthening is performed with RC overlays, correct surface preparation, steel reinforcement detail and selection of new RC grade (compatibility between new and old RC), are enough to assure composite behaviour of the strengthened RC beam.

These procedures can be used in a global strengthening perspective or be oriented to just solve specific problems (such as shear or flexure, for example). Figure 1 illustrates the referred traditional techniques.

These traditional techniques present a major drawback related to the increase of the self-weight of the strengthened element. In addition, cross-section changes will affect not only the aesthetics of the element but also the room space since the strengthened element will have a larger cross-section.

(a) 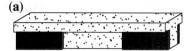   **(b)**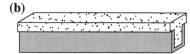

Fig. 1 Examples of traditional strengthening techniques for RC beams: **a** localized intervention with steel plates glued with epoxy to solve shear problem; **b** global intervention with RC overlay to increase load carrying capacity

In the last decades, traditional techniques have been gradually replaced by the use of fibre reinforced polymers. The present section presents some strengthening alternatives for RC beams by the use of FRP.

2.1 Strengthening Procedures

The strengthening of RC beams with FRP was adapted from the concept of jacketing, previously referred. The major difference is related to the material versatility, which allows almost infinite variety of strengthening configurations.

The most common techniques for applying FRP are, in general, based on the use of unidirectional FRP through the: (i) application of fabrics (in situ cured systems) or laminates (pre-cured systems) glued externally on the surface of the element to strengthen (EBR—Externally Bonded Reinforcement); (ii) insertion of laminates (or rods) into grooves opened on the concrete cover (NSM—Near-surface Mounted) [1, 2]. Structural epoxy adhesives are the most used to fix the FRP to concrete. The strengthening performance of these techniques depends significantly on the resistance of the concrete cover, which is normally the most degraded concrete region in the structure due to its greater exposure to environment conditions. As a result, premature failure of FRP reinforcement can occur and, generally, the full mechanical capacity of the FRP is not mobilized, mainly when the EBR technique is adopted. To avoid this premature failure complements have been applied to the aforementioned strengthening techniques, such as the application of anchor systems composed of steel plates bolted in the ends of the FRP (combination with traditional techniques), the use of strapping with FRP fabric or the use of FRP anchor spikes. In addition to the stress concentration that these localized interventions introduce in the elements to strengthen, they require differentiated and time consuming tasks that can compromise the competitiveness of these techniques.

More recently, some FRP-based alternatives for structural strengthening have been proposed. The mechanically fastened fibre reinforced polymer (MF-FRP) technique has been introduced to strengthen concrete structures, and is mainly characterized by the use of hybrid (carbon and glass) FRP strips that are mechanically fixed to concrete using closely spaced fastening pins and, if necessary, anchors at the ends of the strip are applied to prevent debonding [3]. This technique has already been used in some applications, e.g. reinforced concrete, wood and masonry structures, and several benefits have been pointed out, namely, quick installation with relatively simple hand tools, no need for special labour skills, no surface preparation required, and the strengthened structure can be immediately used after the installation of the FRP. From experimental tests an increase of up to 50 % of the load carrying capacity was observed in some cases, when compared with the reference structure. Additionally, the occurrence of a more ductile failure mode for the FRP system is referred [4, 5]. Nevertheless, some notable disadvantages of this technique have been reported, including greater initial cracking induced by the impact of fasteners in high-strength concrete, and less-effective stress transfer between the FRP and concrete due to the discrete attachment points [6].

Based on the MF-FRP technique, the mechanically fastened and externally bonded reinforcement technique (MF-EBR) has been proposed [7, 8]. The MF-EBR combines the fasteners from the MF-FRP technique and the externally glued properties from the EBR. In addition, all the anchors can be pre-stressed which favours the high levels of efficacy that can be observed when this strategy is applied.

In the following paragraphs EBR, NSM and MF-EBR techniques for strengthening RC beams will be presented from the authors' personal experience.

2.1.1 Externally Bonded Reinforcement

Among the strengthening techniques with FRP, EBR was almost naturally the first system to become popular. It is a direct transposition of the principals of jacketing with steel plates. The study of this technique is quite advanced by now and several design codes are already available worldwide, e.g. [1, 2, 9, 10].

The main steps necessary to successfully apply this technique are: (i) roughen the surface where FRP will be applied; (ii) clean the surface and then apply on it epoxy adhesive; (iii) place the FRP on the epoxy layer and press it to create a uniform thickness of 1–2 mm.

Since the performance of this technique is directly influenced by the bond performance between the FRP and cover concrete, some special cares are recommended in order to improve the adherence at that interface. On the other hand, the cover concrete of RC beams that need to be strengthened has normally weaker properties than those from the inner concrete. To take into account these two aspects, it is recommended to completely remove all the surface concrete in bad conditions and to use a primer to regularize the final surface. Nowadays, there are already some complete systems (epoxy adhesive + primer + FRP) that were made compatible and are recommended to use instead of untested combinations of these materials.

Other important detail related to the application of this technique concerns the scenarios of total jacketing. This strategy is normally applied with FRP sheets, which are normally unidirectional orthotropic with low strength in the direction transverse to the fibres. In these cases, beam's edges should be rounded or the strengthening can be jeopardized.

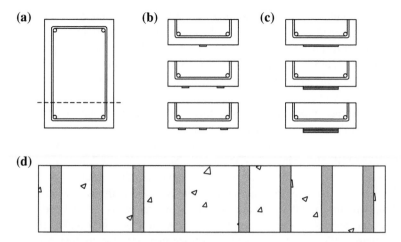

Fig. 2 Configurations used in the strengthening according to the EBR technique: **a** reference specimen; **b** flexural strengthening using different number of CFRP laminates; **c** flexural strengthening using different number of CFRP sheets; **d** shear strengthening using U-shaped CFRP sheets (example of one of the four configurations with different spacing between the CFRP sheets and with different beam depth)

When compared to traditional techniques the major advantages of using EBR are related to the light weight and high strength of the material used. Since FRP are much lighter than steel or concrete, the increase in self-weight in the strengthened beam will be marginal. In addition, as FRP present higher strength, small amounts of strengthening material will achieve the same strengthening level as larger quantities of concrete or steel. This will lead to cost savings in terms of materials and also to minor interference in room space and final appearance of the strengthened element.

Barros et al. [11] conducted a wide experimental bending tests campaign to access the effectiveness of the EBR technique for the flexural and shear strengthening both with unidirectional carbon FRP (CFRP) wet lay-up sheets and precured laminates. The influence of the longitudinal equivalent reinforcement ratio in flexural strengthening and the influence of the longitudinal steel reinforcement ratio and beam depth in shear strengthening, were analysed. Figure 2 shows the configurations tested which were chosen in order to impose a similar $A_f E_f / A_{sl} E_s$ ratio within the beams of the experimental program, where A_f and E_f are the CFRP cross sectional area and the Young's Modulus of the CFRP systems, and A_{sl} and E_s are the cross sectional area and the Young's Modulus of the longitudinal tensile steel bars.

The results, for flexural strengthening showed that, in terms of service load, all the strengthened beams presented increments of more than 40 % when compared to the reference beam with exception for the beam with one wet lay-up CFRP sheet which doubled that value achieving 82 %. In terms of maximum load even though all strengthened beams presented increases when compared to reference one, those were not as regular as service load increases. Their values were 5, 72 and 20 % for beams strengthened with one, two and three laminates, respectively, and 17, 64 and 22 per cent for beams strengthened with one, two and three sheets, respectively.

The most common observed failure modes were debonding of the FRP and delamination of the concrete cover, sometimes after yielding of the longitudinal steel reinforcement.

These findings emphasize the great potential of this technique for strengthening RC beams in flexure. As referred before, one of the biggest problems associated to this technique is related to the premature debonding of the FRP. However, as this work showed, the meaning of premature in this context is related to the ultimate tensile strength of the FRP which is much bigger than that observed at the time of failure. But this happens for load values quite high (sometimes after internal steel yielding) which reveal that the so called premature failure is not that bad.

From the results of shear strengthening tests, it was found that EBR strategy with U shape CFRP sheets achieved an average increase of more than 50 and 70 % in terms of load carrying capacity and maximum load deflection, respectively, when compared to the reference beam which had no shear reinforcement. Other available works confirm these results and emphasize the suitability of EBR technique for shear strengthening of RC beams [12, 13].

2.1.2 Near-Surface Mounted

NSM technique with FRP appeared later compared to EBR one, so the study related to the former is not as advanced as the study about EBR. Nevertheless, some code regulations initially made for the EBR technique have been adapted and nowadays also address NSM [1, 2].

As referred before NSM technique consists on the insertion of laminates or rods into grooves (or slits, where the width is relatively small), thus, the main steps to apply this technique are: (i) cut grooves for inserting the FRP on the concrete cover of the beam; (ii) clean the grooves and fill them with epoxy adhesive; (iii) insert the FRP into the grooves and regularize the outer surface.

When compared to EBR, NSM technique presents the following main advantages: simpler application procedure (no need for surface preparation), better protection against any external damage (since the FRP is inserted into the concrete), minimal changes on the RC beam appearance and high strengthening efficacy (both in flexural and shear strengthening) as will be shown in the following paragraphs (see also Fig. 3).

In the work performed by Barros et al. [11], previously referred, the strengthening technique was also analysed. All the details that had been analysed for the EBR technique were also studied for NSM. In the shear tests, an additional parameter was analysed for NSM technique: the influence of the inclination of the CFRP laminates.

The results, for flexural strengthening showed that, in terms of service load, NSM beams behaved like EBR ones. In terms of maximum load all NSM

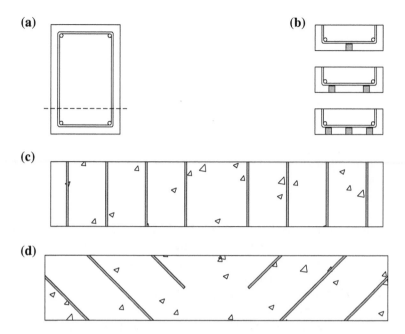

Fig. 3 Configurations used in the strengthening according to the NSM technique: **a** reference specimen; **b** flexural strengthening using different number of CFRP laminates; **c** shear strengthening using CFRP laminates at 90°; **d** shear strengthening using CFRP laminates at 45°. Note: **c** and **d** were tested for four different configurations with different spacing between the CFRP laminates and with different beam depth

strengthened beams behaved better then correspondent EBR's, presenting increases in maximum load of 118, 92 and 35 % for beams strengthened with one, two and three laminates, respectively, when compared to reference beam. In all NSM strengthened beams, the failure occurred by delamination of concrete cover after yielding of the internal steel reinforcement which, once again, highlights the real meaning of premature failure in this context.

In order to compare EBR and NSM techniques, Fig. 4 presents elucidative charts. In Fig. 4a can be seen that, for every level of equivalent reinforcement ratio, NSM technique is more efficient than EBR even though that different is becoming lesser as the equivalent reinforcement ratio increase. In Fig. 4b representative force-displacement relationships are presented for the three strengthening strategies analysed, EBR with laminates (EBR_L), EBR with sheets (EBR_M), and NSM, as well as for reference beam. That last one presents the regular tri-linear behaviour characteristic of RC beams, being the first slope change related to the occurrence of the first crack and the second slope change related to internal steel yielding. Behaviour of strengthened beams is more or less equal to the reference beam until first crack occurs even though it happens for higher load values. After this point, strengthened beams behave stiffer than reference one because of the contribution of the FRP. The yielding initiation of the internal

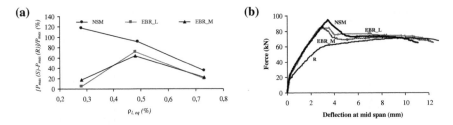

Fig. 4 Comparison between the flexural results obtained with EBR and NSM techniques: **a** strengthening efficacy *versus* longitudinal equivalent reinforcement ratio; **b** Force-deflection relationships

steel reinforcement of the strengthened beams always occurs for a slightly higher displacement level, when compared with the reference beams, which results in a small reduction in the stiffness of the response. Then the beams continued to increase load until the strengthening system failed and the structural response of the beam become similar to that of a regular RC beam.

The behaviour described above for the tested beams is representative of that observed in regular RC beams and RC beams strengthened according to those three techniques, regardless to the strengthening level, internal steel reinforcement or even observed failure mode.

From the shear strengthening tests [11], it was found that NSM technique was more effective, easier and faster to apply then EBR. This efficacy was not only in terms of the beam load carrying capacity but also in terms of deformation capacity at beam failure, presenting average values of more than 80 and 300 % of increase, respectively, when compared to the reference beam which had no shear reinforcement.

Failure modes of the beams strengthened by the NSM technique were not as brittle as those observed in the beams strengthened by the EBR technique.

Other interesting aspect is that when the beam depth increased, laminates at 45° became more effective than vertical laminates.

2.1.3 Mechanically Fastened and Externally Bounded Reinforcement Technique

MF-EBR technique is, at this moment, a recent academic proposal so there are no design codes that can be mentioned at this stage. However, since this technique can be seen as an EBR technique upgraded with discrete anchoring systems and using the principle of superposition effects the appearance of a design code should be straightforward in the next years.

In terms of application procedure the following steps must be done after those referred to EBR have already been performed (see Fig. 5—steps 1 and 2): (3) after

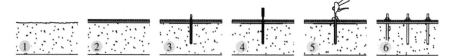

Fig. 5 MF-EBR strengthening system

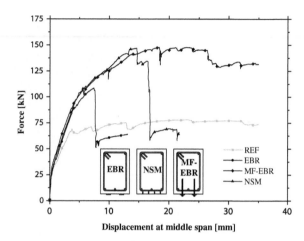

Fig. 6 Force-deflection relationships for tested beams under monotonic loading

the epoxy adhesive cures drill the holes for the anchors directly through the FRP; (4–6) clean the holes, fill them with chemical adhesive and insert the anchors; (7) after the chemical adhesive cures apply the desired pre-stress on the anchors.

This technique uses multidirectional laminates (MDL) in order to be possible the anchorage of these materials with anchors. Sena-Cruz et al. [14] conducted a set of experimental tests where RC beams flexural strengthened by EBR, NSM and MF-EBR techniques were compared for different load actions, namely, monotonic and fatigue loading. Figure 6 presents the appearance of the strengthened beams which had identical amounts of CFRP reinforcement.

The monotonic tests results showed increases in load carrying capacity of almost 40 and more than 80 % for EBR and NSM/MF-EBR techniques, respectively, when compared to the reference beam. Even though MF-EBR achieved the same load carrying capacity that NSM the remarkable advantage of the former was the level of ductility revealed. Figure 6 also presents the force-displacement relationship for all monotonic tests where that important aspect can be verified. Characteristic failure modes were observed in the EBR (debonding) and NSM (concrete cover rip-off) specimens. For the case of the MF-EBR specimen the presence of the fasteners has allowed the development of the highest ductility, justified by the bearing failure of the multidirectional laminates at the anchors vicinity Fig. 7.

Fig. 7 Failure mode of the
MDL-CFRP specimen:
a lateral view; **b** MDL-CFRP
bearing failure

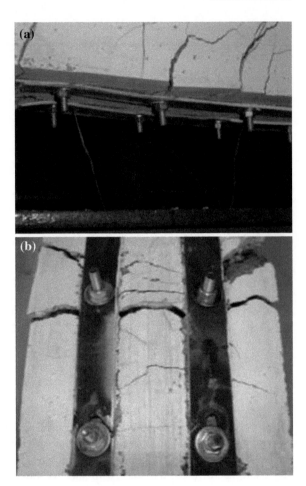

3 Columns

Columns of reinforced concrete framed structures are the most vulnerable
elements since their failure leads to the collapse of the structure. The EBR technique has been used, mainly to increase the axial load carrying capacity, the
ductility and the shear resistance of RC columns of circular or rectangular cross
section of edges of similar dimensions, by applying a FRP jacket around the column [15–17]. The EBR technique is also used to avoid buckling of longitudinal steel bars and to enhance the bond behaviour of the starter bars in the lap
splice zones. However this technique is not especially suitable for the increase
of the flexural capacity of RC elements, since the fibres of the FRP systems have
a predominant orientation orthogonal to the axis of the column. In an attempt

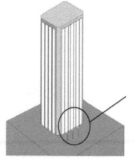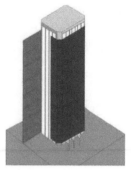

Fig. 8 Technique based on the use of anchored EBR-CFRP laminates and confined by CFRP wet lay-up sheets (adapted from [18])

of also provide flexural strengthening, some authors have used CFRP plates applied according to the EBR technique, and these laminates were externally covered by a CFRP jacket (see Fig. 8) [18]. The laminates were provided with special anchorage configuration with the aim of assuring continuity of the flexural strengthening condition from the column to the adjacent elements (foundations or beams). However, the competitiveness of this technique is arguable since the flexural strengthening improvement provided by these anchorages may not justify their extra-costs.

Combining the NSM for the flexural strengthening with the EBR for the concrete confinement seems to be an effective technique to increase, not only the load carrying capacity for vertical and lateral loads, but also the energy dissipation performance. This technique is described in the following sections.

3.1 NSM Technique for the Flexural Strengthening of RC Columns

To assess the effectiveness of the NSM strengthening technique for concrete columns submitted to static axial compression load and cyclic horizontal increasing load, Barros et al. [19] carried out the three series of reinforced concrete columns indicated in Table 1. Series NON consisted of non-strengthened columns, series PRE was composed of concrete columns strengthened with CFRP laminate strips before testing and series POS consisted of previously tested columns of series NON that were post-strengthened with CFRP. The designation Pnm_s was attributed to tests of series (NON, PRE or POS), where n represents the diameter of the longitudinal steel bars, in mm, (10 or 12), and m can be a or b, since there are two specimens for each series of distinct longitudinal steel reinforcement ratio.

Table 1 Denominations for the RC column specimens

Longitudinal steel reinforcement	Series		
	NON[a]	PRE[b]	POS[c]
4ϕ10	P10a_NON	P10a_PRE	P10a_POS
($A_{sl} = 314$ mm^2)	P10b_NON	P10b_PRE	P10b_POS
4ϕ12	P12a_NON	P12a_PRE	P12a_POS
($A_{sl} = 452$ mm^2)	P12b_NON	P12b_PRE	P12b_POS

[a]Nonstrengthened
[b]Strengthened before testing
[c]Columns of NON series after have been tested and strengthened

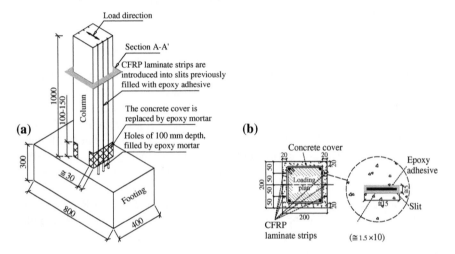

Fig. 9 Strengthening technique in for the RC column: **a** general view; **b** Section A-A' detail (dimensions in mm)

The strengthening technique application procedure was composed of the following steps (see Fig. 9):

- Using a diamond cutter, slits of 4–5 mm width and 12–15 mm depth were cut on the concrete surfaces to be subjected to tensile stresses;
- To anchor the CFRP laminate strips to the footing and to maintain their vertical position, concrete cover of a region having a height of 100–150 mm from the bottom of the column (denoted here by "non-linear hinge region") was removed and, in the alignment of the slits, perforations of about 100 mm depth were made in the footing to anchor the CFRP laminate strips;
- Epoxy adhesive is produced according to supplier recommendations and the slits were filled with it after they were cleaned by compressed air;
- The epoxy adhesive was applied on the faces of the laminate strips of CFRP after they were cleaned by acetone;

Fig. 10 Test set-up
(dimensions in mm)

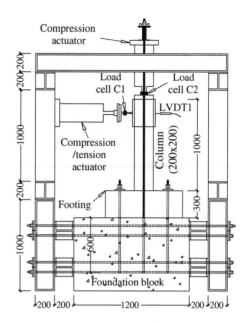

- Laminate strips of CFRP were introduced into the slits and the epoxy adhesive in excess was removed;
- The "non-linear hinge region" and the holes in the footing were filled with epoxy mortar.

The curing/hardening process of the epoxy adhesive lasted for, at least, 5 days prior to testing the strengthened elements. The properties of all the intervening materials are described elsewhere [19].

The test set-up is illustrated in Fig. 10, where it can be seen that each specimen is composed of a column monolithically connected to a footing fixed to a foundation block by four steel bars. A constant vertical load of approximately 150 kN was applied to the column, inducing an axial compression stress of 3.75 N/mm². A history of displacements was imposed for LVDT1, located at the same height as the horizontal actuator, see Fig. 10. The history of horizontal displacements included eight load cycles between ±2.5 and ±20.0 mm, in increments of ±2.5 mm, with a displacement rate of 150 μm/s.

Figure 11a depicts a typical relationship between the horizontal force and the deflection at LVDT1 (see Fig. 10). Since this strengthening technique does not provide significant concrete confinement, the increase on the dissipated energy was marginal. The increment on the load carrying capacity, however, was significant as can be seen in Fig. 11b, where a typical envelope of the maximum values of the relationship between the maximum force registered in the load cycles and its corresponding deflection in the LVDT1 is represented.

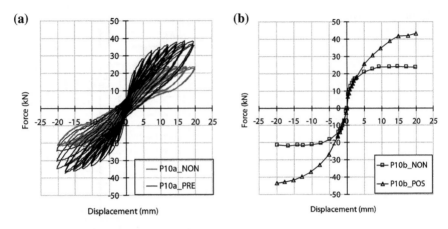

Fig. 11 Cyclic responses in terms of: **a** force-deflection (at LVDT1) relationship for column P10a; **b** Force-deflection (at LVDT1) envelop of all load cycles for column P10b

3.2 NSM/EBR Technique for the Flexural Strengthening and Energy Dissipation Capacity of RC Columns

The available experimental research shows that, for equal percentage of FRP confinement material, full wrapping of the concrete column provides higher load carrying and energy absorption capacities than discrete confinement arrangements. However, taking into account that existing steel hoops assure certain concrete confinement, the effectiveness of discrete confinement arrangements, consisting on applying strips of FRP wet lay-up sheets in between existing steel hoops, can still be significant in terms of load carrying and energy dissipation requirements, and in terms of economy of FRP materials [17]. As verified in previous section, the NSM technique is quite effective for the flexural strengthening of RC slabs, but it is not effective in terms of enhancing the energy dissipation capacity of RC columns. Therefore, combining NSM laminates for the flexural resistance, with strips of wet lay-up FRP sheets located in between existing steel hoops, a high effective technique can be obtained, which was designated as hybrid FRP-based strengthening technique Fig. 12. This figure shows the typical specimen of the experimental program, which is composed of a RC column monolithically connected to a RC footing.

The column was cast in a second phase, three days after the corresponding footing has been cast, in order to reproduce the real practice as much as possible. With the same purpose, starter longitudinal steel bars were used to connect the reinforcement system of the column to the corresponding foot (see Fig. 12). The lap splice of the starter bars had a length of 260 mm. The research program had the purpose of evaluating the influence of the concrete compressive strength, reinforcement ratio of longitudinal steel bars (ρ_{sl}) and number of CFRP layers per each strip, on the load carrying and energy dissipation capacities of RC columns

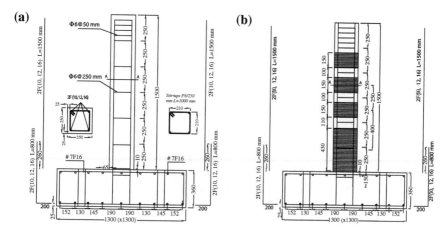

Fig. 12 RC columns tested: **a** Geometry of the columns and steel reinforcement arrangement; **b** disposition of the CFRP wet lay-up strips (dimensions in mm)

Table 2 Groups of tests for RC columns strengthened according to the hybrid technique

Group number	Designation	Specimen reference
1	G1	F10_NON
		F12_NON
		F16_NON
		F10_S2_L2_POS
		F12_S3_L2_POS
		F16_S4_L2_POS
2	G2	F10_S2_L2_PRE
		F12_S3_L2_PRE
		F16_S4_L2_PRE
3	G3	F12_S3_L3_PRE
4	G4	F12_S3_L2_PRE_C25/30

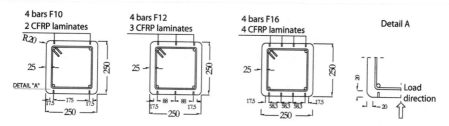

Fig. 13 Arrangement of the steel reinforcement and CFRP laminates in the cross section of the columns (dimensions in mm)

strengthened according to the hybrid technique. The full experimental program is described in Table 2: the first ten columns were built with an 8 MPa concrete compressive strength while the last column had a compressive strength of 29 MPa. The arrangement of the CFRP strips of wet lay-up sheets is represented in Fig. 13. The

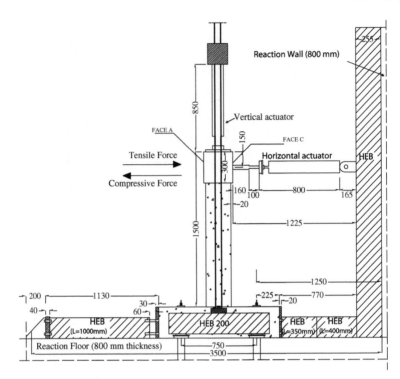

Fig. 14 Test setup [20]

identification of the columns has the format Fa_Sb_Lc_t, where "a" represents the diameter of the longitudinal steel bars, in mm (10, 12 or 16), "b" is the number of CFRP laminates applied in each face of the column subjected to cyclic tension/ compression, in order to increase the column flexural resistance (2, 3 and 4—see Fig. 13), "c" represents the number of CFRP layers in each strip (2 and 3) and "t" is the type of series (NON, PRE and POS). The NON term means a reference column, PRE is a column that was strengthened before having been tested and POS means a column that, after having been tested and strengthened, it was again tested.

The details on the strengthening procedures, as well as the material properties, are described elsewhere [20]. The test set-up is illustrated in Fig. 14. A constant vertical load of approximately 120 kN was applied to the column, introducing an axial compressive stress of about 1.92 MPa.

Figure 15a, c, e represent the force-lateral deflection relationship, F-u_{Act}, (measured by the internal LVDT of the actuator) for the columns of the G1 group of tests. In each graph, the F-u_{Act} curves of the reference column and its corresponding post-strengthened column are superimposed in order to highlight the most relevant features provided by the proposed technique. The relationship between the dissipated energy and the accumulated lateral deflection for the tested

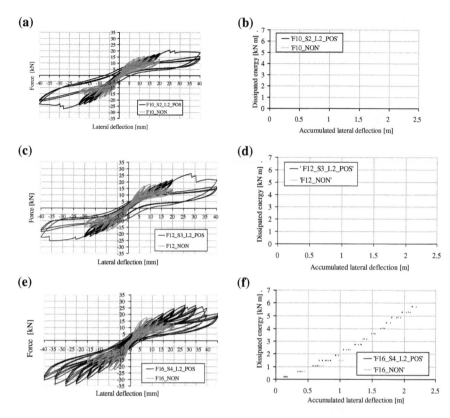

Fig. 15 Force-lateral deflection and dissipated energy versus accumulated lateral deflection for: **a, b** F10_NON and F10_S2_L2_POS columns; **c, d** F12_NON and F12_S3_L2_POS columns; **e, f** F16_NON and F16_S4_L2_POS columns

columns is represented in Fig. 15b, d, f. The accumulated lateral deflection was evaluated adding the absolute values of the recorded displacements in all cycles.

From the obtained results it is verified that the hybrid post-strengthening technique provided an increase of the column load carrying capacity that ranged from 38 to 55 %.

At the end of the tests, the formation of semi-conical fracture surfaces in the concrete surrounding the anchorage zone of the laminates was observed and considered as a possible justification for the load decays that are visible in the Fu_{Act} curve of the post-strengthened columns. Except for F16_S4_L2_POS column, the hybrid technique had no benefit in terms of energy dissipation capacity of the column up to the maximum accumulated lateral deflection of its reference column (see Fig. 15b, d, f), since the very low concrete strength did not allow the strengthening technique to be effective. However, after this deflection, Fig. 15b, d, f reveal that an increase of the ratio between the dissipated energy and the accumulated lateral deflection occurred, indicating that for larger lateral deflections the

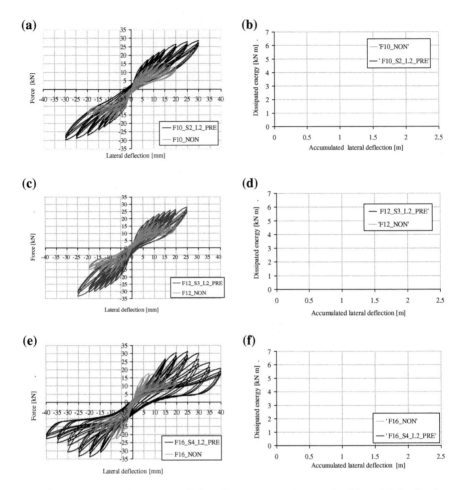

Fig. 16 Force-lateral deflection and dissipated energy versuss accumulated lateral deflection for:
a, b F10_NON and F10_S2_L2_PRE columns; **c, d** F12_NON and F12_S3_L2_PRE columns;
e, f F16_NON and F16_S4_L2_PRE columns

contribution of the CFRP materials becomes more effective, due to the dilation of
the concrete in the confined nonlinear hinge (column bottom zone). In fact, Fig. 15
shows that, above a lateral deflection of ±20 mm, the hysteretic cycles of the
strengthened columns presented a relatively significant level of energy dissipation.

Figure 16a, c, e represent the force-lateral deflection, Fu_{Act}, of the columns of
the G2 group of tests. In each graph, the Fu_{Act} curves of the reference column and
its corresponding pre-strengthened column are superimposed in order to observe
the increase in terms of load carrying capacity and energy dissipation provided
by the strengthening technique. The relationship between the dissipated energy
and the accumulated lateral deflection for the tested columns is represented in
Fig. 16b, d, f.

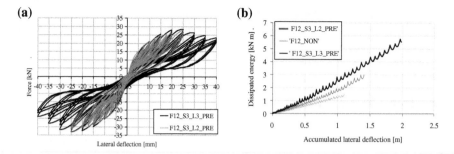

Fig. 17 Force-lateral deflection (**a**) and dissipated energy versus accumulated lateral deflection (**b**) for F12_S3_L2_PRE and F12_S3_L3_PRE columns

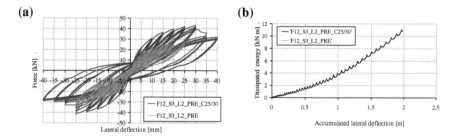

Fig. 18 Force-lateral deflection (**a**) and dissipated energy versus accumulated lateral deflection (**b**) for F12_S3_L2_PRE and F12_S3_L2_PRE_C25/30 columns

From the obtained results it is verified that the hybrid pre-strengthening technique provided an increase of the column load carrying capacity that ranged from 63 to 70 %. In spite of the very low concrete compressive strength of the columns of the G2 group of tests (PRE), the hybrid technique had an evident benefit in terms of energy dissipation capacity of the column up to the maximum accumulated lateral deflection of the reference columns (1.1 m), since an increase of the column energy dissipation capacity ranging from 40–87 % was determined.

To evince the influence of the number of CFRP layers per strip in terms of column load carrying and energy dissipation capacities, Fig. 17a compares the force-lateral deflection of F12_S3_L2_PRE and F12_S3_L3_PRE columns, the first one with two CFRP layers per strip, while the second has three CFRP layers per strip. From the analysis (see also Fig. 17b) it is verified that the increase of the number of CFRP layers had a marginal effect in terms of column carrying capacity, but a significant increase in terms of energy dissipation.

Figure 18 compares the force-lateral deflection and the dissipated energy versus accumulated lateral deflection of the F12_S3_L2_PRE and F12_S3_L2_PRE_ C25/30 columns, from which the favourable effect of the concrete strength in terms of the column load carrying and energy dissipation capacities can be verified. From the obtained results it is verified that an increase of the concrete compressive strength from 8 to 29 MPa provided an increase in terms of load carrying and energy dissipation capacities of 39 % (average of the tensile and compressive maximum force increments) and 109 % (for an accumulated lateral deflection of 1.406 m), respectively.

4 Beam-Column Joints

The main existent techniques for repair and strengthen reinforced concrete beam-column joints can be grouped as follows: repair with epoxy (injection of epoxy resin in the cracks of lightly degraded elements); removal and replacement of concrete in more damaged areas; jacketing with RC layers, masonry blocks or steel plates; use of composite materials. Epoxy repair techniques have exhibited limited success, whereas concrete jacketing of columns and encasing the joint region is an effective but the most labour-intensive strengthening method [21].

Several CFRP (MDL—multidirectional, UD—unidirectional) based configurations for strengthening RC beam-column joints have already been proposed both for interior and exterior joints [22, 23]. In these tests were analysed interior and exterior specimens with same geometry, steel reinforcement details and cyclic load pattern. Table 3 presents the main parameters analysed in each type of joint.

Three major solutions were considered which were designated indirect (i), direct (d) and total (t). The difference between them was related to the face where CFRP was located. In the indirect configuration, CFRP was only placed in the superior and inferior faces of the joint, while in the direct configuration, CFRP

Table 3 Main parameters analysed by Coelho et al. [22, 23]

Interior beam-column joints	Exterior beam-column joints
Interior beam-column joints	**Exterior beam-column joints**

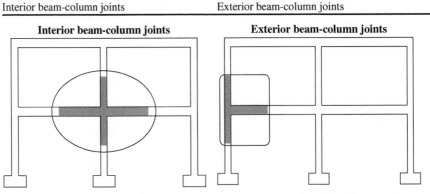

| 3 RC beam-column joints firstly tested until failure and then strengthened with MF-EBR technique | 6 uncracked RC beam-column joints strengthened by MF-EBR (3), MF-FRP (1), NSM (2) |
| Smooth (1 specimen) and ribbed steel (2 specimen) and strengthening configuration were the main analysed variables | Strengthening technique and configuration were the main analysed variables (all specimens had ribbed steel) |

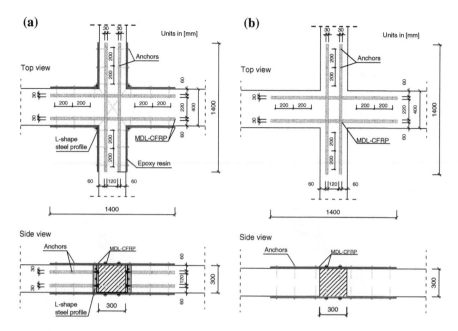

Fig. 19 Details of interior beam-column joints strengthening solutions: **a** MF-EBRt; **b** MF-EBRi

was only placed in the lateral faces of the joint. Finally, the total configuration consisted on the combination of indirect and direct solutions.

4.1 Interior Beam-Column Joints

The preparation of the interior joints [22] involved three main steps: joint reconstruction (removal of deteriorated concrete in the corners area and then reconstruction with grout), cracks sealing (injection of epoxy resin) and CFRP application according to the MF-EBR technique (presented in previous section).

Figure 19 presents an overview on the strengthening configurations used. Indirect configuration (Fig. 19b) was applied on two specimens, one with smooth steel and other with ribbed steel, while total configuration (Fig. 19a) was only applied to one specimen with smooth steel. As Fig. 19 depicts, in the direct configuration L-shape steel profiles were used to assure continuity between the MDL-CFRP in adjacent lateral faces.

Table 4 present a summary of the main results obtained in the interior beam-column joints tests. As can be seen specimen with internal smooth steel reinforcement strengthened with the indirect configuration restored the initial load carrying capacity and dissipated ¼ more energy even though it as lost some initial stiffness and displacement ductility. The identical specimen strengthened with the total

Table 4 Main results obtained for each specimen after strengthening (comparison with the corresponding specimen before the strengthening)

Specimen	Load carrying capacity (%)	Displacement ductility (%)	Initial stiffness (%)	Dissipated energy (%)
Ribbed indirect	+10	−17	−30	−16
Smooth indirect	0	−7	−8	+24
Smooth total	+35	−9	+18	+20

Note (+) means increase and (−) means decrease

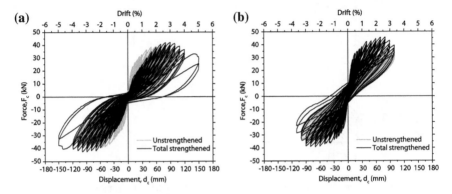

Fig. 20 Example of the relationship between force and horizontal displacement registered at the top of the column (loaded point): **a** specimen with ribbed steel reinforcements; **b** specimen with smooth steel reinforcements (see also Fig. 19)

configuration increased considerably load carrying capacity, initial stiffness and dissipated energy, while displacement ductility was again lower than that of the unstrengthened specimen.

Regarding ribbed steel specimen, only load carrying capacity increased, all other parameters presented considerable decreases. That aspect can be related to the initial damage imposed to each specimen. In fact, while smooth steel specimens presented damage concentration in 4 major cracks (one per each structural element) near the centre of the specimen due to the slippage smooth steel/concrete, in ribbed specimen damage was distributed in several minor cracks along beams and columns length. Since the restoration approach used in the three specimens was the same, it can be said that it is efficient for beam-column joints with smooth steel but not enough or even adequate for ribbed steel beam-column joints. Nevertheless, the strengthening configuration used was able to minimize that aspect and at least load carrying capacity improvement was seen in ribbed specimen.

In Fig. 20 the global response in terms of registered load on the top of the column versus imposed horizontal displacement at the same point for the specimens with worst and best performance. The response of the remaining specimen (smooth steel and indirect strengthening) was almost equal to its original unstrengthened specimen. This figure can be helpful for understanding the results presented above.

4.2 Exterior Beam-Column Joints

The preparation of the exterior joints involved only CFRP application according to the technique analysed in each specimen (procedures previous described), since they were not previously damaged [23]. In these tests for MF-EBR technique one specimen was strengthened with a direct configuration (MF-EBRd), other with an indirect configuration using mechanical anchors with a limited depth (MF-EBRi) and other with an indirect configuration using threaded rods through the entire thickness to anchor the MDL-CFRP (MF-EBRib). MF-FRP was only applied in a specimen with direct configuration for comparison (MF-FRPd). NSM was applied in two specimens, on with direct configuration (NSMd) and other with indirect configuration (NSMi). Figure 21 presents elucidative pictures of the strengthening solutions adopted.

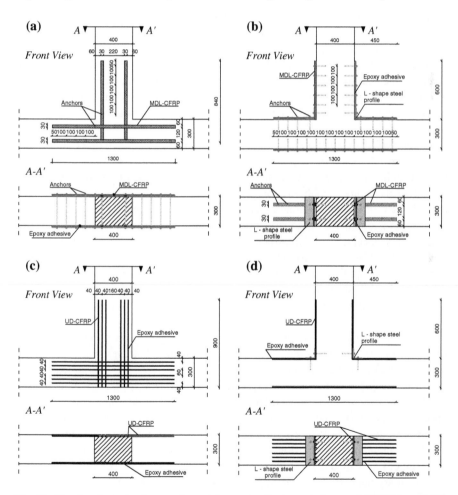

Fig. 21 Details of exterior beam-column joints strengthening solutions: **a** MF-EBRi and MF-EBRib; **b** MF-EBRd and MF-FRPd; **c** NSMi; **d** NSMd

Table 5 Main results obtained for each specimen after strengthening (comparison with the corresponding specimen before the strengthening)

Specimen	Load carrying capacity (%)	Displacement ductility (%)	Initial stiffness (%)	Dissipated energy (%)
MF-EBRi	16	−29	13	−1
MF-EBRib	21	−10	14	4
MF-EBRd	40	−23	26	43
MF-FRPd	52	−36	33	34
NSMd	52	−45	23	35
NSMi	75	−8	16	101

Note: (+) means increase and (−) means decrease

Table 5 present a summary of the main results obtained in the exterior beam-column joints tests. In general all strengthened specimens presented increases in terms of load carrying capacity, initial stiffness and dissipated energy, when compared with the unstrengthened specimen. This corroborates the conclusion taken on the interior joints tests that performance of strengthened specimens was affected by the repair strategy which did not allow full exploration of the strengthening strategies. In these tests, since specimens were not pre-damaged, strengthening strategies impact was clearly seen.

Other similar conclusion in both types of joint tests is that indirect configurations achieved lower increments in each parameter analysed than direct solutions, which was expected since the FRP was not being used in its best direction. Nevertheless, indirect configurations were considered for comparison with direct ones. In fact, in real in situ conditions of residential buildings, sometimes the indirect strengthening is the only viable option for applying these techniques.

The only exception to this inferior performance of indirect configurations was obtained in NSM specimens. This fact is related to an extra 5 mm cover epoxy layer that was applied in NSMi specimen because there was not enough cover thickness in this specimen due to a construction error. That aspect can be the reason for the high increases presented in this specimen for almost all measured parameters.

For all the proposed solutions, only one drawback point was observed in terms of displacement ductility, which was always lower in all strengthened specimens when compared to reference one.

Figure 22 presents the global response in terms of the registered applied load on the top of the beam *versus* horizontal displacement at the same point for each specimen. As can be seen, until the maximum load point all strengthened specimens presented higher stiffness and maximum load than reference specimen.

Other remarkable conclusions that can be drawn from this results is that the confinement provided by solution MF-EBRib compared to solution MF-EBRi added a marginal extra load carrying capacity and MF-FRPd performed better than MF-EBRd, which was not expected since the later was supposed to be an upgrade of the former.

Fig. 22 Relationship between force and horizontal displacement registered at the top of the beam (loaded point) for each specimen

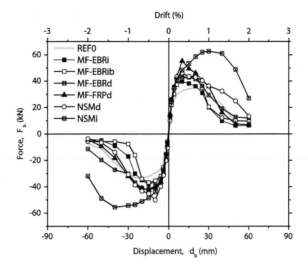

5 Conclusions

Reinforced concrete is undoubtedly one of the major solutions used worldwide in buildings construction. Many of them were built in times where regulations weren't fully developed and several problems have sometimes come to surface when earthquakes occurred. This fact has called attention to the need of preventing measures for those constructions, in one hand, and the need for repairing and strengthening solutions, on the other hand. This subject has been widely addressed by the civil engineering community in last decades. This chapter intended to present a compilation of the most recent common repairing and strengthening solutions for reinforced concrete buildings developed by the authors.

References

1. ACI (2008) Guide for the Design and Construction of Externally Bonded FRP Systems for Strengthening Concrete Structures. Report by ACI Committee 440.2R-08, American Concrete Institute, Farmington Hills, MI, USA
2. SA, Design handbook for RC structures retrofitted with FRP and metal plates: beams and slabs. HB 305-2008, Standards Australia GPO Box 476, Sydney, NSW 2001, Australia (2008)
3. Bank, L.C.: Strips for strengthening RC structures—a viable alternative. In: Proceedings of 2nd International Conference on FRP Composites in Civil Engineering: CICE, Adelaide, Australia 12 (2004)
4. Lamanna, A.J., Bank, L.C., Scott, D.W.: Flexural strengthening of reinforced concrete beams by mechanically attaching fiber-reinforced polymer strips. J. Compos. Constr. **8**(3), 204–209 (2004)
5. Quattlebaum, J.B., Harries, K.A., Petrou, M.F.: Comparison of three flexural retrofit systems under monotonic and fatigue loads. J. Bridge Eng. **10**(6), 731–740 (2005)

6. Ray, J.C., Scott, D.W., Lamanna, A.J., Bank, L.C.: Flexural behavior of reinforced concrete members strengthened using mechanically fastened fiber reinforced polymer plates. In: Proceedings of the 22nd Army Science Conference, United States Army, Washington, DC, pp. 556–60 (2000)
7. Micelli, F., Rizzo, A., Galati, D.: Anchorage of composite laminates in RC flexural beams. Struct. Concr. 11(3), 117–126 (2010)
8. Sena-Cruz, J.M., Barros, J.A.O., Coelho, M.R.F.: Bond between concrete and multidirectional CFRP laminates. Adv. Mater. Res. 133–134, 917–922 (2010)
9. Fib. Bulletin 14 (2001) Externally bonded FRP reinforcement for RC structures. Technical report by Task Group 9.3 FRP (fiber reinforced polymer) reinforcement for concrete structures. Féderation Internationale du Béton – fib 130
10. CNR-DT 200/2004, Guide for the design and construction of externally bonded FRP systems for strengthening existing structures. Consiglio Nazionale delle Richerche, Roma, Italy (2004)
11. Barros, J.A.O., Dias, S.J.E., Lima, J.L.T.: Efficacy of CFRP-based techniques for the flexural and shear strengthening of concrete beams. Cement Concr. Compos. 29(3), 203–217 (2007)
12. Pellegrino, C., Modena, C.: FRP shear strengthening of RC beams with transverse steel reinforcement. ASCE J. Compos. Constr. 6(2), 104–111 (2002)
13. Pellegrino, C., Modena, C.: FRP shear strengthening of RC beams: experimental study and analytical modeling. ACI Struct. J. 103(5), 720–728 (2006)
14. Sena-Cruz, J.M., Barros, J.A.O., Coelho, M.R.F., Silva, L.F.F.T.: Efficiency of different techniques in flexural strengthening of RC beams under monotonic and fatigue loading. Constr. Build. Mater. 29, 175–182 (2012)
15. Berthet, J.F., Ferrier, E., Hamelin, P.: Compressive behaviour of concrete externally confined by composite jackets. J. Constr. Build. Mater. 19, 223–232 (2005)
16. Harajli, M.H., Hantouche, E., Soudki, K.: Stress-strain model for fiber-reinforced polymer jacketed concrete columns. ACI Struct. J. 105(5), 672–682 (2006)
17. Barros, J.A.O., Ferreira, D.R.S.M.: Assessing the efficiency of CFRP discrete confinement systems for concrete column elements. ASCE Compos. Constr. J. 12(2), 134–148 (2008)
18. Sadone, R.: Comportement de poteaux en béton armé reinforces par matériaux composites, soumis à des sollicitations de type sismique, et analyse d'éléments de dimensionnement. Ph.D. thesis, IFSTTAR (2011)
19. Barros, J.A.O., Ferreira, D.R.S.M., Fortes, A.S., Dias, S.J.E.: Assessing the effectiveness of embedding CFRP laminates in the near surface for structural strengthening. Constr. Build. Mater. 20, 478–491 (2006)
20. Perrone, M., Barros, J.A.O., Aprile, A.: CFRP-based strengthening technique to increase the flexural and energy dissipation capacities of RC columns. ASCE Compos. Constr. J. 13(5), 372–383 (2009)
21. Engindeniz, M., Kahn, L.F., Zureick, A.: Repair and strengthening of non-seismically designed RC beam-column joints: state-of-the-art. Georgia Institute of Technology, School of Civil and Environmental Engineering, Structural Engineering, Mechanics and Materials, Research Report No. 04-4:59 (2004)
22. Coelho, M.R.F., Fernandes, P.M.G., Melo, J., Sena-Cruz, J.M., Varum, H., Barros, J.A.O., Costa, A.: Seismic retrofit of RC beam-column joints using the MF-EBR strengthening technique. Adv. Mater. Res. 452–453, 1110–1115 (2012)
23. Coelho, M.R.F., Fernandes P.M.G., Sena-Cruz, J.M., Barros, J.A.O.: Efficiency of different techniques in seismic strengthening of RC beam-column joints. In: 15th World Conference of Earthquake Engineering, Lisbon (2012)

Numerical Modelling Approaches for Existing Masonry and RC Structures

Alexandre A. Costa, Bruno Quelhas and João P. Almeida

Abstract Assessment of existing buildings making use of numerical simulation methods, even under the hypothesis of full knowledge of current conditions and materials, it is not an easy and straightforward task due to the limitations and complexities of such analysis tools. In this chapter, a discussion of different approaches for the simulation of structural response is introduced and applied to two of the most common building typologies: masonry structures and reinforced concrete frames. Following a brief introduction of the problematic, an overview of different modelling possibilities for masonry structures is presented. Afterwards, choices made during numerical modelling are discussed, based mainly on the finite element method. Moreover, the problematic of different modelling techniques is addressed, where some paths and best practices are suggested. The last section is devoted to the response simulation of reinforced concrete structures. Efficient frame elements and sectional models, which allow capturing an extended range of elastic and inelastic response, are analysed first. Strut-and-tie modelling is then recalled as a powerful analysis tool, and its application to the assessment of old buildings is studied.

Keywords Masonry • Modelling strategies • Finite elements • Strut-and-tie • Frame elements • Sectional fibre analysis

A. A. Costa (✉)
Department of Civil Engineering, Polytechnic of Porto, School of Engineering,
Porto 4200-072, Portugal
e-mail: alc@isep.ipp.pt

B. Quelhas
Department of Civil, Environmental and Architectural Engineering, University of Padova,
35131 Padova, Italy
e-mail: bruno.silva@dicea.unipd.it

J. P. Almeida
School of Architecture, Civil and Environmental Engineering, Ecole Polytechnique Fédérale
de Lausanne (EPFL), 1015 Lausanne, Switzerland
e-mail: joao.almeida@epfl.ch

A. Costa et al. (eds.), *Structural Rehabilitation of Old Buildings*,
Building Pathology and Rehabilitation 2, DOI: 10.1007/978-3-642-39686-1_10,
© Springer-Verlag Berlin Heidelberg 2014

1 On the Structural Response of Existing Constructions

The study of existing constructions and their actual behaviour, even for gravity loads, is a difficult task, being however a key point on current engineering practices and scientific research.

When dealing with existing constructions and their numerical simulation, the next points should be respected and taken into consideration, in order to minimize uncertainties and increase the confidence on the developed numerical models, namely:

(i)   Overall characterization: all the structural elements must be surveyed (vertical and horizontal ones, including foundations and roof) for a correct definition of the elements' geometry and constituent materials, as well as existing loads applied on the structure;
(ii) Detailed survey of the structural elements: quantification of reinforcement on reinforced concrete (RC) elements and corresponding distribution (amount of longitudinal reinforcement, stirrups diameter and spacing, etc.), detection of design and/or construction errors, eventual cracking pattern and apparent causes (overburden, design errors, foundation settlement, etc.);
(iii) Construction detailing (especially for seismic assessment): connection between horizontal and vertical elements, structural detailing at beam-column joints, connection between perpendicular load-bearing walls and existence of out-of-plane devices (for masonry structures only) such as tie-rods, floor connections to the walls, RC ring beams, etc.

After the first stage of inspection and diagnosis, which is essential for the development of the numerical model on the actual conditions of the structure, a first problematic may arise related to material characterization. Current codes are in general not very helpful since they mainly focus on the design of new structures, while poor guidance is offered to the analysis of existing buildings.

Since the numerical simulation of structural behaviour is directly related to the material properties, the second stage should cope with in situ material testing to estimate boundaries of their mechanical characteristics. Despite some current attempts made towards this purpose (e.g. [1–4]), the amount of information and experimental tests required to obtain a good confidence level may lead to an unrealistic number of experiments (e.g. core drilling) to perform on single elements or the whole structure, disrespecting its integrity or even its future usage.

On the other hand, the material characterization of existing masonry structures is a major problem because current non-destructive techniques (NDT, as sonic tests or tomography) and even minor-destructive (as the flat-jack technique) may not give relevant results (usually only the Young's modulus) when applied to multi-leaf stone masonry walls or irregular masonry. Only destructive testing methods, as in situ vertical or diagonal compression tests, may give pertinent results (as maximum compressive strength) for the numerical model. Great care should be taken when dealing with existing constructions and unknown materials, belonging to the engineer the decision on material properties and material safety factors to be used in the model. Improper preparation of technicians for a correct assessment of existing structures leads usually to erroneous and/or too intrusive interventions, also correlated with the analytical and/or numerical approach followed.

Modal analysis is an easy and straightforward technique which may reduce the uncertainties of numerical modelling within the elastic range of the structure. Indeed, a good numerical model should be always calibrated based on the eigen-frequencies and mode shapes resulting from experimental modal identification, which can be carried out using simple ambient vibration tests.

On RC structures, this permits to infer the current load conditions of the structure, the influence of the surrounding constructions on its dynamic properties (if dealing with adjoining buildings), but more important than that, its current elastic modulus and current dynamic mass, as well as the influence of masonry infills in the global behaviour.

Modal analysis is even more important for masonry structures because it allows, in addition to the calibration of Young's modulus and current masses, the perception of existing connections between all horizontal and vertical elements. In this manner, it is possible to understand the current loading conditions of the structure, and also to simulate a strengthening intervention to cope with eventual deficiencies. Concerning the seismic behaviour, the evaluation of the existence of connections between horizontal and vertical elements (yielding the so-called "box behaviour", which modal analysis can help to identify), and despite the connections' unknown efficiency for higher excitation levels, is mandatory for a thorough structural assessment.

In the work presented by Ilharco et al. [5], the modal identification permitted to observe the most vulnerable part of an existing three-storey building, Fig. 1a, as well as to identify the absence of an efficient connection between horizontal (timber floors, not represented in the figures) and vertical elements (load-bearing masonry walls), Fig. 1b. Local mode shapes were observed in both parallel façades without frequency or phase correlation, meaning that the floor was not

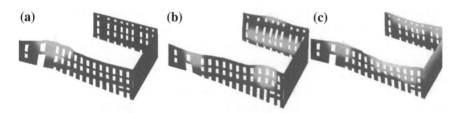

Fig. 1 Local mode shapes of an existing building [5]: **a** major vulnerable area; **b** lack of connection between timber floors and exterior masonry wall; **c** good connection and in-plane stiffness of timber roof

connecting both façades at that level, and mobilized only portions of the façades. On the other hand, good connection at the roof was observed as well as a mode shape governed by its in-plane behaviour, Fig. 1c, connecting parallel façades of the building. It is possible to conclude the above because both façades are mobilized in the same direction and with equal frequency of vibration, meaning that the roof was acting as an effective connection. This information was important for the design of the strengthening intervention. In general, the design of rehabilitation and/or strengthening techniques should thus be based on a comprehensive and exhaustive characterization of the structures' current conditions.

Despite experimental advances on material characterization and assessment of actual structural conditions, as expressed in the previous lines, the engineering experience and judgment plays the main role when dealing with rehabilitation interventions, criticizing and eventually rejecting the results obtained with the numerical models.

As an attempt to minimize the influence of engineering judgment and experience in the final decisions, the following paragraphs present some recommendations and modelling strategies to perform better and more efficient numerical simulations of existing constructions.

2 Numerical Assessment of Existing Masonry Structures

Masonry is one of the oldest structural material still in use; it has been applied on a huge diversity of constructions throughout the world. It is by nature a heterogeneous material whose components present a quite unknown geometry and a high mechanical variability. The structural behaviour of masonry depends on several factors such as member geometry; the characteristics of its texture; the physical, chemical and mechanical properties of its components and finally, the characteristics of masonry as a composite material [6].

All of the above mentioned factors make the analysis of the stone masonry mechanical behaviour a very complex matter. This is why it is of great interest the development and calibration of effective modelling and analysis strategies capable of predicting the behaviour of stone masonry structures, in particular under cyclic loads for seismic assessments. For this, it is necessary to accurately characterize this type of material through experimental testing. However, achieving a good characterization of masonry structures, detailed enough to be used confidently on the simulation, is, most of the times, a very demanding task, both in terms of cost and time [7].

The mechanical in situ characterization through non- and minor-destructive tests (sonic tests, flat jacks, etc.) gives precious information, almost without damaging the buildings, but with arguable global representation, while performing destructive tests on existing structures, either in situ or by removing samples large enough to be representative, is most often not possible especially when the structures have a high cultural value. As so, laboratory tests on masonry specimens representative of real constructions appear as a feasible alternative.

All the previous testing techniques are very useful to characterize masonry within its linear or non-linear range, to be used with different analysis methods.

2.1 Analysis Methods

To analyse the behaviour of existing masonry structures, there are nowadays several methods and computational tools that are based on different theories and strategies, resulting in different levels of complexity, different calculation times and, of course, different costs.

Simple analytical models may be used to assess the existing conditions of masonry structures, such as the simple rule of thumb, limit equilibrium analysis, arch theory, kinematic analysis, among others.

On the other hand, the analysis of existing masonry structures concerning seismic behaviour can be performed mainly through four different methods: (i) linear static (or simplified modal), (ii) linear dynamic (typically multimodal with response spectrum), (iii) non-linear static ("pushover") and (iv) non-linear dynamic.

When opting for one method of analysis one must have clearly defined the desired type of analysis, its objectives and also the knowledge of the advantages and limitations of the available tools, bearing in mind that more complex analysis are not necessarily synonymous of better results [8]. Above all, an analysis must be informed and planned in order to maximize its simplicity. A practical analysis of existing masonry structures implies great simplifications in the creation of the model geometry; the technician responsible for the analysis has to assess what is or is not important for a given analysis.

Nowadays the technological and scientific development permits the execution of increasingly complex analysis in increasingly shorter times, often enhancing the detail and size of a model. However, this may cause loss of objectivity and

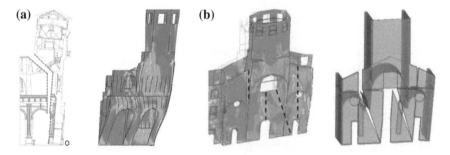

Fig. 2 Comparative analyses, using different methods, of the Cathedral of Santa Maria Assunta, Reggio Emilia (Italy) [9]: **a** Out-of-plane mechanism of the frontal façade; **b** In-plane mechanism of the frontal façade

result on a huge amount of information, rendering more difficult the analysis of the structural behaviour.

Given the large number of parameters of which depends the analysis of existing masonry structures and the degree of uncertainty and lack of knowledge that surrounds them, it is not appropriate and/or correct to propose a single method of analysis. In addition, on the process of choosing a method of analysis one must take into consideration the limitations and advantages of each method and the objectives of the analysis. As such, it is correct to propose analyses of a same problem using different methods, and the comparison between the results of each method can increase the degree of confidence in the obtained results. Several authors combined different modelling strategies, on the analysis of this type of structures, such as limit analysis and numerical modelling (Fig. 2). The numerical models allow individualizing macro-elements and their respective lines of collapse, as well as the formation of mechanisms on a structure, whose vulnerability and safety can be analysed through simplified limit analysis.

2.2 Modelling Strategies

Historically, structures began to be designed using simple rules of thumb based on the workers' experience. This method, although quite basic, was used in the construction of big and important structures such as bridges. After this, static graphics started being used; it is a quite simple method, which allows solving graphically the structural problems.

Later, methods based on the concept of limit analysis began to be employed. These methods assume that a structure is collapsing and compares the state of collapse with the actual condition of the structure, thus defining its structural safety. The main advantage of this type of analysis is the combination of the knowledge on the mechanisms and failure loads with the simplified implementation on practical computational tools. The number of required material parameters is thus reduced to a minimum, which is convenient as this type of parameters have a high degree of uncertainty, being very difficult to find reliable information [10].

Only recently, due to the high (and ever increasing) capacity of modern computers to solve numerical problems, it became possible to simulate the response of materials, such as masonry, considering their non-linear structural behaviour. Several methods, such as the Finite Element Method (FEM), were then implemented. The basic unit that characterizes this method, the finite element, usually does not represent a structural member but rather one of its sub-parts. Thus, the FEM can be applied to simulate separately the behaviour of various materials of the elements that compose the assembly (micro-modelling), or to simulate in a homogenized and continuous way the global behaviour of a composite material (macro-modelling).

Apart from the micro- and macro-modelling techniques, there is also the so-called homogenized modelling. In this type of approach the structure is composed by a finite repetition of an elementary cell, and the masonry is seen as a continuum whose constitutive relations are derived from the characteristics of its individual components (blocks and mortar), and from the geometry of the elementary cell. In recent years, some of these homogenization techniques have been developed and applied by several researchers (e.g. [11, 12]).

Other modelling methods widely used are the Structural Element Models (SEM), the Discrete Element Method (DEM) and its new formulation, and the Discontinuous Deformation Analysis (DDA). These last two methods are extensively used on rock mechanics, which in many cases is quite similar to masonry problems. They are also quite useful on the analysis of the failure mechanisms of masonry structures.

2.3 Finite Element Method

The Finite Element Method (FEM) is one of the most used approaches for the modelling of structures. It offers a widespread variety of possibilities concerning the description of the masonry structures within the frame of detailed non-linear analysis. Most of modern possibilities based on the FEM fall within two main approaches: modelling at the micro level, considering the material as discontinuous, or at the macro level. Hybrid models can also be created, which have considerable interest when, for example, one intends to analyse in detail a specific structural element within a more complex structure.

2.3.1 Micro-modelling

Some authors, such as Costa et al. in [13] on the analysis of stone masonry structures through the FEM, used a detailed micro-modelling approach reducing the masonry to its basic components (joints, blocks and infill), Fig. 3. The units and the mortar at joints are described using continuum finite elements, whereas the

Fig. 3 Micro-modelling using finite elements—replica of an existing wall from Azores [13]

unit-mortar interface is represented by discontinuous elements accounting for potential crack or slip planes.

This type of detailed modelling has a greater degree of accuracy; however, it is inevitably accompanied by an increase in the calculation time and modelling effort, which makes this simulation strategy unpractical for common engineering practice. Nonetheless, it is suitable for the study of localized areas and effects, provided that there is detailed knowledge of the geometry and composing elements and materials. It is particularly adequate to describe the local response of the material. Elastic and inelastic properties of both unit and mortar should be realistically taken into account.

The detailed micro-modelling strategy leads to very accurate results, but requires an intensive computational effort. This drawback is partially overcome by the simplified micro-models. Some authors (e.g. [14, 15]) opt for this simplified micro-modelling strategy, which is characterized by the combination, or omission, of certain constituents, allowing to drastically reducing the computation time without a great loss of accuracy.

The primary aim of the micro-modelling approaches is to closely represent masonry based on the knowledge of the properties of each constituent and the interface. The necessary experimental data for calibration must be obtained from laboratory tests on the constituents and small masonry samples.

2.3.2 Macro-modelling

Still in the field of finite element modelling, some authors opt for the macro-modelling using macro mechanical models, also known as homogeneous or continuous, in which all elements of an assembly of materials are incorporated into a continuum, wherein it is established a relation between the average extensions and stresses of the masonry. These relations are obtained by adopting a phenomenological point of view or by using homogenization techniques.

Macro-modelling is probably the most popular and common approach due to its lesser calculation demands. In practice-oriented analyses on large structural members or full structures, a detailed description of the interaction between units and mortar may not be necessary. In these cases, macro-modelling, which does not make any distinction between units and joints, may offer an adequate approach to the characterization of the structural response.

The macro-models have been extensively used with the aim of analysing the seismic response of complex masonry structures, such as arch bridges (e.g. [16]), historical buildings (e.g. [17]), and mosques and cathedrals (e.g. [18, 19]).

The smeared crack scalar damage models or other similar models, such as those presented in [20, 21], are often used in macro-modelling of masonry. This type of models, where the damage is defined in a given point by a scalar value which defines the level of material degradation (ranging from the elastic state until collapse), and the cracking is considered as distributed along the structure, are commonly used in the modelling/analysis of reinforced concrete structures (e.g. [22]) or large volumes of

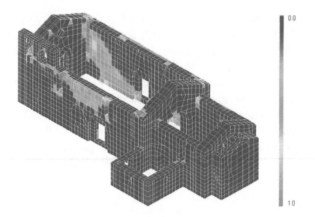

Fig. 4 Macro-modelling of the Gondar church using Finite Elements based on a damage model [25], Tensile damage map (d^+)

concrete (e.g. [20]). Such models were later adapted for the use in masonry buildings through the work of different authors, such as [23, 24] or [25] (Fig. 4).

In terms of applicability, it is a type of modelling clearly suitable when factors such as time, simplicity of modelling and computational capacity are crucial. Further, it is oriented for the everyday use in the analysis of real structures and when there is the need of maintaining a balance between accuracy and speed/efficiency [8].

2.4 Modelling Problematic

For what concerns the available analysis methods, the linear elastic analysis is the most commonly used in current practice, due to its advantages in terms of computational time. However, its application to masonry structures is, in principle, inadequate because it does not take into account the non-tension response or non-linear sliding-shear, among other essential features of masonry behaviour. It must be noted that, due to its very limited capacity in tension, masonry shows a complex non-linear response even at low or moderate stress levels. Moreover, simple linear elastic analysis cannot be used to simulate masonry strength responses, typically observed in arches and vaults, characterized by the development of sub-systems working in compression. Attempts to use linear elastic analysis to design arches may result in very conservative or inaccurate approaches. Linear elastic analysis is not useful, in particular, to estimate the ultimate response of masonry structures and should not be used to conclude on their strength and structural safety.

Notwithstanding, linear elastic analysis has been used, with partial success, as an auxiliary tool assisting in the diagnosis of large masonry structures. Easy availability and reduced computer costs have promoted its use, in spite of the mentioned limitations, before the development and popularization of more powerful computer applications. Some examples are the studies of San Marco's Basilica in

Venice by [26], the Metropolitan Cathedral of Mexico [27], the Tower of Pisa [28], the Colosseum of Rome [29] and the Church of the Güell Colony in Barcelona [30], among many others. In all these cases, the limitations of the method were counterbalanced by the very large expertise and deep insight of the analysts.

Whenever possible, it is preferable to individually analyse the elements of a structure, simplifying its geometry and reducing the computation effort. In this case it may be advisable to use 2D models instead of the 3D ones. In cases where it is considered important to use the complete model of the structure, it is advised the use of mixed complexity, detailing (from the geometrical and/or material behaviour point of view) the areas of major interest, with higher influence on the overall behaviour of the structure or critical areas that present particular problems to be analysed.

For what regards modelling with finite elements, in the case of continuous elements such as masonry walls, it is frequent to use shell elements, since these present advantages in terms of creation of the geometrical model and of computational effort. However, certain precautions need to be taken into account and several difficulties arise when using this type of element, such as:

- the use of this type of element does not allow considering, in a direct manner, the load or support eccentricity effect, since the shells are aligned with the axis. In cases where there are large variations in thickness on a structure, it is important to take into account this phenomenon, in particular, when the floors are supported only in one of the leaves of a multiple-leaves wall panel;
- the use of such elements makes it impossible to consider directly the phenomenon of leaves' separation, i.e. it can only be used when this type of phenomenon is not important to analyse the problem, or when the structural element in question is monolithic;
- in finite elements, the stress distribution along the thickness of a wall is linear, which may deviate from reality, at a local level;
- in such elements, the stiffness in the areas of intersection of two shells (e.g. facade angles) is not always well represented.

The use of volumetric elements allows reproducing, in a more realistic way, the intersection zones of structural elements. By using this approach it is possible to evaluate the stresses in the thickness of a wall; however, more than one element to discrete the mesh along the thickness has to be used, otherwise the errors will be large, and the greater the thickness of the actual element the greater the error. By contrast, the use of volumetric elements makes the creation of the geometric model more complex and time consuming.

The modelling of the links between elements of a masonry structure is almost always one of the most important numerical modelling problems of historical structures. This difficulty arises not only from the difficulty in defining them appropriately, but also from the choice of numerical models capable of realistically characterizing these links. For a question of simplicity, in most cases the connections are considered as continuous and fixed, but this type of assumption is only valid in certain cases and it is the responsibility of the technician in charge of the analysis to evaluate this assumption in each case.

In the particular case of the support conditions of a structure on the soil, most of the times the structure is considered as fixed at the base. However, this assumption is only valid in the cases where the soil presents good quality and the wall foundation can be considered as properly fixed to the soil. Otherwise, the soil should be modelled as well, for example with springs at the base featuring equivalent stiffness characteristics to the ones presented by the considered type of soil.

In the modelling of wooden roofs reinforced with metal rods, only the tensile resistance must be considered on the ties, using appropriate models in order to render the analysis more realistic.

3 Reinforced Concrete Structures

Today, most (if not all) structural engineering offices use software—typically based on the Finite Element Method as mentioned in the previous section, but not only—to assess the structural condition of old RC buildings and rehabilitation interventions that may be required. The present section deals with some of the numerical models that are often used for such intent.

The number of non-linear mechanisms that describe the behaviour of RC in its cracked state is countless. It ranges from compression and tension softening to creep, shrinkage, core confinement effects, aggregate interlock, rebar bond slip, dowel action, tension stiffening, scale effects, and so on. The role of each of those mechanisms in the strength and deformation capacity of the structure is highly dependent on the type and magnitude of the applied loads (in the static, cyclic or dynamic regimes), as well as on the geometrical and mechanical member properties. Despite the current fast development of software for RC analysis, it is still virtually impossible to build a numerical model that simultaneously accounts for all the aforementioned mechanisms and load case scenarios. Consequently, the engineer is faced with the choice of selecting one (or a few) models that can provide a satisfactory insight into the problem at hand. That decision becomes yet more delicate in the context of assessment of old RC buildings, wherein additional physical phenomena—associated to material deterioration and other damage sources—may have to be simulated too.

Likewise, the selection of the modelling methods to present in these few pages is far from straightforward. Nevertheless, the very fact that RC buildings, from the force flow viewpoint, are essentially frame structures justifies addressing beam-column elements in the next section. Not only they are indispensable in almost every RC building model, but quite often also the only component used in assembling large models. The latter element can be suitably supplemented by a multipurpose analysis tool that, although applicable to the whole RC structure, is typically used near statistical or geometrical discontinuities: the strut-and-tie method, discussed in the subsequent section. Its versatility implies successful adaptation to the specificities of old buildings, in addition to modelling of strengthening systems. Finally, some closing remarks on other relevant physical phenomena and applicable simulation techniques are briefly addressed.

3.1 The Fundamental Role of Frame Elements

Among the major sources of deformation on reinforced concrete structures, member bending is typically the most important and should thus be always considered in the corresponding models. Other mechanisms of deformation, including member shear, second-order effects, anchorage slip, etc., are briefly referred in Sect. 3.3.

RC simulation should provide tools for reliable conventional assessment of existing buildings (at the serviceability and ultimate states). However, in earthquake-prone regions, their seismic vulnerability is often the conditioning factor of the overall assessment. Therefore, an ideal member model should be resourceful enough to capture the flexural behaviour of reinforced and prestressed concrete members in its various stages of linear and non-linear behaviour, both under static and cyclic loading. The present section recalls efficient models for sectional and element analysis that achieve the previous goal.

Due to the inherent 1D character of any beam theory, its displacement field can be decomposed into a component along the beam longitudinal axis, and another in the cross-section. The latter include the assumptions that make each beam theory unique, leading to a specific set of beam local compatibility and equilibrium equations, as well as their boundary counterparts.

In particular, the well-known Euler–Bernoulli hypothesis of plane sections remaining plane and perpendicular to the longitudinal axis gives rise to the most widely used beam theory for simulation of member flexural behaviour. Therein, the distribution of uniaxial normal strains is uniquely defined throughout the cross-section by the strain at the reference axis and the curvature (or curvatures about the two axes, if bi-axial bending is considered). The sectional behaviour can be best simulated by discretising the section into layers (or fibres—s in Fig. 5, for bi-axial bending), wherein it is assumed that the strain in each layer (fibre)

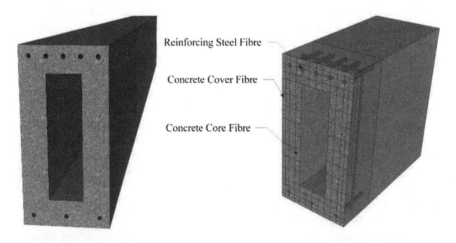

Fig. 5 Rectangular hollow RC section (*left*) and discretisation into concrete/steel fibres (*right*)

is uniform and equal to the strain at the centre of that layer (fibre). Making use of appropriate concrete and steel strain–stress relations, the corresponding normal stress is obtained. Their sectional integration is obtained by summation throughout the layers (fibres), yielding the axial force and moment(s) [31]. An advantage of the latter method is that the axial force-bending moment(s) interaction in both directions is directly simulated, as well as the post-peak softening response.

Alongside the sectional response, the other crucial point in the simulation of member behaviour is the choice of the element model. Shortly, the solution of the previously mentioned compatibility or equilibrium equations leads to two distinct types of frame finite element formulations: the classical displacement-based (DB), for long the most widely employed in structural analysis software, and the force-based (FB) approach, which now has stable and efficient state determination algorithms [32]. The former enforce a linear curvature and constant axial strain along the element. On the other hand, the latter assume a constant approximating polynomial for the axial force and a linear shape function for the bending moment. Under nodal element loading, it is thus apparent that the features of the FB formulation allow for a strict verification of equilibrium along the length, which holds irrespective of material constitutive behaviour. The DB element, on the other hand, only provides exact response for linear elastic behaviour. Furthermore, it can also be shown that span loading can be directly and exactly accounted for in FB formulations, unlike DB, and that no shear-locking phenomena exists. The first sensible consequence of the abovementioned features is that practitioners should give preference to FB over DB elements.

The strict verification of equilibrium is an advantage of fundamental relevance whenever the member response simulation involves highly non-linear material behaviour, such as under seismic loads (where cyclic uniaxial material laws must be utilized) or blast loads. It also implies that only one FB element is required to model each structural member, whilst a refined mesh of DB elements is called for to attain results of a comparable level of accuracy. Nevertheless, FB elements can also be very effectively employed to model structural behaviour in earlier stages of post-cracking response. The model's exact consideration of distributed loading, for instance, proves of the utmost significance for the computation of deflections for service load checks, wherein gravity and other span live loads control.

Consider a simply supported, uniformly loaded, eccentrically prestressed concrete beam. While the load is low, the beam will be curved upward. A single FB element with a layered section approach can be used to estimate the short-term camber. The long-term upward deflection—accounting for creep, shrinkage and relaxation—can also be straightforwardly predicted with the same model through adequate modification of the strain–stress relationships of the prestressing strands and concrete. For an increasing value of the uniformly distributed load, failure will eventually occur at mid-span, and once again the previous simple finite element scheme can be used to compute the corresponding deflection. Many strengthening techniques can also be modelled with the current method.

However, the element response also depends on the integration scheme and number of integration points where the sectional response is computed. The use

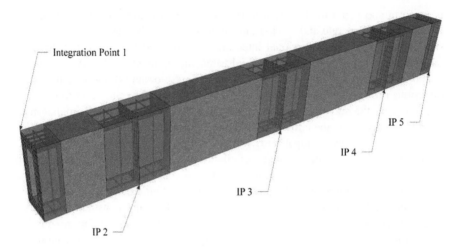

Fig. 6 Recommended lower bound for FB elements: five-point Gauss–Lobatto integration

of a Gauss–Lobatto integration scheme is advised as it controls the end sections of the element, which are privileged locations for the occurrence of inelastic behaviour. The number of integration points to consider is a question of required numerical accuracy: five can be roughly considered as the lower bound—see Fig. 6, but seven or even more may be advisable on occasion.

Finally, it is noted that if softening branches of the moment–curvature relation are attained (in one or more integration points), a pathological numerical behaviour known as strain-localisation takes place. As a consequence, the model results begin losing their physical meaningfulness and regularization techniques should be applied to restore it [33].

3.2 Strut-and-Tie in Assessment and Rehabilitation

It is most likely that the standards on which the design of an 'old' RC building was based no longer comply with current state-of-the-art principles, in terms of modelling and analysis methods, as well as detailing rules. Furthermore, it is also very probable that all computations were carried out by hand. Therefore, the influence of computer-aided methods of structural simulation—fundamentally connected to the development of the Finite Element Method—cannot be found.

Although many regular RC buildings can be entirely modelled by frame elements (henceforth denoted as B-regions, from Bernoulli or beam), there are also, in general, statical or geometrical discontinuity zones depicting two or three-dimensional stress states that require further special-purpose analysis and verification. In those so-called D-regions (from disturbed or discontinuity), the assumptions of the previously studied layered analysis do no longer apply since plane sections do not remain plane, transverse strains may not be negligible, etc. Additionally, old structures often evidence distinct effects of damage, such as cracking patterns, deterioration of concrete, rebar corrosion, etc., and it would thus be most helpful to bring such information inside the modelling and analysis framework process. Finally, within the context of rehabilitation, simulation of possible strengthening solutions is also frequently required.

The so-called strut-and-tie modelling (STM) is a versatile tool that rationally tackles the aforementioned points (D-regions, damage and strengthening) and whose relevance has been recognized by incorporation into major international codes for structural concrete. Since the famous works by Marti [34] and Schlaich [35], almost 30 years ago, the STM has progressively established as a powerful analysis and design method—now increasingly computerized—that has replaced the empirical approaches historically applied in the design of D-regions.

Figure 7a depicts an example of division between B-regions—analysed in the previous section—and D-regions in a frame, assumed under in-plane loading. Typically, D-regions are areas under point or reaction loads, openings, re-entrant corners, frame joints, etc. Based on Saint–Venant's principle, the dimensions of a D-region can be taken within a distance corresponding to the largest of the depth or width of the member to either side of the disturbance, see Fig. 7b. Once D-regions are identified they can be isolated for analysis purposes.

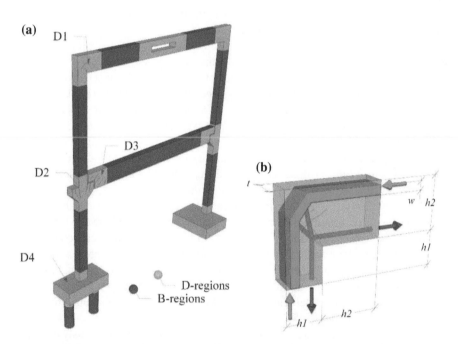

Fig. 7 **a** Illustrative identification of *B*-and *D*- regions in cracked frame structure; **b** STM for region *D1*, under gravity loads, with indication of geometrical properties

STM is based on the theory of plasticity and requires sketching, for each D-region, an internal truss (which consists of struts, ties and nodes) representing a statically admissible field. The latter has to be in equilibrium: (i) externally, with the applied loading and reactions (boundary forces from supports or adjacent B-regions); (ii) internally, at each node. The method produces a lower bound solution and therefore conservative predictions of the ultimate capacity of D-regions are obtained.

In general, ties represent one or multiple layers of reinforcing or prestressing steel. A strut represents a concrete compressive stress field whose centreline is in the same direction of the predominant principal stresses. It can have distinct shapes (prismatic, bottle-shaped, fan-shaped) [35], but in STM a uniform distribution is typically assumed. The effective cross-sectional area of the strut is given by $A_c = w \times t$, where t is the member thickness and w is the critical effective width, both indicated in Fig. 7b. Finally, nodes are regions where forces are transferred between struts and ties.

The stress limit of a strut (also known as effective strength), f_{cu}, is obtained from the uniaxial concrete compressive strength, f_c', and the so-called effectiveness factor, ν (≤ 1.0), as follows: $f_{cu} = \nu f_c'$. The purpose of the previous effectiveness factor ν is to account for the limited deformation capacity of concrete, its post-peak softening behaviour, the strength reduction caused by the shape of the strut's stress field and disturbances due to cracks and tensile strains, and still the strength degradation that may arise during cyclic loading. The effectiveness factor can also consider strength enhancements from confinement, as that provided by transversal reinforcement. Assuming a uniformly distributed stress field, the capacity of the strut is given simply by $F_{cu} = A_c f_{cu}$.

On the other hand, the stress in the ties is limited by the yield strength of ordinary reinforcing steel, f_y, or prestressing steel, f_{py}. Hence, the capacity of the tie is computed by $F_{tu} = f_y A_s + \Delta f_p A_{ps}$, where A_s is the area of rebars, A_{ps} is the area of prestressing steel, and Δf_p is the difference between f_{py} and the installed prestressing steel stress.

It should be noted that more than one STM may be envisaged for each load case, as long as equilibrium is satisfied. However, attention is required in order to respect the limited deformation capacity of concrete. This is the most time-consuming phase of the process, often requiring iterations to optimize the model and satisfy stress limit criteria. The devised truss should stand as an idealization of the actual flow of tensile and compressive forces in the region. Whilst experience is advantageous, the following tips can also prove useful: (i) loads tend to reach the support through the shortest path, (ii) the nodal angle between struts and ties should be reasonably large to minimize strain incompatibilities caused by strut shortening and tie lengthening in almost the same direction, (iii) based on the principle of minimum strain energy after cracking and the smaller deformability of concrete struts in comparison to steel ties, Schaich et al. [35] proposed to select STMs that minimize the total length of the ties. However, other less subjective aids can be used in sketching the STM, as follows.

To start with, the engineer can and should take advantage of previously existing STMs associated with similar regions (under analogous loading conditions) that can be found in the literature, guidelines and standards. A common alternative is to arrange the STM according to the elastic principal stresses produced by a finite element model, which will arguably satisfy both serviceability and ultimate limit states of D-regions; the deviation between the STM and the elastic solution should be kept within 15 degrees [35]. More recently, techniques for automatic generation of STMs, based on system performance criteria and topology optimization of continuum structures, have been developed [36, 37]. They are a few examples of the great increase in STM software, which is generalizing.

One should also look at the particulars of analysing existing members and at how damage inspection of old buildings can be used to calibrate appropriate STMs. The review of the building construction plans is invaluable, since they contain clear indications on the physical placement of the steel reinforcement and hence the centroid of the modelled ties can be accurately determined. In their absence, monitoring of crack patterns assumes extra pertinence. It is known that principal compressive stresses are parallel to cracks, which are caused by the orthogonal principal tensile stresses; therefore, struts should be set along the cracks' direction, and possibly centred in the cracked region width. Figure 7a and b, as well as Fig. 8a and c, help understanding the aforementioned relation. On the other hand, clear signs of concrete degradation, evidence of rebar corrosion or steel–concrete bonding problems, should also be duly accounted for in the STM. That can be achieved by readjustment of the truss or by an appropriate reduction (or even elimination) of the capacity of the concerned struts and ties.

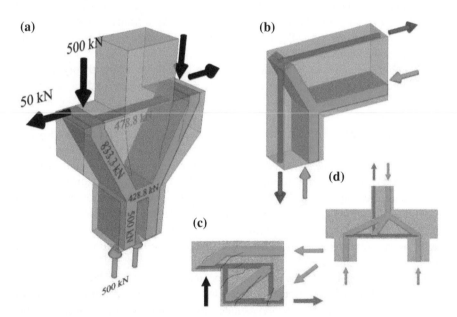

Fig. 8 **a** STM for region *D2* of Fig. 7, with sample calculations, **b** Alternative STM for region *D1*, under seismic loads, **c** STM for region *D3*, **d** STM for region *D4*

The selection of the effective width *w* of a strut is usually computed so that the strut's capacity is larger than the force it carries. On the other hand, there is also a tie effective width that can be computed during the analysis of the node in order to satisfy the nodal bearing capacity. Several methods for construction and verification of nodes, which won't be addressed, have been proposed up to now, including the hydrostatic approach [34] and the modified hydrostatic approach [38].

Figure 8a shows a straightforward computation of sample truss member forces; it is noted that other more complex STMs may be challenging in that they may imply solving statically indeterminate systems. Figure 8b illustrates the necessity that can arise, on occasion, of sketching a different truss when a distinct load case is considered.

Finally, it is highlighted that the STM can also be used in the rehabilitation context to assess the efficiency of a specific strengthening technique. Using the same underlying principles of static admissibility and flow of forces, STM—eventually combined with frame fibre analysis—can readily simulate common retrofitting solutions such as RC or steel member jacketing, steel bracing, infill strengthening, use of fibre reinforced polymers, etc.

3.3 Other Relevant Modelling Tools

From the major sources of deformation on reinforced concrete buildings, member bending is typically the most important and is therefore always considered by linear and non-linear beam-column models, such as those of Sect 3.1. However, besides normal stresses, shear stresses are also present in frame members that may result in diagonal cracking. The latter can lead to premature member failure unless adequate reinforcement has been provided.

In frame elements, a Timoshenko or a higher-order beam theory must be considered to simulate shear deformation, and the corresponding governing equations solved either with a displacement-based or force-based formulation. At the sectional level, the most simplified approaches use approximate shear force-shear deformation relationships. More detailed frame models employs 2D or 3D material models for reinforced concrete to characterize the layer (fibre) behaviour; however, the assumed imposition of a sectional shear-strain profile is unable to respect local equilibrium throughout the whole range of non-linear analysis. Therefore, the correct coupling between normal and shear stresses cannot be reproduced in a direct way, which impairs the theoretical soundness of such models. Several iterative procedures with simplifying assumptions have been proposed by different authors [39], but further developments are still required in this complex field of research.

On the opposite end of the modelling spectrum one can find detailed 2D or 3D finite element models. Advanced concrete models are required, for instance based on non-linear elasticity, theory of plasticity, fracture, microplane models, etc. They have been used by the research community to reproduce the behaviour of RC walls and piers with relevant shear deformations, but are not a practical tool for practitioners. Additionally, they call for a demanding combination of expertise in numerical modelling, as well as powerful processing capabilities.

Second-order effects can also be relevant in assessing serviceability and failure limit states. For each type of structural element (beam, plate, shell, solid…) there is a wealth of theoretical models to account for such effect. In view of the discussion in Sect. 0, it is worth mentioning the appropriateness of the co rotational approach for non-linear geometrical analysis of force-based elements [40].

Fixed end rotations, caused by slip of longitudinal steel reinforcing bars along their embedment length in beam-column joints, or footings, can be another important mechanism of member deformation. In terms of contribution to the total lateral member displacement, it may represent, on occasion, almost as much as the flexural component during inelastic response [41]. It is possible to find in the literature different simulation methods and pseudo-empirical expressions addressing this issue, but again there is room for more insightful research on the mechanics of its non-linear performance.

Finally, a thorough assessment of the condition of old buildings may require assessing the influence of additional physical phenomena, for which adequate models can in general be found. One may wish to consider, for instance, creep and shrinkage in concrete, corrosion of reinforcement (which affects not just the rebars, but bond, cracking and spalling of concrete), the deformability of beam-column joints and foundations, crack widths and propagation, among others.

4 Final Remarks

Some guidance and paths were shown in the previous sections, as well as state-of-the-practice tools and solutions for current engineering problems that researchers and engineers may face while developing numerical models for masonry and RC structures. It is thus expected that some uncertainties related to modelling issues were mitigated, contributing to an improvement of the overall quality of the numerical model and subsequent results.

Nevertheless, one of the foremost remarks is that there is no unique solution for the modelling of an existing structure. The usage of different modelling approaches, such as sub-assemblages and detailed finite element analysis of components of a structure, together with macro analysis of the complete construction, increases the global knowledge of structural performance and confidence on the developed numerical model.

The selection of the adequate tools, methods and elements is a crucial step, to the success of the model, as thoroughly discussed for masonry and RC structures. Moreover, the inspection of the current conditions of the structure is of meaningful importance for tuning the numerical model, and therefore to obtain reliable final assessment results. For masonry structures, materials and connections should be checked with particular care, while for RC buildings, concrete cracking, rebar corrosion, and other damage sources should be identified for the definition of appropriate material properties of beam-column models, as well as configuration and parameters of strut-and-tie approaches.

For the overall interpretation of the results, the engineering judgment and experience will play a decisive role, yet the guidelines presented herein are expected to reduce modelling uncertainties and assist in the final decision making process.

References

1. CEN (2007) EN 13791: assessment of in situ compressive strength in structures and precast concrete components
2. NTC (2008) Norme Tecniche per le Costruzioni, D. M. 14 gennaio 2008, Suppl. ord. n° 30 alla G.U. n. 29 del 4/02/2008, Consiglio Superiore dei Lavori Pubblici
3. CEN (2005) EN 1998-3: Eurocode 8-Part 3: Assessment and Retrofitting of Buildings
4. ACI 214.4R-10 (2010) Guide for obtaining Cores and interpreting compressive strength results. ACI Committee 214
5. Ilharco, T., Costa, A.A., Lopes, V., Costa, A., Guedes, J.: Assessment and intervention on the timber structure of a XVII century building in Lisbon: an example of seismic retrofitting. Revista Portuguesa de Engenharia de Estruturas **II**(11), 25–38 (2012)
6. Binda, L., Saisi, A.: Research on historic structures in seismic areas in Italy. Prog. Struct. Mat. Eng. **7**(2), 71–85 (2005)
7. Oliveira, D.: Experimental and numerical analysis of blocky masonry structures under cyclic loading. PhD Thesis (2003)
8. Lourenço, P.B.: Computations on historic masonry structures. Prog. Struct. Mat. Eng. **4**(3), 301–319 (2002)
9. Casarin, F.: Structural assessment and vulnerability analysis of a complex historical Building. PhD Thesis, University of Trento, Trento, Italy (2006)
10. Orduña, A., Lourenço, P.B.: Limit Analysis as a tool for the simplified assessment of ancient masonry structures. In: International Conference on Structural Analysis of Historical Constructions—SAHC01, Guimarães, Portugal, pp. 511–520 (2001)
11. Lourenço, P.B., Milani, G., Tralli, A., Zucchini, A.: Analysis of masonry structures: review of and recent trends of homogenisation techniques. Can. J. Civil Eng. **34**, 1443–1457 (2007)
12. Sacco, E.: A non-linear homogenization procedure for periodic masonry. Eur. J. Mech. A/Solids, Elsevier Masson SAS **28**(2), 209–222 (2009)
13. Costa, A.A., Silva, B., Arêde, A., Guedes, J., Arêde, A., Costa, A.: Experimental assessment, numerical modelling and strengthening of a stone masonry wall. Bull. Earthq. Eng. **10**(1), 135–159 (2012)
14. Lourenço, P.B., Rots, J.G.: A multi-surface interface model for the analysis of masonry structures. J. Eng. Mech. ASCE **123**(7), 660–668 (1997)
15. Lagomarsino, S., Gambarotta, L.: Damage models for the seismic response of brick masonry shear walls. Part I: The mortar joint model and its applications. Earthquake Eng. Struct. Dynam. **26**, 423–439 (1997)
16. Pelá, L., Aprile, A., Benedetti, A.: Seismic assessment of masonry arch bridges. Eng. Struct. **31**(8), 1777–1788 (2009)
17. Mallardo, V., Malvezzi, R., Milani, E., Milani, G.: Seismic vulnerability of historical masonry buildings: a case study in Ferrara. Eng. Struct. **30**(8), 2223–2241 (2007)
18. Roca, P., Massanas, M., Cervera, M., Arun, G.: Structural analysis of Küçük Ayasofya Mosque in Istanbul. IV International Seminar on Structural Analysis of Historical Constructions—SAHC04, Padova, Italy, pp. 679–686, (2004)
19. Martínez, G., Roca, P., Caselles, O., Clapés, J.: Characterization of the dynamic response for the structure of Mallorca Cathedral. V International Seminar on Structural Analysis of Historical Constructions—SAHC06, New Delhi, India, (2006)
20. Faria, R., Oliver, J., Cervera, M.: A strain-based plastic viscous-damage model for massive concrete structures. Int. J. Solids Struct. **35**(14), 1533–1558 (1998)

21. Cervera, M.: Visco-elasticity and rate-dependent continuum damage models. Centre for Numerical Methods in Engineering—CIMNE, Monograph M79, Barcelona, Spain, (2003)
22. Faria, R., Pouca, N.V., Delgado, R.: Simulation of the cyclic behaviour of R/C rectangular hollow section bridge piers via a detailed numerical model. J. Earthquake Eng. **8**(5), 725–748 (2004)
23. Clemente, R., Roca, P., Cervera, M.: Damage model with crack localization—Application to historical buildings. Structural V International Seminar on Structural Analysis of Historical Constructions—SAHC06, New Delhi, India, pp. 1125–1135, (2006)
24. Pelá, L., Cervera, M., Roca, P., Benedetti, A.: An orthotropic damage model for the analysis of masonry structures. VIII International Seminar on Structural Masonry—ISSM 08, pp. 175–178 (2008)
25. Silva, B., Guedes, J., Arêde, A., Costa, A.: Calibration and application of a continuum damage model on the simulation of stone masonry structures: Gondar church as a case study. Bull. Earthq. Eng. **10**(1), 211–234 (2012)
26. Mola, F., Vitaliani, R.: Analysis, diagnosis and preservation of ancient monuments: the St. Mark's Basilica in Venice. Structural analysis of historical constructions I—CIMNE, Barcelona, Spain, pp. 166–188 (1995)
27. Meli, R., Sánchez-Ramíre, A.R.: Structural aspects of the rehabilitation of the Mexico City Cathedral. Structural analysis of historical constructions I—CIMNE, Barcelona, Spain, pp. 123–140 (1995)
28. Macchi, G., Ruggeri, M., Eusebio, M., Moncecchi, M.: Structural assessment of the leaning tower of Pisa. Structural preservation of the architectural heritage, IABSE, Zürich, Switzerland, pp. 401–408 (1993)
29. Croci, G.: The Colosseum: safety evaluation and preliminary criteria of intervention. International Seminar on Structural Analysis of Historical Constructions—SAHC, Barcelona, Spain (1995)
30. Gonzalez, A., Casals, A., Roca, P., Gonzalez, J.L.: Studies of Gaudi's Cripta de la Colonia Güell. International Conference on Composite Construction—Conventional and Innovative—IABSE, Rome, Italy, pp. 457–464, (1993)
31. Collins, M.P., Mitchell, D.: Pre-stressed concrete structures. Response Publications, Canada (1997)
32. Neuenhofer, A., Filippou, F.C.: Evaluation of non-linear frame finite-element models. J. Struct. Eng. **123**(7), 958–966 (1997)
33. Calabrese, A., Almeida, J.P., Pinho, R.: Numerical issues in distributed inelasticity modeling of RC frame elements for seismic analysis. J. Earthquake Eng. **14**(S1), 38–68 (2010)
34. Marti, P.: Basic tools of reinforced concrete beam design. ACI J. Proc. **82**(1), 45–56 (1985)
35. Schlaich, J., Schafer, K., Jennewein, M.: Toward a consistent design of structural concrete. J. Prestress Concr. Inst. **32**(3), 74–150 (1987)
36. Ali, M.A., White, R.N.: Automatic generation of truss model for optimal design of reinforced concrete structures. ACI Struct. J. **98**(4), 431–442 (2001)
37. Liang, Q.Q., Uy, B., Steven, G.P.: Performance-based optimization for strut-tie modeling of structural concrete. J. Struct. Eng. **128**(6), 815–823 (2002)
38. Schlaich, M., Anagnostou, G.: Stress fields for nodes of strut-and-tie models. J. Struct. Eng. **116**(1), 13–23 (1990)
39. Bentz, E.C.: Section analysis of RC members. PhD Thesis. University of Toronto, Canada, (2000)
40. de Souza, R.M.: Force-based finite element for large displacements inelastic analysis of frames. PhD Thesis. University of California, Berkeley, US, (2000)
41. Sezen, H.: Seismic behavior and modeling of reinforced concrete building columns. PhD Thesis. University of California, Berkeley, US, (2002)

Seismic Vulnerability and Risk Assessment of Historic Masonry Buildings

Romeu Vicente, Dina D'Ayala, Tiago Miguel Ferreira, Humberto Varum, Aníbal Costa, J. A. R. Mendes da Silva and Sergio Lagomarsino

Abstract Seismic risk evaluation of built-up areas involves analysis of the level of earthquake hazard of the region, building vulnerability and exposure. Within this approach that defines seismic risk, building vulnerability assessment assumes great importance, not only because of the obvious physical consequences in the eventual occurrence of a seismic event, but also because it is the one of the few potential aspects in which engineering research can intervene. In fact, rigorous vulnerability assessment of existing buildings and the implementation of appropriate retrofitting solutions can help to reduce the levels of physical damage, loss of life and the economic impact of future seismic events. Vulnerability studies of urban centres

R. Vicente (✉) · T. M. Ferreira · H. Varum · A. Costa
Civil Engineering Department, Aveiro University, 3810-193 Aveiro, Portugal
e-mail: romvic@ua.pt

T. M. Ferreira
e-mail: tmferreira@ua.pt

H. Varum
e-mail: hvarum@ua.pt

A. Costa
e-mail: agc@ua.pt

D. D'Ayala
Civil Environmental and Geomatic Engineering, University College London,
WC1E 6BT, London, UK
e-mail: d.d'ayala@ucl.ac.uk

S. Lagomarsino
Department of Structural and Geotechnical Engineering,
University of Genova, 16145 Genova, Italy
e-mail: sergio.lagomarsino@unige.it

J. A. R. M. da Silva
Civil Engineering Department, Coimbra University, 3030-788 Coimbra, Portugal
e-mail: raimundo@dec.uc.pt

A. Costa et al. (eds.), *Structural Rehabilitation of Old Buildings*,
Building Pathology and Rehabilitation 2, DOI: 10.1007/978-3-642-39686-1_11,
© Springer-Verlag Berlin Heidelberg 2014

should be developed with the aim of identifying building fragilities and reducing seismic risk. As part of the rehabilitation of the historic city centre of Coimbra, a complete identification and inspection survey of old masonry buildings has been carried out. The main purpose of this research is to discuss vulnerability assessment methodologies, particularly those of the first level, through the proposal and development of a method previously used to determine the level of vulnerability, in the assessment of physical damage and its relationship with seismic intensity.

Keywords Vulnerability • Risk • Masonry • Fragility curves • Damage scenarios

1 Vulnerability Assessment and Risk Evaluation

The assessment of the vulnerability of the building stock of an urban centre is an essential prerequisite to its seismic risk assessment. The other two ingredients are the expected hazard over given return periods and the distribution and values of the assets constituting the building stock. All three elements of the seismic risk assessment are affected by uncertainties of aleatory nature, related to the spatial variability of the parameters involved in the assessment, and epistemic, related to the limited capacity of the models used to capture all aspects of the seismic behaviour of buildings and of describing them in simple terms, suitable for this type of analysis. Hence it should always be kept in mind that the computation of a risk level is highly probabilistic, and that to accurately represent the risk the expected values should always be accompanied by a measure of the associated dispersion. A very preliminary estimate of the seismic capacity of the local building stock can be obtained by consulting the requirement included in the seismic standards and code of practices in force at the time of construction of such buildings. This information together with a temporal and spatial record of the growth of the urban centre can provide a first definition of classes of buildings assumed to have different capacity class by class. This information can be obtained by looking at past and present cadastral maps with ages of buildings and knowing the historical development and enforcement of codes at the site. In general however for a correct assessment of the seismic risk a more detailed inventory and classification should be considered, the extent of which is a function of the economic and technical resources available and of the extent of the area under investigation and the diversity within the building stock.

In the case of historic masonry buildings constituting the core of city centres data on their structural layout and lateral capacity cannot generally be obtained from seismic standard, as this do not include these buildings typologies. However in the last twenty years extensive historical studies on the development of so-called non engineered structural typologies and documentation of the associated local construction techniques have been produced in many region of Europe exposed to significant earthquake hazard. These studies tend to provide construction details and qualitative assessment that can constitute some of the ingredients of a more structured analytical vulnerability assessment, based on engineering principles. For instance a study at the urban scale can provide insight on the shape of single buildings and aggregate and hence an understanding of the interactions among buildings. Details on floor construction and layout, type and layout of masonry, presence of connections among walls, can lead the seismic assessor to a qualitative judgement of relative robustness and resilience of different construction solutions. It is however only when these details are interpreted within a mechanical framework and the relations among the parts expressed in mathematical terms that the relevance of the various parameters to the overall seismic behaviour can be established and the relative vulnerability of different objects quantified with a measured level of reliability. To achieve so, such information cannot be simply descriptive, but needs to be collected in a systematic way to be used in mathematical models. Moreover in order to correctly measure the level of uncertainty and hence reliability of the risk assessment of a particular urban centre, the sampling and data collection needs to follow some consistent rules.

The appropriate approach to a seismic risk assessment at territorial scale needs to address diverse issues, to balance the relative simplicity of the analysis vis-à-vis the variability in the building stock, so as to properly represent the diverse typologies present and hence accurately characterise the global vulnerability and cumulative fragility, while explicitly accounting for the uncertainties related to modelling limitation, the inherent randomness of the sample, and the randomness of the response.

For ordinary buildings seismic risk assessment is typically carried out for the performance condition of life safety and collapse prevention, related to a seismic hazard scenario related to a 10 % probability of exceedence in 50 year or 475 year return period. For historic buildings in city centre and in case of assets of particular value, it might be more appropriate to consider the performance condition of damage limitation or significant damage associated to lower-intensity and shorter return period seismic hazard. Recently has been argued by the author that for specific studies of high value historic buildings, such as the ISMEP project [1], and where sufficient information on the seismicity of the region is available, such as the case of Istanbul, a deterministic analysis can be used to define the hazard, rather than the probabilistic one, and consider the most credible seismic scenario within the set timeframe of assessment.

In the following sections of the chapter, after a review of earlier approaches to seismic vulnerability, the derivation of fragility functions is illustrated for three different methodologies: an empirical approach based on a modified version of the Vulnerability Index [2], an analytical approach based on mechanical simulation called FaMIVE [3] and a similar analytical approach for aggregates.

2 Vulnerability Assessment Methodologies

As stated in the introduction, when performing vulnerability assessment of large numbers of buildings and over an urban centre or a region, the resources and quantity of information required is large and thus the use of less sophisticated and onerous inspection and recording tools is more practical. Methodologies for vulnerability assessment at the national scale should hence be based on few parameters, some of an empirical nature based on knowledge of the effects of past earthquakes, which can then be treated statistically.

In the recent past, European partnerships [4–6] constituting various work-groups on different aspects of vulnerability assessment and earthquake risk mitigation have defined, particularly for the former, methodologies that are grouped into essentially three categories in terms of their level of detail, scale of evaluation and use of data (first, second and third level approaches). First level approaches use a considerable amount of qualitative information and are ideal for the development of seismic vulnerability assessment for large scale analysis. Second level approaches are based on mechanical models and rely on a higher quality of information (geometrical and mechanical) regarding building stock. The third level involves the use of numerical modelling techniques that require a complete and rigorous survey of individual buildings. The definition and nature of the approach (qualitative and quantitative) naturally condition the formulation of the methodologies and the level at which the evaluation is conducted, from the expedite evaluation of buildings based on visual observation to the most complex numerical modelling of single buildings (see Fig. 1).

A most important criterion of distinguishing vulnerability approaches for historic buildings, is whether the method is purely empirical, i.e. based on obser-vation and record of damage in past earthquake, from which a correlation between

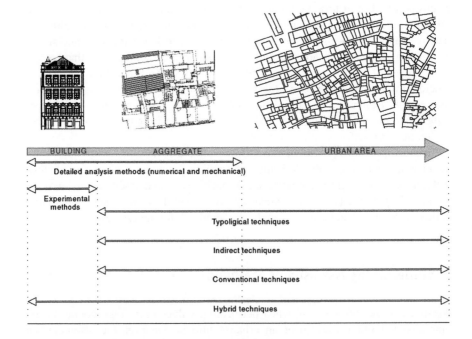

Fig. 1 Analytical techniques used at different evaluation scales

building typologies and damage level given a seismic intensity level can be derived, or analytical, where a model of a representative building for a typology is defined, and the response of such model to expected shaking intensities is computed. The first approach is particularly suited to historic city centres where a record of past earthquakes is available and damage to the building has been collected systematically over a number of events. This is for instance the case of the GNDT-AEDES approach developed in Italy over several decades from the earthquake in Friuli onwards [7]. The second approach is suitable to areas for which construction details are recorded and well understood, there might be some experimental work available to characterise their mechanical behaviour, there is some record of damage to calibrate the procedure, but most importantly they are suitable to be used to produce scenarios for future event and help define strengthening strategies, at the level of the single building, urban block, district or entire city [4].

A third approach is the heuristic or expert opinion approach by which vulnerability is attributed to building typologies by a panel of experts elicited to perform an assessment based on a common set of information and their previous knowledge. An example of such approach is the development of the vulnerability classes defined within the European Macroseismic Scale EMS-98 [8]. To the above three approaches a fourth, hybrid, can be added.

Following the first example of such classification developed by [8] and refined by [9], vulnerability approaches can also be grouped in direct and indirect. A brief review of the most significant approaches in each group is included in the reminder of this section.

Direct techniques use only one step to estimate the damage caused to a structure by an earthquake, employing two types of methods; *typological* and *mechanical*:

Typological methods—classify buildings into classes depending on materials, construction techniques, structural features and other factors influencing building response. Vulnerability is defined as the probability of a structure to suffer a certain level of damage for a defined seismic intensity. Evaluation of damage probability is based on observed and recorded damage after previous earthquakes and also on expert knowledge. Results obtained using this method must be considered in terms of their statistical accuracy, since they are based on simple field investigation. In effect the results are valid only for the area assessed, or for other areas of similar construction typology and equal level of seismic hazard. Examples of this method are the vulnerability functions or Damage Probability Matrices (DPM) developed by [9], in which a matrix for each building type or vulnerability class is defined that directly correlates seismic intensity with probable level of damage suffered.

Mechanical methods—predict the seismic effect on the structure through the use of an appropriate mechanical model, which may be more or less complex, of the whole building or of an individual structural element. Methods based on simplified mechanical models are more suitable for the analysis of a large number of buildings as require only a few input parameters, modest computing burden and can lead to reliable quantitative evaluations. A commonly used method belonging to this group is the limit state method, based on limit state analysis (displacement capacity and demand) [10] applied this method to the analysis of the historic city centre of Catania considering only in plane mechanisms. The FAMIVE method [11] is a more holistic and reliable mechanically-based method, considering a suite of different mechanisms directly correlated to structural and constructional features. More sophisticated methods are generally used to evaluate single structures at a higher level of detail (in terms of building structure and construction) and are based on more refined modelling techniques. The analytical procedure for this type of method can involve non-linear static push over analysis such as the methodologies at [12, 13] and Capacity Spectrum Method (CSM) [14]. Examples of CSM application are provided in Sects. 3.2 and 3.3.

Indirect techniques initially involve the determination of a vulnerability index, followed by establishment of the relationships between damage and seismic intensity, supported by statistical studies of post-earthquake damage data. This form of evaluation is used extensively in the analysis of vulnerability on a wide scale. Of the various techniques currently available, the methodology initially developed by GNDT in the 1980s has undergone various modification and applications, for example Catania in 1999 and Molise in 2001 [7]. The method involves the determination of a building vulnerability classification system (vulnerability index) based on observation of physical construction and structural characteristics. Each building is classified in terms of a vulnerability index related to a damage grade determined via the use of vulnerability functions. These functions enable the formulation, of the damage suffered by buildings for each level of seismic intensity (or peak ground acceleration, PGA) and vulnerability index. These types of methods use extensive databases of building characteristics (typological and mechanical properties) and rely on observed damage after previous earthquakes to classify vulnerability, based on a score assignment. The rapid screening ATC-21 technique (1988) is extensively used in the U.S. to obtain such a vulnerability score [15]. An example of application of GNDT approach is shown in Sect. 3.1

Conventional techniques are essentially heuristic, introducing a vulnerability index for the prediction of the level of damage. There are essentially two types of approach: those that qualify the different physical characteristics of structures empirically and those based on the criteria defined in seismic design standards for structures, evaluating the capacity-demand relationship of buildings. ATC-13 [16], the best known of the first type, defines damage probability matrices for 78 classes of structure, 40 of which refer to buildings. Uncertainty is treated explicitly through a probabilistic approach. The HAZUS methods [17] belongs to the second type, providing parameters for capacity curves and damage through the CSM approach. Damage level are derived heuristically for 36 building classes [18]. For each construction type and level of earthquake-resistant design, the capacity of the structure, spectral displacement and inter-story drift limit are defined for different levels of damage.

Hybrid techniques combine features of the methods described previously, such as vulnerability functions based on observed vulnerability and expert judgment, in

DB Managing System   GIS Application

Fig. 2 Database and GIS framework (from [20])

which vulnerability is based on the vulnerability classes defined in the European Macroseismic Scale, EMS–98 [8]. This is the case in the Macroseimic Method devised by [19], which combines the characteristics of typological and indirect methods using the vulnerability classes defined in the EMS-98 scale and a vulnerability index improved by the use of modification factors.

For a robust decision making process following a risk analysis of a region it is essential to visualise and interpret the results considering their spatial distribution. The use of relational database within a GIS environment, allows to manage data regarding historic building stock characteristics, conservation requirements, seismic vulnerability, damage and loss scenarios, cost estimation and conduct risk-impact assessment.

Figure 2 represents such an application. Such platforms allow visualising both collected data and damage distributions for different hazard scenarios, and depending on the resolutions results can be mapped down to a single building.

3 Vulnerability of Historic Masonry Buildings

3.1 Empirical Approach

Historic masonry buildings do not have adequate seismic capacity and consequently require special attention due to their incalculable historical, cultural and architectural value. The amount of resources spent on their vulnerability assessment and structural safety evaluation is justifiable, since not only does a first level assessment [21, 22] include building inspection, but also can help in the identification of building for which a more detailed assessment is required, as well as the definition of priorities for both retrofitting and in support of earthquake risk management [23]. The definition and validation of a scoring method for the urban scale assessment of historic building is described in this section.

The methodology presented here can be classified as a hybrid technique. The vulnerability index formulation proposed is based essentially on the GNDT Level II approach [7] based on post-seismic damage observation and survey data covering a vast area, focussing on the most important parameters affecting building damage which must be surveyed individually.

Overall vulnerability is calculated as the weighted sum of 14 parameters (see Table 1) used in the formulation of the seismic vulnerability index. These parameters are related to four classes of increasing vulnerability: A, B, C and D. Each parameter represents a building feature influencing building response to seismic activity. A weight pi is assigned to each parameter, ranging from 0.50 for the less important parameters (in terms of structural vulnerability) up to 1.5 for the most important (for example parameter P3 represents conventional strength) as shown in Table 1. The vulnerability index obtained as the weighted sum of the 14 parameters initially ranges between 0 and 650, with the value then normalised to fall within the range $0 < I_v < 100$. The calculated vulnerability index can then be used to estimate building damage due to a seismic event of given intensity.

This procedure has been used in Italy for the last 25 years and was later adapted by [24] for Portuguese masonry buildings and improved by: (i) introducing a more detailed analysis based on better data on the building stock; (ii) clarifying the definition of some of the most important parameters; and (iii) introducing new parameters that take into account the interaction between buildings (structural aggregates) and other overlooked building features. The addition of parameters P5, P7 and P10 provides: the height of the building (P5); the interaction between contiguous buildings (P7)—a very important feature when assessing buildings in urban areas; and the alignment of wall façade openings which affects the load path and load bearing capacity (P10).

The 14 parameters are arranged into four groups, as shown in Table 1, in order to emphasise their differences and relative importance (see [24]). The first group includes parameters P1 and P2 characterising the building resisting system, the type and quality of masonry, through the material (size, shape and stone type), masonry fabric, arrangement and quality of connections between walls; P3 roughly estimates the shear strength capacity; P4 evaluates the potential risk of out-of-plane collapse, P5 evaluate the height and P6 the foundation soil. The second group of parameters is mainly focused on the relative location of a building in the area as a whole and on its interaction with other buildings (parameter P7). This feature, not considered in other methodologies, is extremely important, since the seismic response of a group of buildings is rather different to the response of a single building. Parameters P8 and P9 evaluate irregularity in plan and height, while parameter P10 identifies the relative location of openings, which is important in terms of the load path. The third group of parameters, which includes P11 and P12, evaluates horizontal structural systems, namely the type of connection of the timber floors and the thrust of pitched roofing systems. Finally, P13 evaluates structural fragilities and conservation level of the building, while P14 measures the negative influence of non-structural elements with poor connections to the main structural system. As can be seen in Table 1, among all parameters, P3, P5

Table 1 Vulnerability index (I_v)

Parameters	Class C_{vi}				Weight p_i	Vulnerability index
	A	B	C	D		
1. Structural building system						
P1 Type of resisting system	0	5	20	50	0.75	
P2 Quality of the resisting system	0	5	20	50	1.00	
P3 Conventional strength	0	5	20	50	1.50	$I_v^* = \sum_{i=1}^{14} C_{vi} \times p_i$
P4 Maximum distance between walls	0	5	20	50	0.50	
P5 Number of floors	0	5	20	50	1.50	
P6 Location and soil conditions	0	5	20	50	0.75	
2. Irregularities and interaction						
P7 Aggregate position and interaction	0	5	20	50	1.50	
P8 Plan configuration	0	5	20	50	0.75	
P9 Regularity in height	0	5	20	50	0.75	
P10 Wall facade openings and alignments	0		20	50	0.50	
3. Floor slabs and roofs						Normalised index
P11 Horizontal diaphragms	0	5	20	50	1.00	$0 \leq I_v \leq 100$
P12 Roofing system	0	5	20	50	1.00	
4. Conservation status and other elements						
P13 Fragilities and conservation state	0	5	20	50	1.00	
P14 Non-structural elements	0	5	20	50	0.50	

and P7 have the highest weight values (p_i) in the vulnerability index. On the other hand, parameter P2, P11, 12 and P13 are those whose increase could be defined as representing a strengthening action (masonry consolidation, timber floor stiffening, retrofitting of trussed roofing systems, effective connection between horizontal and vertical structural elements and building maintenance strategy).

The definition of each parameter weight is a major source of uncertainty as it is based on expert opinion. Consequently in order for the results to be accurately interpreted statistically, upper and lower bounds of the vulnerability index, I_v were defined. This method can be considered robust when two conditions are verified: (i) the inspection of the majority of buildings under analysis was carried out in detail; and (ii) accurate geometrical information was available. A confidence level indicator is associated with each parameter, so that the vulnerability index is also coupled to a confidence level rating.

To resolve the conflict of a detailed inspection versus a large number of building to be inspected in an urban area a strategy is chosen to undertake a vulnerability assessment in two phases: in the first phase, an evaluation of vulnerability index, I_v, is made for those buildings for which detailed information is available—geometrical and morphological information, blue prints, survey sheets, etc. -; in the second phase a more expeditious approach is adopted, based on the mean values obtained from the first phase. The underlying assumption is that masonry building characteristics are homogeneous in the region under study. The mean vulnerability index value obtained for all masonry buildings in the first detailed evaluation is used as vulnerability index for a typology, to be weighed by modifiers for each building. Classification of these modifiers will affect the total vulnerability index computed in Table 1 as sum of all the weighed parameters, some of which act as modifiers of the mean score.

Table 2 presents the seven modifier parameters and their scores in relation to the average vulnerability value for each parameter. The vulnerability index, I_v, is defined according to the sum of the modifier parameter scores for each non-detailed assessment.

Table 2 Vulnerability modifier factors and scores

Vulnerability modifiers	Vulnerability classes, c_{vi}				Modified score:
	A	B	C	D	$\dfrac{p_i}{\sum_{i=1}^{7} p_i} \times (c_{vi} - \bar{c}_{vi})$
P5 Number of floors	−4.1	−3.1	0.0	6.2	
P6 Location and soil conditions	−0.5	0.0	1.6	4.7	p_i: parameter, i,
P7 Aggregate position and interaction	−1.0	0.0	3.1	9.3	weight assigned
P8 Plan configuration	−2.1	−1.6	0.0	3.1	$\sum_{i=1}^{7} p_i$: sum of
P9 Regularity in height	−2.1	−1.8	0.0	3.1	parameter weights
P12 Roofing system	−2.8	−2.1	0.0	4.1	
P13 Fragilities and conservation state	−2.8	−2.1	0.0	4.1	c_{vi}: modifier factor vulnerability class
Maximum modifier range, $\sum \Delta I_v$	−15.3	−10.3	4.7	34.7	$\bar{c}_{vi}$: average vulnerability class of parameter i

The scores for each parameter are defined with respect to the average value of that parameter obtained for the mean value of the vulnerability index and the weight of each parameter in the overall definition. For example, as the mean vulnerability class value for parameter P8 (plan configuration) obtained by the detailed assessment is taken as that of class C, the modifier scores are computed with respect to this average value. The final vulnerability is defined as:

$$\overline{\overline{I_v}} = \overline{I_v} + \sum \Delta I_v \tag{1}$$

where $\overline{\overline{I_v}}$ is the final vulnerability index, $\overline{I_v}$ is the average vulnerability index from the detailed assessment and $\sum \Delta I_v$ is the sum of the modified scores.

It is then possible to estimate damage associated with a certain level of seismic intensity, I, described in terms of macroseismic intensity [8]. The validation of this vulnerability index method was carried out by [21] through correlation between the GNDT II method [2] and the EMS-98 Macroseismic Scale, as indicated in [20]. On the basis of the EMS-98 scale damage definitions it is possible to derive damage probability matrices for each of the defined vulnerability classes (A–F). Through numerical interpretation of the linguistic definitions, Few, Many and Most, complete Damage Probability Matrices (DPM) for every vulnerability class may be obtained. Having solved the incompleteness using probability theory, the ambiguity and overlap of the linguistic definitions is then tackled using fuzzy set theory [25], by deriving, for each building typology and vulnerability class, upper and lower limits for the correlation between macroseismic intensity and mean damage grade.

For the operational implementation of the methodology, an analytical expression is proposed [26] which correlates hazard with the mean damage grade ($0 < \mu_D < 5$) of the damage distribution (discrete beta distribution) in terms of the vulnerability value, as shown in Eq. 2.

$$\mu_D = 2.5 + 3 \times tanh \left(\frac{I + 6.25 \times V - 12.7}{Q} \right) \times f(V,I) \tag{2}$$

where I is the seismic hazard described in terms of macroseismic intensity, V the vulnerability index as calculated by [20], Q a ductility factor and $f(V, I)$ is a function of the vulnerability index and intensity. The latter is introduced in order to understand the trend of numerical vulnerability curves derived from EMS-98 DPMs for lower values of the intensity grades ($I = V$ and VI) where:

$$f(V,I) = \begin{cases} e^{\frac{V}{2} \times (I-7)} & I \leq 7 \\ 1 & I > 7 \end{cases} \tag{3}$$

This analytical expression derives from the interpolation of vulnerability curves calculated from the completed DPMs, as suggested in the EMS-98 scale. Used to estimate physical damage, this mathematical formulation is based on work previously proposed by [27]. The vulnerability index, V, determines the position of the curve, while the ductility factor, Q, determines the slope of the vulnerability function (rate of damage increases with rising intensity). Regression analysis and parametric studies performed by [27] lead to a mean value of $Q = 3.0$ being suggested for masonry buildings of fairly ductile behaviour.

Based on the comparison between both the methods (see [21]), the following analytical expression for the vulnerability index, V, was derived:

$$V = 0.56 + 0.0064 \times I_v \tag{4}$$

Via this relationship, the vulnerability index, I_v, can be transformed into the vulnerability index, V (used in the Macroseismic Method), enabling the calculation of the mean damage grade through Eq. 2 and subsequently the estimation of damage and loss. For those buildings where detailed evaluation was not carried out, the mean vulnerability index can be defined as a function of the vulnerability classes defined in terms of the EMS-98 scale. In this case, the modifier parameters can also be expressed in vulnerability index V format, taking into account the I_v values re-defined in Eq. 4.

Once vulnerability has been defined, the mean damage grade, μ_D, can be calculated for different macroseismic intensities, using Eq. 2. Figure 3 shows one example of vulnerability curves for a mean value of vulnerability index, $I_{v,mean}$, as well as for the upper and lower bound ranges ($I_{v,mean} - 2\sigma_{Iv}$; $I_{v,mean} - 1\sigma_{Iv}$; $I_{v,mean} + 1\sigma_{Iv}$; $I_{v,mean} + 2\sigma_{Iv}$). From these mean damage grade values, μ_D, different damage distribution histograms for events of varying seismic intensity and their respective vulnerability index values can be defined, using a probabilistic approach. The most commonly-applied methods are based on the binomial probability mass function and the beta probability density function.

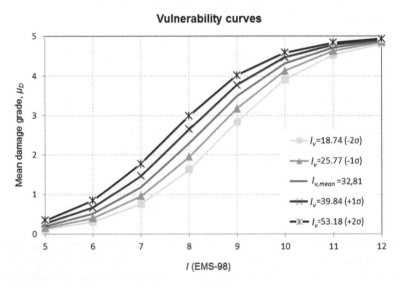

Fig. 3 Example of vulnerability curves for an old building stock

$$\text{PMF} : p_k = \frac{n!}{k!\,(n-k)!} \times d^k \times (1-d)^{n-k} \, n \geq 0;\, 0 \leq p \leq 1 \qquad (5)$$

The damage distribution fits to a beta distribution function, where t and r are geometric parameters associated with the damage distribution. Research carried out by [25] shows that the beta distribution is the most versatile, as variation of t and r enables the fitting of both very narrow and broad damage distributions. This continuous beta probability density function in which Γ is the known gamma function is expressed as:

$$\text{PMF} : p_\beta(x) = \frac{\Gamma(t)}{\Gamma(r)\,\Gamma(t-r)}\,(x-a)^{r-1}\,(b-x)^{t-r-1}$$
$$;\, a \leq x \leq b;\, a = 0;\, b = 5 \qquad (6)$$

Assuming that a $= 0$ and b $= 5$, it can then be simplified to:

$$\text{PMF} : p_\beta(x) = k(t,r) \times x^{r-1} \times (5-x)^{t-r-1} \qquad (7)$$

where for a continuous variable x, the variance (σ_x^2) and the mean value (μ_x) of the values are related to t and r as defined below:

$$t = \frac{\mu_x(5-\mu_x)}{\sigma_x^2}, \qquad (8)$$

$$r = t.\frac{\mu_x}{5} \qquad (9)$$

The discrete distribution of the probability associated with each damage grade, D_k, with $k \in [0,5]$, is defined as:

$$P(D_0) = p(0) = \int_0^{0.5} k(t,r) \cdot x^{r-1}(5-x)^{t-r-1}\,dx$$

$$P(D_k) = p(k) = \int_{k-0}^{k-0.5} k(t,r) \cdot x^{r-1}(5-x)^{t-r-1}\,dx \qquad (10)$$

$$P(D_5) = p(5) = \int_{4.5}^{5} k(t,r) \cdot x^{r-1}(5-x)^{t-r-1}\,dx$$

For the definition of parameters t and r in the beta discrete distribution, the numerical damage distributions derived from the EMS-98 scale [26] can be used. The reduced variation obtained for parameter t in the numerical damage distributions justifies the adoption of a unique value of t (equal to 8) with which to represent the variance of all possible damage distributions. Based on this assumption, it is then possible to define the damage distributions exclusively through use of the average value μ_D, characterized by variance coherent with that found via completion of the EMS-98 DPM's.

$$r = 8 \cdot \frac{\mu_D}{5} \tag{11}$$

Figure 4 presents examples of damage distributions obtained through use of the beta probability distribution ($t = 8$; $a = 0$; $b = 5$) for events of different seismic intensity and the mean value of the building vulnerability index ($I_v = 32.88$).

Another method of representing damage using damage distribution histograms involves the use of fragility curves. Here the probability of exceeding a certain damage grade or state, D_k ($k \in [0,5]$) is obtained directly from the physical building damage distributions derived from the beta probability function for a determined building typology. Just like the vulnerability curves, fragility curves define the relationship between earthquake intensity and damage in terms of the conditional cumulative probability of reaching a certain damage state. Probability histograms of a certain damage grade, $P(D_k = d)$, are derived from the difference of cumulative probabilities:

$$P\left(D_k = d\right) = P_D\left[D_k \geq d\right] - P_D\left[D_{k+1} \geq d\right] \tag{12}$$

Fragility curves are influenced by the parameters of the beta distribution function and allow for the estimation of damage as a continuous probability function. Figure 5 shows fragility curves corresponding to the damage distribution histograms of the mean vulnerability index value (I_v) as well as of the mean value plus one standard deviation ($I_v + \sigma_{I_v}$).

The next step in a risk assessment process is the estimation of losses. Loss estimation models can also be based on damage grades and involve correlating the probability of the occurrence of a certain damage level with the probability of building collapse and loss of functionality. The most frequently employed approaches are those based on observed damage data, such as the one proposed in [17] or that of the Italian National Seismic Survey. The latter was based on work by [28] which involved the analysis of data associated with the probability of

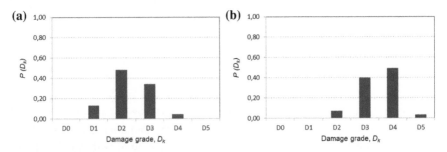

Fig. 4 Discrete damage distribution histograms for $I_v = 32.88$: **a** I(EMS-98) = VIII; **b** I(EMS-98) = IX

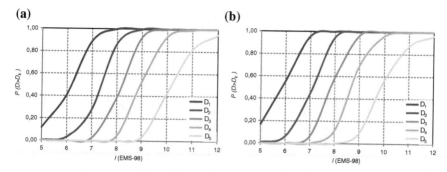

Fig. 5 From *left* to *right*: examples of fragility curves for I_v and $I_v + \sigma_{Iv}$

buildings to be deemed unusable after minor and moderate earthquakes. Although such events produce lower levels of structural and non-structural damage, higher mean damage values may occur which are associated with a higher probability of building collapse.

The probability of the occurrence of each damage grade is multiplied by a factor. This range from 0 to 1 and differs from proposal to proposal, based on statistical correlation. In Italy, data processing undertaken by [28] enabled the establishment of these weighted factors and respective expressions for their use in the estimation of building loss. For the analysis of collapsed and unusable buildings the following equations have been derived:

$$P_{collapse} = P(D_5) \qquad (13)$$

$$P_{unusable\ buildings} = P(D_3) \times W_{ub,3} + P(D_4) \times W_{ub,4} \qquad (14)$$

where $P(D_k)$ is the probability of the occurrence of a certain level of damage (D_1 to D_5) and $W_{ub,3}$, $W_{ub,4}$ are weights indicating the percentage of buildings associated with the damage level D_k, that have suffered collapse or that are considered unusable. The values of the weighting factors presented in the SSN [28] and [17] are slightly different. The weights: $W_{ub,3} = 0.4$; $W_{ub,4} = 0.6$; can be used for the evaluation of stone masonry buildings.

Figure 6 shows an example of probability curves which describe the results of building collapse and unusable building estimations for the mean value of the vulnerability index (I_v) as well as for other values of vulnerability, namely: ($I_{v,mean} - 2\sigma_{Iv}$; $I_{v,mean} - 1\sigma_{Iv}$; $I_{v,mean} + 1\sigma_{Iv}$; $I_{v,mean} + 2\sigma_{Iv}$).

One of the most serious consequences of an earthquake is the loss of human life and thus one of the major goals of risk mitigation strategies is ensuring human safety. Over the last hundred years the world has been struck by more than 1,250 strong earthquakes and over 1.5 million people have died as a consequence [29]. However official numbers are not always accurate and the actual totals may be much higher. Of the various casualty rate analyses and correlation laws found in the literature, those developed by [1, 29–31] are the most frequently cited.

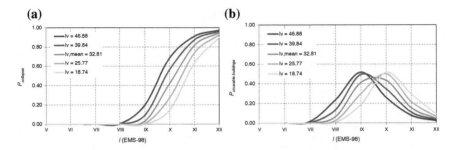

Fig. 6 Estimate of the collapsed and unusable buildings for different vulnerability values

Once again the Italian proposal [28] is presented here for consistency with the loss assessment procedure. The rate of dead and severely injured is projected as being 30 % of the residents living in collapsed and unusable buildings, with the survivors assumed to require short term shelter. Casualty (dead and severely injured) and homelessness rates are determined via Eqs. 15 and 16 respectively.

$$P_{dead\ and\ severely\ injured} = 0.3 \times P(D_5) \tag{15}$$

$$P_{hom\,eless} = P(D_3) \times W_{ub,3} + P(D_4) \times W_{ub,4} + P(D_5) \times 0.7 \tag{16}$$

These two indicators are of great interest for risk management. Following the same logic, Fig. 7 shows an estimation of the numbers of dead, severely injured and homeless for the mean value of the vulnerability index (I_v), as well as for other vulnerability values ($I_{v,mean} - 2\sigma_{Iv}$; $I_{v,mean} - 1\sigma_{Iv}$; $I_{v,mean} + 1\sigma_{Iv}$; $I_{v,mean} + 2\sigma_{Iv}$).

Finally, the estimated damage grade can be interpreted economically, as defined by [2], i.e. the ratio between the repair cost and the replacement cost (building value). The correlation between damage grades and the repair and rebuilding costs are obtained by processing of post-earthquake damage data. As shown in Table 3, a variety of correlations are found in literature.

The most reasonable relationship, as confirmed by the post-seismic investigation of [32], is that which assumes a similar value of the damage index for damage

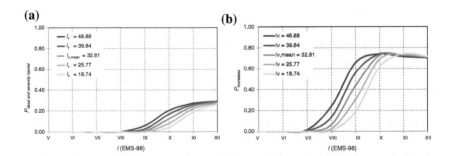

Fig. 7 Estimation of homeless and casualty rate for different values of vulnerability

Table 3 Correlation between damage levels and damage index

Damage grade, D_k	0	1	2	3	4	5
[28]	0.000	0.010	0.100	0.350	0.750	1.000
[16]	0.000	0.050	0.200	0.550	0.900	1.000
[31]	0.005	0.035	0.145	0.305	0.800	0.950

grade 4 and 5 and a greater difference between the damage index for the lower damage grades of 1 and 2. The values obtained by [16] and [31] are in agreement with these criteria. The statistical values obtained by these authors were derived from analysis of the data collected, using the GNDT-SSN procedure, after the Umbria-Marche (1997) and the Pollino (1998) earthquakes [31], and based on the estimated cost of typical repairs for more than 50,000 buildings.

The probabilities of the repair costs are defined as the product of the following two probabilities: The conditional probability of the repair cost for each damage level, $P[R|D_k]$, expressed by the values presented in Table 3, and the known conditional probability of the damage condition for each level of building vulnerability and seismic intensity, $P[D_k |v, I]$, given by:

$$Prob[R|I] = \sum_{D_k=1}^{5} \sum_{I_v=0}^{100} Prob[R|D_k] \times Prob[D_k|I_v,I] \tag{17}$$

These values should be calculated for both the mean vulnerability index value and the lower and upper bound values ($I_{v,mean} - 2\sigma_{Iv}$; $I_{v,mean} - 1\sigma_{Iv}$; $I_{v,mean} + 1\sigma_{Iv}$; $I_{v,mean} + 2\sigma_{Iv}$). Note that according to this methodology, for seismic events of intensity in the range of V–IX the variation between estimated minimum and maximum repair cost is significant. For higher earthquake intensities, the difference is much smaller as a result of the high damage levels caused by severe seismic events.

3.2 Analytical Mechanical Approach: FaMIVE

The seismic vulnerability assessment of unreinforced masonry or adobe historic buildings can be performed with the Failure Mechanisms Identification and Vulnerability Evaluation (FaMIVE) analytical method, developed in [3, 33]. The FaMIVE method uses a nonlinear pseudo-static structural analysis with a degrading pushover curve to estimate the performance points by way of a variant of the N2 method [14], included in EC8 part 3 [34]. It yields as output collapse multipliers which identify the occurrence of possible different mechanisms for a given masonry construction typology, given certain structural characteristics.

Developed over the last decade, it is based on a suite of 12 possible failure mechanisms directly correlated to in situ observed damage [33, 35, 36] and laboratory experimental validation [37] as shown in Fig. 8.

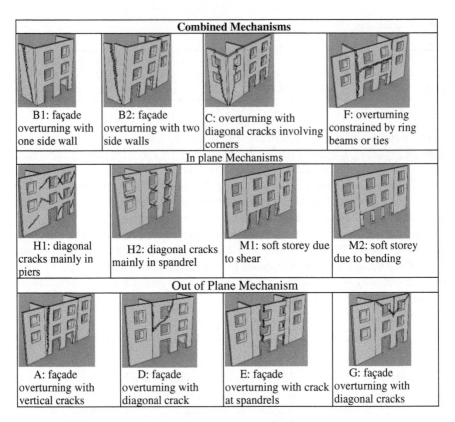

Fig. 8 Mechanisms for computation of limit lateral capacity of masonry façades

Each mode of failure corresponds to different constraints conditions between the façade and the rest of the structure, hence a collapse mechanism can be univocally defined and its collapse load factor computed. As shown in the flowchart of Fig. 9 the programme FaMIVE, first calculates the collapse load factor for each façade in a building, then taking into account geometric and structural characteristics and constraints, identifies the one which is most likely to occur considering the combination of the largest portion mobilised with the lowest collapse load factor at building level.

The FaMIVE algorithm produces vulnerability functions in terms of ultimate lateral capacity for different building typologies and quantifies the effect of strengthening and repair intervention on reduction of vulnerability. In its latest version it also computes capacity curves, performance points and outputs fragility curves for different seismic scenarios in terms of intermediate and ultimate displacements or ultimate acceleration. Within the FaMIVE database capacity curves and fragility functions are available for various unreinforced masonry typologies, from adobe to concrete blocks, for a number of reference typologies studied at sites in Italy [33, 36], Spain, Slovenia [38], Turkey [1], Nepal, India, Iran and Iraq. The procedure has been validated against the EMS-98 vulnerability classes [8, 26]

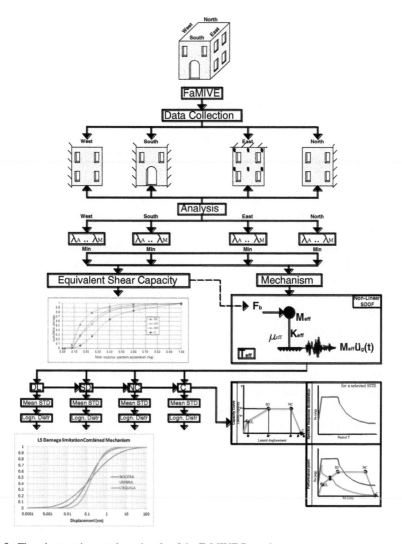

Fig. 9 Flowchart setting out the rationale of the FaMIVE Procedure

and recently used to produce capacity curves and fragility curves for use in the USGS PAGER environment [39, 40].

The mechanism's characteristics are used to derive an equivalent non-linear single degree of freedom capacity curve to be compared to a spectrum demand curve, and eventually define performance points as illustrated in the flowchart in.

3.2.1 Definition of Damage Limit States and Damage Thresholds

In order to derive fragility curves the next step consist of defining limit state performance criteria to be correlated to damage states. This step is fraught with uncertainties, as very limited consolidate evidence exist to perform such correlation over a wide range of building typologies and shaking levels. While robust database of damage states exist in literature no attempt has been so far made to record permanent drift and corresponding ground shaking in a consistent way, so as to provide empirical evidence for capacity curves. As an alternative, a number of authors have worked on correlating performance indicator and damage indicator on experimentally obtained capacity curves, by way of shaking table tests or push-over tests [41, 42]. The major limitation of these tests have been carried out focusing only on the capacity of in-plane walls, while very limited experimental work has been conducted on the characterisation of out-of-plane capacity for URM [43] have considered the out-of-plane failure of URM bearing walls constrained by flexible diaphragm, however the support conditions predefine the failure mode with three horizontal cylindrical hinges, already highlighted by [44], and rather different from on site and laboratory observation collected by [45]. A testing scheme more informed by observation of post-earthquake damage in existing masonry structures is the one devised by [46], however by predefining a state of damage the mechanism is also predefined.

Table 4 compares ranges for drift limit states as average from experimental literature, with the EC8 [34] provision for URM for the damage limit states of Significant damage and Near Collapse. The EC8 values relate to the in-plane failure of single pier elements, either with prevalent shear or flexural behaviour, while there is no indication for out of plane behaviour. In Table 4 are also included the range of values of performance drift obtained with the FaMIVE simulations for over 1000 cases as obtained from ten different sites for any type of masonry fabric and floor structure. The next section explains in detail how in the FaMIVE procedure the capacity curves are derived and the drift limit states computed.

Table 4 Performance drift value for damage limit states

	Limit state	Damage limitation (%)	Significant damage (%)	Near Collapse (%)	Collapse (%)
In-plane prevalent behaviour	EC 8 Part 3		0.4–0.6	0.53–0.8	
	Experimental	0.18–0.23	0.65–0.90	1.23–1.92	2.1–2.8
	FaMIVE	0.023–0.132	0.069–0.679	0.990–1.579	1.801–2.547
Out-of-plane prevalent behaviour	EC8 Part 3		0.8–1.2 (H_0/D)	1.06–1.60 (H_0/D)	
	Experimental	0.33	0.88	2.3	4.8
	FaMIVE	0.263–0.691	0.841–1.580	1.266–1.961	2.167–5.562
Combined prevalent behaviour	FaMIVE	0.030–0.168	0.181–0.582	0.724–1.401	1.114–3.307

3.2.2 Derivation of Capacity Curves

Capacity curves can be derived for each façade on the basis of the following steps. The first step is to calculate the lateral effective stiffness for each wall and its tributary mass. The effective stiffness for a wall is calculated on the basis of the type of mechanism attained, the geometry of the wall and layout of opening, the constraints to other walls and floors and the portion of other walls involved in the mechanism:

$$K_{eff} = K_1 \frac{E_t I_{eff}}{H_{eff}^3} + K_2 \frac{E_t A_{eff}}{H_{eff}} \tag{18}$$

where H_{eff} is the height of the portion involved in the mechanism, E_t is the estimated modulus of the masonry as it can be obtained from experimental literature for different masonry typologies, I_{eff} and A_{eff} are the second moment of area and the cross sectional area, calculated taking into account extent and position of openings and variation of thickness over height, k_1 and k_2 are constants which assume different values depending on edge constraints and whether shear and flexural stiffness are relevant for the specific mechanism.

The tributary mass Ω_{eff} is calculated following the same approach and it includes the portion of the elevation activated by the mechanisms plus the mass of the horizontal structures involved in the mechanism:

$$\Omega_{eff} = V_{eff}\delta_m + \Omega_f + \Omega_r \tag{19}$$

where V_{eff} is the solid volume of the portion of wall involved in the mechanism, δ_m is the density of the masonry Ω_f, Ω_r are the masses of the horizontal structures involved in the mechanism. Effective mass and effective stiffness are used to calculate a natural period T_{eff}, which characterise an equivalent single degree of freedom (SDoF) oscillator:

$$T_{eff} = 2\pi \sqrt{\frac{\Omega_{eff}}{K_{eff}}} \tag{20}$$

The mass is applied at the height of the centre of gravity of the collapsing portion with respect to the ground and a linear acceleration distribution over the wall height is assumed. The elastic limit acceleration A_y is identified as the combination of lateral and gravitational load that will cause a triangular distribution of compression stresses at the base of the overturning portion, just before the onset of partialisation:

$A_y = \frac{t_b^2}{6h_0} g$ with corresponding displacement

$$\Delta_y = \frac{A_y}{4\pi^2} T_{eff} \tag{21}$$

where, t_b is the effective thickness of the wall at the base of the overturning portion, h_o is the height to the ground of the centre of mass of the overturning portion, and T the natural period of the equivalent single degree of freedom (SDF) oscillator. The maximum lateral capacity A_u is defined as:

$$A_u = \frac{\lambda_c}{\alpha_1} \qquad (22)$$

where λ_c is the load factor of the collapse mechanism chosen, calculated by FaMIVE, and α_1 is the proportion of total mass participating to the mechanism. This is calculated as the ratio of the mass of the façade and sides or internal walls and floor involved in the mechanism Ω_{eff}, to the total mass of the involved macroelements (walls, floors, and roof). The displacement corresponding to the peak lateral force, Δ_u is

$$3\Delta_y \leq \Delta_u \leq 6\Delta_y \qquad (23)$$

as suggested by [47]. The range in Eq. (22) is useful to characterize masonry fabric of variable regularity and its integrity at ultimate conditions, with the lower bound better describing the behavior of adobe, rubble stone and brickwork in mud mortar, while the upper bound can be used for massive stone, brickwork set in lime or cement mortar and concrete blockwork.

Finally the near collapse condition is determined by the displacement Δ_{nc} identified by the condition of loss of vertical equilibrium which, for overturning mechanisms, can be computed as a lateral displacement at the top or for in plane mechanism by the loss of overlap of two units in successive courses:

$$\Delta_{nc} = t_b/3 \quad or \quad \Delta_{nc} = l/3 \qquad (24)$$

where t_b is the thickness at the base of the overturning portion and l is the typical length of units forming the wall. In the case of in-plane mechanism the geometric parameter used for the elastic limit is, rather than the wall thickness, the width of the slender pier.

The thresholds points identified by Eqs. (20)–(23) can be associated to corresponding states of damage. Specifically **DL**, *damage limitation*, corresponds to the elastic lateral capacity threshold (D_y, A_y) defined by Eq. (20), **SD**, *significant damage*, corresponds to the peak capacity threshold (Δ_u, A_u) defined by Eqs. (21) and (22), and **NC**, *near collapse*, corresponds to incipient or partial collapse threshold $(\Delta_{nc} A_u)$ defined by Eq. (23).

The procedure's approach also allows a direct analysis of the influence of different parameters on the resulting capacity curves, whether these are geometrical, mechanical or structural. By way of example Fig. 10 shows a comparison of average capacity curves grouping the results by different criteria for the same sample of buildings. In Fig. 10 the average curves are obtained by considering whether failure occurs by out-of-plane, in-plane or combined mechanism involving both sets of walls as presented in Fig. 8. In Fig. 11 the capacity curves are obtained by considering different structural typologies, as classified by the WHE-PAGER project [48] and shown in Table 5. It can be seen that the correlation between mode

Fig. 10 Average capacity *curves* for sample grouped by collapse mechanism classes

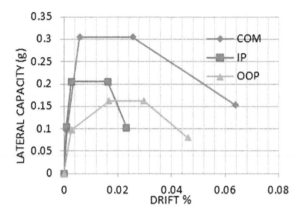

Fig. 11 Average capacity *curves* for sample grouped by structural typology

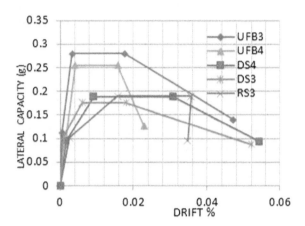

Table 5 Structural typologies classification according to PAGER [48]

Load bearing material	PAGER structure code	Description
Stone Masonry	RS3	Local field stones with lime mortar
	RS4	Local field stones with cement mortar, vaulted brick roof and floors
	DS2	Rectangular cut stone masonry block with lime mortar
	DS3	Rectangular cut stone masonry block with cement mortar
	DS4	Rectangular cut stone masonry block with reinforced concrete floors and roof
	MS	Massive stone masonry in lime or cement mortar
Brickwork or blockwork	UFB1	Unreinforced brick masonry in mud mortar without timber posts
	UFB3	Unreinforced brick masonry in lime mortar. Timber flooring
	UFB4	Unreinforced fired brick masonry, cement mortar. Timber flooring.
	UFB5	Unreinforced fired brick masonry, cement mortar, but with reinforced concrete floor and roof slabs
	UCB	Unreinforced concrete block masonry with lime or cement mortar

of failure and structural typology is qualitatively good but not univocal, and the grouping affects both ultimate lateral capacity and drift.

Substantial differences also exist for nominally the same structural typology from different regional setting. In Fig. 12 average capacity curves for structural typologies based on unreinforced brickwork with different mortars and horizontal structures are compared from different locations, one in Italy, one in Turkey, one in Nepal.

The results in Fig. 12 show that the parameter location, and hence construction details, layout and local tradition, might have a greater influence on the resulting curves, than the nominal structural typology class, usually considered of universal reference in many general purpose databases (such as HAZUS 99 [17], RISK-EU [5], LESS-LOSS [6], etc.). Such results bring in sharper focus the limitation and inaccuracy of using idealised models and average curves without adequately considering the inherent aleatoric variation associated with any given site where the assessment is conducted, and the importance of a detailed knowledge of the local construction characteristics when sampling the buildings representative of the building stock. A substantial variation in the drift associated with the various limit states can be also observed.

3.2.3 Performance Points and Their Correlation with Damage States

The lateral acceleration capacity and the relative proportion of drift for the three limit states identified in the previous section are essential indicators of the seismic performance. A method for assessing the overall behaviour by use of a global

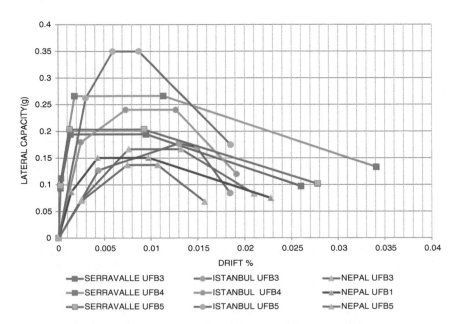

Fig. 12 Average capacity *curves* for different location and masonry typologies

performance indicator is the computation of the performance point. In order to calculate the performance point it is necessary to intersect the capacity curve derived above with the demand spectra for different return periods in relation to the performance criteria considered. Two broadly equivalent approaches for the derivation of the non-linear demand spectra exist: the N2 method [14] included in the EC8 [34] and the Capacity Spectrum method (CSP) [14]. The two methods differ essentially in the way the non-linear demand spectrum is arrived at: the N2 method uses a reduction factor R, function of the structure expected ductility μ, while the CSP uses a fictitious damping factor derived from the hysteresis loop of the structure. There exists a rich literature that compares the benefits of the two approaches [49]. In the following the N2 method will be used to illustrate the derivation of performance points.

To calculate the coordinates of the performance point in the displacement-acceleration space, the intersection of the capacity curve with the nonlinear demand spectrum for an appropriate level of ductility μ can be determined as shown in Eq. (24), given the value of A_u:

$$if\ T < T_c$$

$$if\ A_u \geq A_{nl}(T) \Rightarrow SD_{nl}(\mu) = \frac{T_c^2(\beta A_{el}(0) - A_{nl}(T))^2}{(\mu - 1)^2} * \frac{g\mu}{4\pi^2 A_{nl}(T)}$$

$$if\ A_{nl}(T_c) < A_u < A_{nl}(T) \Rightarrow SD_{nl}(\mu) = \frac{T_c^2(\beta A_{el}(0) - A_u)^2}{(\mu - 1)^2} * \frac{g\mu}{4\pi^2 A_u}$$

$$if\ A_u \leq A_{nl}(T_c) \Rightarrow SD_{nl}(\mu) = \frac{gT_c^2(\beta A_{el}(0))^2}{4\pi^2 \mu A_u} \tag{25}$$

$$if\ T \geq T_c$$

$$if\ A_u \geq A_{nl}(T) \Rightarrow SD_{nl}(\mu) = \frac{gT_c^2(\beta A_{el}(0))^2}{4\pi^2 \mu A_{nl}(T)}$$

$$if\ A_u < A_{nl}(T) \Rightarrow SD_{nl}(\mu) = \frac{gT_c^2(\beta A_{el}(0))^2}{4\pi^2 \mu A_u}$$

where two different formulations are provided for values of ultimate capacity A_u greater or smaller than the nonlinear spectral acceleration $A_{nl}(T_c)$ associated with the corner period T_c marking the transition from constant acceleration to constant velocity section of the parent elastic spectrum. In (24) SD_{nl} is the non-linear spectral displacement, function of the chosen target ductility μ; β is the acceleration amplification factor calculated as the ratio of the elastic maximum spectral acceleration and the peak ground acceleration $A_{el}(0)$; $A_{nl}(T)$ is the non-linear spectral acceleration for the value of natural period that defines the elastic branch of the capacity curve; g is the gravity constant. Note that in Eq. (24) $A_{el}(T)$, $A_{nl}(T)$ and A_u are dimensionless quantities, expressed as proportion of g.

In Fig. 13 the damage thresholds for the limit state of near collapse for each building in the sample of Nocera Umbra, Italy, are compared with the regional response spectrum for 475 year return period (or 10 % of exceedance in 50 years)

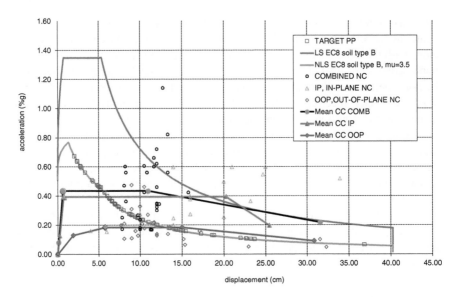

Fig. 13 Representation of target performance points and damage thresholds for Near Collapse limit states in the acceleration/displacement space. The PGA is the value recorded for the 1997 Umbria Marche earthquake in Nocera Umbra. On the mean capacity curve for the three mechanism classes, the mean significant damage thresholds are marked in *red*

anchored to the PGA of the second shock of the Umbria-Marche September 1997 sequence. For the non-linear spectrum obtained with the N2 method approach a ductility $\mu = 3.5$ has been chosen in agreement with experimental evidence provided by [47] and [50] and to match the performance point of NC for the mean capacity curve for the combined mechanism. It can be seen that there is quite a significant scatter of performance and most of the out-of plane mean curves lies below the nonlinear spectrum, meaning that a higher level of ductility is required to meet the performance.

It should also be noted that a consistent proportion of the representative points of Near Collapse lies under the nonlinear response spectrum, equally deficient in terms of acceleration and displacement, especially for the out of plane behaviour. Such outliers should not be overlooked as they usually point out to inherent construction deficiency in a regional context, inhibiting seismic resilience.

3.2.4 Derivation of Fragility Curves

Advanced uncertainty modelling and probability of occurrence of given phenomena is usually confined to the hazard component of the risk equations, while when probabilistic models are developed for vulnerability components, these usually relate to simplified modelling of the structure seismic response and assumption of pre-determined dispersion as might be found in literature [17, 51].

Usually it is also assumed that fragility curves for different limit states can be obtained by using mean values of the performance point displacement and deriving lognormal distributions by either computing the associated standard deviation if some form of random sampling has been considered, or by assuming values of β from empirical distribution or literature. To this end the average displacement for each limit state can be calculated as:

$$\bar{\Delta}_{LS} = e^{\mu} \quad \text{with} \quad \mu = \frac{1}{n} \sum (\ln x) \tag{26}$$

and the corresponding standard deviation as:

$$\beta_{LS} = e^{\mu + \frac{1}{2}\sigma^2} \sqrt{e^{\sigma^2} - 1} \quad \text{with} \quad \sigma = \sqrt{\frac{\sum (\ln x - \ln \bar{x})^2}{n}} \tag{27}$$

Figures 14 and 15 show the set of fragility curves obtained for each of the damage limit states of DL and SD as computed for the two Italian sites of Nocera Umbra and Serravalle considering separately the three types of structural behaviour. As, once a structural typology has been assigned, the values of the mechanical characteristics are the same across the two samples, while the structural details are accounted for directly in the three classes of mechanisms, the variability observed in each chart between samples can be related directly to geometric

Fig. 14 Fragility distribution for limit state of damage limitation for the three classes of collapse mechanisms for two different Italian sites

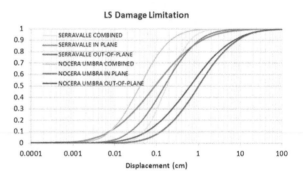

Fig. 15 Fragility *curves* for limit state of significant damage for two Italian samples for the three classes of collapse mechanism

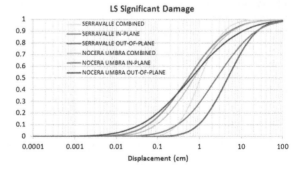

differences and masonry fabric, i.e. to the local aspects of the construction practice and architectural layout. Hence curves on the left of the diagrams are stiffer in the case of damage limitation or have lesser ductility in the case of significant damage. However the distribution does not bare consistency across the three classes of mechanism for the two sites.

Figures 16 and 17 show the fragility distribution at ultimate conditions in terms of near collapse displacement and ultimate lateral capacity for the three failure behaviour. While there is little difference among the two locations for the out-of-plane behaviour both the in-plane and the combined behaviour show high variability. The higher deformability of Serravalle for the in-plane behaviour is related to a higher proportion in this sample of facades with porticoes at ground level, resulting in possible soft storeys, while the lower value of limit displacement for the Nocera Umbra sample is dependent on a high proportion of masonry fabric of poorly hewed stone classified as RS3. On the other end the lower lateral capacity of the Serravalle sample for the combined mechanism is to be associated with slenderer façades. Moreover Nocera Umbra has a greater lateral capacity both for combined mechanisms and for in-plane mechanism than Serravalle (see Fig. 17) while ultimate capacity for the out-of-plane mechanism provides similar fragility curves.

The reliability of the results obtained in the previous section can be considered within the framework set out in the Eurocode 8 [34], whereby the reliability associated to the results of a seismic assessment of a structure is expressed as a

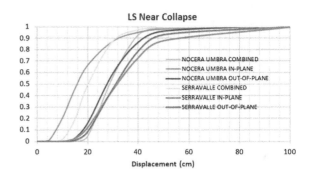

Fig. 16 Fragility *curves* for the limit state of near collapse for two Italian sites and three classes of mechanisms

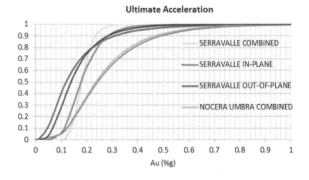

Fig. 17 Fragility *curves* for the ultimate lateral capacity for two Italian sites and three classes of mechanisms

function of the level of knowledge and quantified by means of the confidence factor. Hence this can be considered a measure of the epistemic uncertainty. Eurocode 8 recognises three levels of knowledge: limited, normal and full; and three fields of knowledge: geometry, construction details and materials. As data used in the FaMIVE approach are collected by on site visual inspection with some measurement and in situ accurate observation of construction details, while only very limited in situ non-destructive test on materials are performed and material characteristics are otherwise assigned based on literature or surveyor experience, then the level of knowledge is superior to KL1, *limited*, but not quite equal to KL2, *normal*. For this level of knowledge, a static nonlinear analysis, such as the limit state mechanism approach, leading to a capacity curve is deemed appropriate. Hence according to the recommended values the confidence factor CF should be in the range 1.2–1.35 depending on how closer the actual knowledge can be considered to the reference level identified by KL2. The confidence factor is then used in EC8 to reduce the capacity values as obtained from the assessment.

Although the EC8 approach recognise the importance of treating epistemic uncertainties, the level of knowledge is translated in a safety factor value rather than a probability or possibility of a specific value to occur. While this approach can be considered acceptable for the assessment of single buildings, it does not account explicitly for aleatoric variation.

The FaMIVE procedure uses a measure of reliability of the input data to determine the reliability of the output. Depending on whether data, in each section of the data collection form, has been collected and measured directly on site, or collected on site and confirmed by existing drawings or photograph, or collected from photographic evidence only, three level of reliability are considered, as high, medium and low, respectively, to which three confidence ranges of the value given for a parameter can be considered corresponding to 10 % variation, 20 % variation and 30 % variation. The parameter value attributed during the survey is considered central to the confidence range so that the interval of existence of each parameter is defined as $\mu \pm 5$ %, $\mu \pm 10$ %, $\mu \pm 15$ %, depending on highest or lowest reliability. The reliability applied to the output parameters, specifically lateral acceleration and limit states' displacement, is calculated as a weighted average of the reliability of each section of the data form, with minimum 5 % confidence range to maximum 15 % confidence range.

To quantify the effect of the level of the epistemic uncertainty on the fragility curves obtained with the FaMIVE procedure, the samples from three different locations in Italy and for the three failure behaviours introduced in the previous section, are analysed together. For each entry in the sample a separate reliability parameter is computed as indicated above, then two new sets of values representing the lower bound and upper bound for each entry are computed. For these two sets logarithmic mean and standard deviation are calculated using Eqs. (25) and (26) and the lognormal distributions obtained. These are presented in

Figure 18 for the three displacement limit states and for the ultimate acceleration, respectively. The reliability indicator for the overall sample is ±11 %, showing that the data reliability is medium–low, i.e. no availability of drawings in most

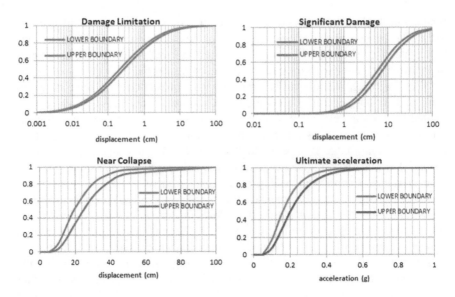

Fig. 18 Effect of epistemic uncertainty on fragility distribution for limit states

cases and onsite measurement on a modest number of cases. This is a typical situation in the aftermath of an earthquake, such as the conditions in which both the Nocera Umbra sample and the L'Aquila sample were collected.

3.3 Building Aggregates

In historic centres the evolution of the urban layout is a critical factor. The diachronic process of construction means that in some cases adjacent buildings share load-bearing masonry walls and their façades are aligned. In this case, buildings do not constitute independent units, resulting in their structural interaction, particularly critical for horizontal actions. Hence the structural performance should be studied at the level of the aggregate and not only for each isolated building.

This chapter presents an extension of the mechanical methods introduced in the previous chapter, to undertake vulnerability assessment, evaluate seismic risk and estimate loss at the urban scale for historic city centres in which the building stock is structurally linked. It is assumed that collapse or ultimate limit state of the structure is due to shear-type failure.

A building aggregate can be considered as a unit, for which it is fundamental, the knowledge on building typology, conservation state and connection scheme between buildings, as a consequence of the evolution of the urban layout (see Fig. 19). The building interaction does not only change the load paths, but also the global and local seismic response, as a consequence of the quality of the connections. The vulnerability assessment for single buildings overlooks the integrity

Fig. 19 Diachronic construction process and building interaction (adapted from [50])

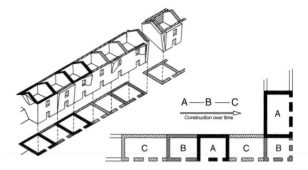

of the aggregate weather it is small or large aggregate, the irregularity created by confining buildings, connection to neighbouring buildings, etc. [50].

The interaction of buildings is first of all very dependent on irregularity raised by differences in height and stiffness of neighbouring buildings. Since the aggregate is constituted by single buildings, which have different level of vulnerability when considered individually, the position and layout of these can increase or reduce the vulnerability of the aggregate as a whole. In this sense the aggregate is a structural unit and should be evaluated as a global structure and from its collective behaviour and response to seismic action more vulnerable buildings can benefit from this confinement, however the interaction of the buildings can worsen the global vulnerability of an aggregate due to changes in height or stiffness. In general the global behaviour is beneficial for the more vulnerable buildings while for the stiffer units the level of damage suffered during a seismic event is greater, due to the interaction of strong building-weak building.

Building aggregates can take a number of shapes, as shown in Fig. 20, although buildings in a row are very characteristic of the eighteenth century urban layout for many European historic city centres. Whatever the aggregate shape, the seismic behaviour is evaluated in two main directions: parallel to the building façades development and perpendicular to them. More complex aggregate shapes can be sub-divided in smaller aggregates of simpler shape.

For the case of a row of buildings, many situations can arise from the interaction among buildings. Normally flexural failure is expected for buildings with slender masonry piers at ground floor due to big openings and shear failure for buildings with thick masonry piers between openings, but these kind of failure modes are altered because of the group response. The misalignments of building front, misalignments of window openings of adjacent buildings, big differences in wall area and stiffness of aligned buildings may change completely the load paths for the horizontal forces and the resulting failure mechanism.

Figure 21 shows an example of the influence of aggregate's layout on building failure mechanisms

It is often noted that end buildings are very vulnerable due to their position and normally suffer most damage by rotation and sliding phenomenon's induced by inertial forces of the whole aggregate in one direction. Furthermore the rigidity of timber diaphragms of masonry buildings do not oppose to the global behaviour

Fig. 20 Building aggregate shapes

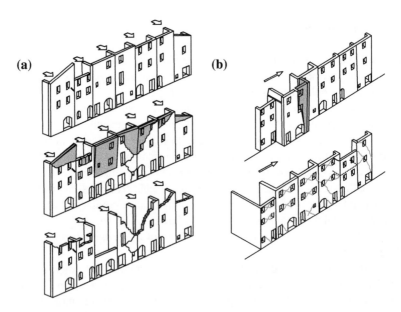

Fig. 21 Building interaction: **a** Out-of-plane; **b** In-plane

since they are flexible diaphragms but are important in the horizontal load distribution among masonry shear walls. In this direction the global response is proven to be of great importance, however in the perpendicular direction, the building response is substantially self-ruling. The masonry mid-walls of adjacent buildings, lacking openings, charged by floor structures leading to high values of normal stress appear to have high shear strength in the in-plane response and do not condition building failure. A critical issue for the facades of the aggregates, often observed in post seismic survey, is the out of plane collapse of walls. The weak connections to orthogonal walls, due to the building process of buildings in-between existent ones or to the addition of extra floor on the other may compromise the quality of connections among orthogonal walls. Out of plane collapses at roof level are also common due to the combined effect of weak connections and low values of normal stress reducing the shear capacity.

3.4 Mechanical Method for Building Aggregate

The vulnerability assessment procedure is based on the use of a simplified capacity curve for each building. To better understand the assessment process, it has been broken down into steps following the same logic as in Sect. 3.

3.4.1 Identification of Building Typology

A subdivision into two different typologies relating to two different wall arrangements are identified as A and B. This division is necessary to identify primarily the more vulnerable direction of the masonry building and define a more probable collapse mechanism as shown in Fig. 22:

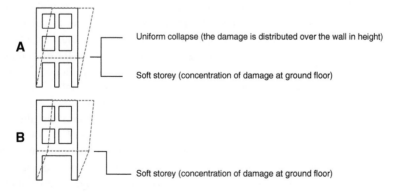

Fig. 22 Building typology and collapse mechanism

Type A—Masonry walls that have regular openings in height or few or no openings whatsoever (midwalls, gable end walls)

Type B—Masonry walls with big openings at ground floor level: This situation is a frequent characteristic in the refurbishment and transformation of historic masonry buildings where wall are suppressed to create larger open spaces.

3.4.2 Collapse Mechanism

The building aggregate is analysed considering two possible mechanisms: uniform collapse and soft-storey collapse. For each of the building typologies identified and relative to the direction considered, the analysis of a building or a group of buildings is undertaken considering the collapse mechanism and the typology. The following situation can be identified (see Fig. 22):

- For buildings of typology A, two collapse mechanisms are possible: the uniform collapse considers that the damage is distributed over the height of the wall and for the soft storey mechanism damage is concentrated at ground floor.
- For buildings of typology B only one collapse mechanism is considered because of its increased vulnerability at ground floor level.

3.4.3 Vulnerability Assessment

To evaluate the response of building aggregate with a bulky or array shape, in both principal directions (X, Y) it is assumed that the X-direction is the weaker direction of the building aggregate for which the occurrence of a soft storey mechanism is prevalent, for both building types A and B. For the other direction, Y, both collapsed mechanisms are considered in the assessment.

In an array of buildings the YY direction assumed as the stronger, is usually the direction of the majority of the party walls between buildings within the aggregate. These walls are assumed to have individual response. This hypothesis is fairly acceptable, because in this direction buildings do not interact as strongly as in the other direction (façade walls). In this direction a very straightforward vulnerability assessment is attained for each building using the mechanical model in which the simplified bilinear capacity curve (SDOF system) is constructed for each building [51, 53], limit states and the level of seismic action are defined, hence the performance point is retrieved through the capacity spectrum methodology (see [24]). Once the fragility curves for the four damage states are obtained, the evaluation of the probabilistic damage distribution is performed. The damage distribution of the aggregate in this direction is evaluated by the average value of the single damage distribution for each building for both collapse mechanisms (uniform collapse and soft storey mechanism), defining in this way a damage range for the building aggregate in this direction, without considering the damage of each building within the aggregate.

For the *XX* Direction, considered the weaker direction as mentioned, usually building façades are aligned and the interaction between buildings in this direction is much more important. The procedure adopted in this case is as follows:

(i) Construction of each simplified bilinear capacity curve corresponding to a single degree of freedom system for each building in this direction. Once obtained these simplified capacity curves, they can be transformed into force–displacement curves and summed to obtain a global push-over curve for the aggregate. But since aggregates are formed by buildings with different height, horizontal displacements should be normalized in such a way that $\phi_n = 1$ (modal shape vector), where n is the control node. This must be done because buildings that compose and aggregate have different number of floors and consequently different height and therefore top-displacement at roof level that is normally considered cannot be the selected control node. To achieve this, the displacements are divided by the number of floors, therefore the control node is the displacement at ground floor and the curves can be summed (see Fig. 23). Each simplified capacity curve (*Ay*, *dy*, *du*) is then normalized by transformation of coordinates into the force–displacement using the following expressions:

$$\text{Force}: \quad F = Ay \times m^* \Gamma \tag{28}$$

$$Displacement: \quad d = \frac{dy \times \Gamma}{N}; \quad N : numt \tag{29}$$

(ii) The force displacement curves are summed and the global pushover curve of the building aggregate is obtained in this direction (see Fig. 23)

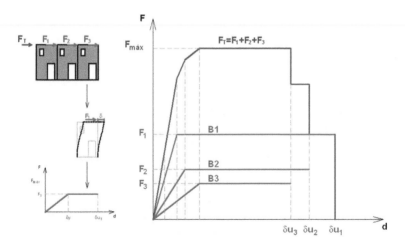

Fig. 23 Construction of the global push-over curve

(iii) The determination of an equivalent elasto-perfectly plastic force–displacement relationship for the building aggregate is constructed (non-linear static analysis through a simplified mechanical model) that the elastic stiffness of an equivalent bilinear system is found by marking the secant to the push-over curve at the point corresponding to a shear base 70 % of the maximum value (maximum base shear). The horizontal section of the bilinear curve shall be found by equalizing the areas underneath the two curves up to the ultimate displacement of the system. The value of the ultimate displacement which is considered equal to the ultimate limit state corresponds to a force degradation of not more than 20 % of the maximum. The construction of the equivalent global pushover curve, an equivalent capacity curve to evaluate the response of the aggregate structure must take into account two possible situations:

(a) There is no building within the aggregate that collapses of a value of shear base 70 % of the maximum shear of the global pushover curve and in this case the equivalent bilinear curve is defined analytically as followed in Fig. 24.

(b) If a building within the aggregate collapses before attaining the 70 % of the maximum shear value defined for the global push-over curve, it will drop of a value of the shear capacity of the building that prematurely failed. In this case, the equivalent stiffness is found by marking the secant to the unfailed push-over curve and the horizontal section is defined as

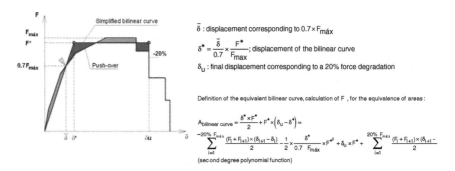

Fig. 24 Construction of the equivalent bilinear curve—case a)

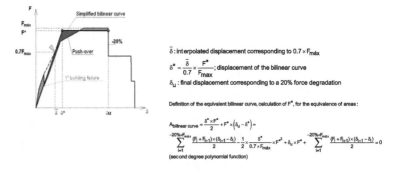

Fig. 25 Construction of the equivalent bilinear curve—case b)

defined in the normal procedure. For this case in Fig. 25 is shown the steps to construct the equivalent elasto-perfectly plastic force–displacement relationship.

(iv) The construction of the equivalent bilinear capacity curve of an equivalent single degree of freedom is attained by a global transformation factor, Γ_{global} considering the number of floors of each building and the singular transformation factors of each building, to return to a system coordinates of (S_a, S_d). The transformation factor is given by:

$$\Gamma = \frac{M^*}{\sum\limits_{j=1}^{N} \frac{m_j^*}{\Gamma_j}} = \frac{\sum\limits_{i}^{N} n_{pj} \times m_j^*}{\sum\limits_{i=1}^{N} \frac{n_{pj}^2 \times m_j^*}{\Gamma_j}}; \ \Gamma \times m^* = \frac{M^*}{\sum\limits_{j=1}^{N} \frac{m_j^*}{\Gamma_j}} = \frac{\left(\sum\limits_{i}^{N} n_{pj} \times m_j^*\right)}{\sum\limits_{i=1}^{N} \frac{n_{pj}^2 \times m_j^*}{\Gamma_j}} \quad (30)$$

in which:

i = 1...., N buildings;

m^*_j: $\sum\limits_{i} m_i \times \Phi_i; equivalant\ mass$;

n_{pj}: number of floors of building;

Γ_j: transformation factor

(v) Once computed the equivalent bilinear curve, it is possible to evaluate the performance point by using the capacity spectrum method (see Fig. 26). After identifying the final performance point by doing the reverse process the evaluation of the damage state of each building is possible by identifying on each curve the target displacement. In order to access the damage state suffered by

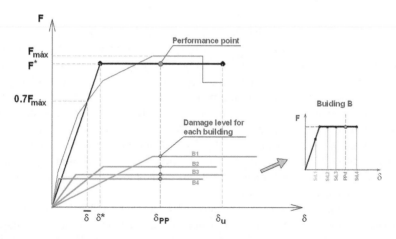

Fig. 26 F-δ curves for each building and performance level identification and limit state definition

Table 6 Thresholds for damage states

Spectral displacement threshold	Damage state
$S_{d,1} = 0.7 \times D_y$	Slight damage
$S_{d,1} = 1.5 \times D_y$	Moderate damage
$S_{d,1} = 0.5 \times (D_y + D_u)$	Severe damage
$S_{d,1} = D_u$	Heavy damage

each building in the aggregate under the defined seismic action, the displacement corresponding to the performance point can then evaluated the performance level of each building, by defining the damage threshold states, the values used for the damage state definition have been widely discussed in [51] and are based on expert judgment and for this case are defined as (Table 6):

Once defined the equivalent bilinear curve of the aggregate the performance can be retrieved by applying known procedure for the CSM (see [24]). Then the correspondent displacement is evaluated over the push-over curves for each building, evaluating individually the probabilistic damage distribution for each building and the global response in the direction evaluated. Finally the damage distribution of the aggregate in this direction is evaluated by the average value of the single distribution for each building for only each collapse mode mechanism (soft storey mechanism or uniform collapse) or the global response depending on the direction evaluated, defining in this way a damage range for the building aggregate in this direction, without losing the perception of the damage for each building with the aggregate.

4 Final Remarks

The chapter offers a review and classification of the most commonly adopted procedures for carrying out a seismic vulnerability assessment at territorial scale of large number of historic masonry buildings. By way of exemplification of each of the classes of methods identified, three procedures are presented in greater details. The first one relies on empirical data only and it is an extension of the Vulnerability Index method. By combining this procedure with the vulnerability classes and damage states proposed by EMS'98, is possible to derive fragility curves, cumulative losses and casualty for building pertaining to diverse vulnerability classes. A simple treatment of the uncertainty is proposed, by using the standard deviation of the Vulnerability Index. This does not account for the uncertainty associated with the hazard.

However, the uncertainties associated with the empirical vulnerability curves and the quality of vulnerability classification data are still issues that must be studied further with respect to post-seismic data collection. For risk mitigation, a reduction in building vulnerability is a priority and therefore the development of more reliable vulnerability assessment models which combine statistical and mechanical methods should lead to better results.

The second procedure proposed, FaMIVE, represent a robust attempt to meet these requirements. It moves from a survey of the local structural and vulnerability characteristics of the building stock in an historic centre, and uses the collected data within the framework capacity spectrum method and performance base design to derive performance points and fragility curves, for classes of buildings of same typology. Damage thresholds are defined on the basis of observation, numerical analysis and comparison with existing experimental results. The results show that, by considering diverse types of mechanisms, construction details and resilient features, it is possible to tune, capacity curves, first, and then fragility curves, to specific construction typologies and local building characteristics. The aleatoric uncertainty is dealt by considering variability in construction as obtained through the direct survey. The epistemic uncertainty associated with the methodology is accounted for by developing a reliability framework.

Buildings in historic centres are usually built adjacent to each other and their vulnerability is highly affected by the connections to neighbouring buildings. The third procedure shows a first attempt to interpret the overall behaviour of an aggregate by considering in detail the interaction of buildings' facades in plane. This allows deriving capacity curves at the level of the aggregate and captures the global response of the aggregate opening the possibility of defining vulnerability functions at the level of the aggregate based on mechanical behaviour. Out-of-plane failures, although classified, have not been considered and this will be a feature extension of the method.

The three procedures illustrated here lend themselves to the use of a GIS platform and database management system to best communicate the information collected about building feature and geometry, the output of seismic vulnerability assessment and the development of damage and other risks scenarios. Such tools, depending on the scale and type of procedure used can be very helpful for the development of strengthening strategies, cost-benefit analyses, civil protection and emergency planning.

References

1. D'ayala, D., Ansal, A.: Non linear push over assessment of heritage buildings in Istanbul to define seismic risk. Bull. Earthq. Eng. **10**(1), 285–306 (2012)
2. Benedetti, D., Petrini, V.: Sulla Vulnerabilità Di Edifici in Muratura: Proposta Di Un Metodo Di Valutazione. L'industria delle Costruzioni **149**(1), 66–74 (1984)
3. D'ayala, D.: Force and displacement based vulnerability assessment for traditional buildings. Bull. Earthq. Eng. **3**(3), 235–265 (2005)
4. D'Ayala, D., Spence, R., Oliveira, C.S., Pomonis, A.: Earthquake loss estimation for Europe's historic town centres. Earthq. Spectra **13**(4), 773–793 (1997)
5. Mouroux, P., Le Brun, B.: Risk-Ue project: an advanced approach to earthquake risk scenarios with application to different European towns. In: Oliveira, C., Roca, A., Goula, X. (eds.). Assessing and Managing Earthquake Risk, vol. 2, pp. 479–508. Springer Netherlands (2006)

6. Calvi, G.M., Pinho, R.: LESSLOSS: A European Integrated Project on Risk Mitigation for Earthquakes and Landslides. University of Pavia, Structural Mechanics Department, Pavia (2004)
7. GNDT-SSN. Scheda di esposizione e vulnerabilità e di rilevamento danni di primo e secondo livello (murata e cemento armato)," *Gruppo Nazionale per la Difesa dai Terremoti, Roma* (1994)
8. Grünthal, G.: European Macroseismic Scale 1998 (EMS-98), European Seismological Commission, Subcommission on Engineering Seismology, Working Group Macroseismic Scales," *Cahiers du Centre Européen de Géodynamique et de Séismologie*, vol. 15, (1998)
9. Whitman, R.V., Reed, J.W., Hong, S.T.: Earthquake damage probability matrices. In: Proceedings of the 5th World Conference on Earthquake Engineering, Rome, vol. 2, pp. 2531–2540, (1973)
10. Calvi, G.M.: A displacement-based approach for vulnerability evaluation of classes of buildings. J. Earthq. Eng. **3**(03), 411–438 (1999)
11. Speranza, E.: An Integrated Method for the Assessment of the Seismic Vulnerability of Historic Buildings. University of Bath (2003)
12. ATC- 40. Seismic evaluation and retrofit of concrete buildings. Technical report, ATC-40. Applied Technology Council, Redwood City, California (1996)
13. FEMA 273. NEHRP Guidelines for the Seismic Rehabilitation of Buildings: Second Ballot Version. The Council (Federal Emergency Management Agency), Washington DC (1997)
14. Fajfar, P.: Capacity spectrum method based on inelastic demand spectra. Earthq. Eng. Struct. Dynam. **28**(9), 979–994 (1999)
15. ATC-21. Rapid Visual Screening of Buildings for Potential Seismic Hazards: A Handbook. Redwood City, California (1988)
16. ATC-13. Earthquake damage estimation data for California. Report ATC-13. Applied Technology Council, Redwood City, California (1985)
17. HAZUS. Earthquake loss estimation methodology—technical and user manual. Federal Emergency Management Agency, Washington, D.C. (1999)
18. FEMA 178. NEHRP Handbook for the Seismic Evaluation of Existing Buildings. The Council (Federal Emergency Management Agency), Washington, DC (1992)
19. Lagomarsino, S., Giovinazzi, S.: Macroseismic and mechanical models for the vulnerability and damage assessment of current buildings. Bull. Earthq. Eng. **4**(4), 415–443 (2006)
20. Santos, C., Ferreira, T.M., Vicente, R., Mendes da Silva, J.A.R.: Building typologies identification to support risk mitigation at the urban scale—case study of the old city centre of Seixal, Portugal. J. Cult. Heritage (2012)
21. Vicente, R., Parodi, S., Lagomarsino, S., Varum, H., Mendes da Silva, J.A.R.: Seismic vulnerability and risk assessment: case study of the historic city centre of Coimbra, Portugal. Bull. Earthq. Eng. **9**(4), 1067–1096 (2011)
22. Ferreira, T.M., Vicente, R., Mendes da Silva, J.A.R., Varum, H., Costa, A.: Seismic vulnerability assessment of historical urban centres: case study of the old city centre in Seixal, Portugal. Bull. Earthq. Eng. 1–21 (LA—English) (2013)
23. Ferreira, T., Vicente, R., Varum, H., Mendes da Silva, J.A.R., Costa, A.: Vulnerability assessment of urban building stock: a hierarchic approach. In: International Disaster and Risk Conference IDRC, pp. 245–248 (2012)
24. Vicente, R.: Estratégias e metodologias para intervenções de reabilitação urbana. Avaliação da vulnerabilidade e do rísco sísmico do edificado da Baixa de Coimbra. PhD THesis. Universidade de Aveiro, Portugal. (in Portuguese), (2008)
25. Giovinazzi, S.: The vulnerability assessment and the damage scenario in seismic risk analysis. Technical University Carolo-Wilhelmina at Braunschweig, Braunschweig, Germany and University of Florence, Florence (2005)
26. Bernardini, A., Giovinazzi, S., Lagomarsino, S., Parodi, S.: Vulnerabilità e previsione di danno a scala territoriale secondo una metodologia macrosismica coerente con la scala EMS-98. In: *ANIDIS, XII Convegno Nazionale l'ingegneria sismica in Italia* (2007)

27. Sandi, H., Floricel, I.: Analysis of seismic risk affecting the existing building stock. In: Proceedings of the 10th European Conference on Earthquake Engineering, vol. 3, pp. 1105–1110 (1995)
28. Bramerini, F., Di Pasquale, G., Orsini, A., Pugliese, A., Romeo, R., Sabetta, F.: Rischio sismico del territorio italiano. Proposta per una metodologia e risultati preliminari. Servizio Sismico Nazionale, Rapporto Tecnico, SSN/RT/95/01, Roma, (1995)
29. Coburn, A., Spence, R., Comerio, M.: Earthquake protection. Earthq. Spectra **19**, 731 (2003)
30. Tiedemann, H.: Casualties as a function of building quality and earthquake intensity. In: Proceedings of the International Workshop on Earthquake Injury Epidemiology for Mitigation and Response, pp. 10–12 (1989)
31. Dolce, M., Kappos, A., Masi, A., Penelis, G., Vona, M.: Vulnerability assessment and earthquake damage scenarios of the building stock of Potenza (Southern Italy) using Italian and Greek methodologies. Eng. Struct. **28**(3), 357–371 (2006)
32. Di Pasquale, G., Goretti, A.: Economic and functional vulnerability of residential buildings stricken by Italian recent seismic events. In: Proceedings of the 10th Italian National Conference on Earthquake Engineering (2001)
33. D'Ayala, D., Speranza, E.: Definition of collapse mechanisms and seismic vulnerability of historic masonry buildings. Earthq. Spectra **19**(3), 479–509 (2003)
34. CEN. Eurocode 8: design of structures for earthquake resistance. Part 3: General Rules, Seismic Actions and Rules for Buildings, prEN 1998-1. CEN, Brussels (2005)
35. D'Ayala, D., Kansal, A.: Analysis of the seismic vulnerability of the architectural Heritage in Buhj, Gujarat, India. In: Proceedings of IV International Seminar on Structural Analysis of Historical Constructions, pp. 1069–1078 (2004)
36. D'Ayala, D., Paganoni, S.: Assessment and analysis of damage in L'Aquila historic city centre after 6th April 2009. Bull. Earthq. Eng. **9**(1), 81–104 (2011)
37. D'Ayala, D., Shi, Y.: Modeling Masonry Historic Buildings by Multi-Body Dynamics. Int. J. Architect. Heritage **5**(4–5), 483–512 (2011)
38. Bosiljkov, V., Kržan, M., D'Ayala, D.: Vulnerability study of Urban and rural heritage masonry in Slovenia through the assessment of local and global seismic response of buildings. In: Proceedings of the 15th World Conference on Earthquake Engineering (2012)
39. EERI. Final Technical Report- Providing building vulnerability data and analytical fragility functions for PAGER (2012)
40. D'ayala, D., Kishali, E.: Analytically derived fragility curves for unreinforced masonry buildings in urban contexts. (2012)
41. Paquette, J., Bruneau, M.: Pseudo-dynamic testing of unreinforced masonry building with flexible diaphragm and comparison with existing procedures. Constr. Build. Mater. **20**(4), 220–228 (2006)
42. Magenes, G., Penna, A., Rota, M., Galasco, A., Senaldi, I.: Shaking table test of a full scale stone masonry building with stiffened floor and roof diaphragms. In: Proceedings of the 15th World Conference on Earthquake Engineering (2012)
43. Meisl, C., Elwood, K., Ventura, C.: Shake table tests on the out-of-plane response of unreinforced masonry walls this article is one of a selection of papers published in this special Issue on Masonry. Can. J. Civ. Eng. **34**(11), 1381–1392 (2007)
44. Griffith, M.C., Lam, N.T.K., Wilson, J.L., Doherty, K.: Experimental investigation of unreinforced brick masonry walls in flexure. J. Struct. Eng. ASCE **130**(3), 423–432 (2004)
45. D'Ayala, D., Shi, Y., Stammers, C.W.: Dynamic multibody behaviour of Historic masonry building models. In: Proceedings of the VI International Conference on Structural Analysis of Historic Construction, SAHC08, vol. I, pp. 489–496 (2008)
46. Al Shawa, O., De Felice, G., Mauro, A., Sorrentino, L.: Out-of-plane seismic behaviour of rocking masonry walls. Earthq. Eng. Struct. Dynam. **41**(5), 949–968 (2012)
47. Tomaževič, M.: Damage as a measure for earthquake-resistant design of masonry structures: slovenian experience this article is one of a selection of papers published in this special Issue on masonry. Can. J. Civ. Eng. **34**(11), 1403–1412 (2007)

48. Jaiswal, K., Wald, D., D'Ayala, D.: Developing empirical collapse fragility functions for global building types. Earthq. Spectra **27**(3), 775–795 (2011)
49. Freeman, S.A.: Review of the development of the capacity spectrum method. ISET J. Earthq. Technol. **41**(1), 1–13 (2004)
50. Griffith, M.C., Vaculik, J., Lam, N.T.K., Wilson, J., Lumantarna, E.: Cyclic testing of unreinforced masonry walls in two-way bending. Earthq. Eng. Struct. Dynam. **36**(6), 801–821 (2007)
51. Kappos, A.J., Panagopoulos, G., Penelis, G.G.: Development of a seismic damage and loss scenario for contemporary and historical buildings in Thessaloniki, Greece. Soil Dynam. Earthq. Eng. **28**(10–11), 836–850 (2008)
52. Giuffrè, A.: Mechanics of historical masonry and strengthening criteria. In: Proceedings of the XV Regional Seminar on Earthquake Engineering, pp. 60–122 (1989)
53. Pagnini, L.C., Vicente, R., Lagomarsino, S., Varum, H.: A mechanical model for the seismic vulnerability assessment of old masonry buildings. Earthq. Struct. **2**(1), 25–42 (2011)

CPI Antony Rowe
Chippenham, UK
2018-04-07 04:53